High Time Resolution Astrophysics

Astrophysics and Space Science Library

EDITORIAL BOARD

Chairman

W. B. BURTON, *National Radio Astronomy Observatory, Charlottesville, Virginia, U.S.A.*
(bburton@nrao.edu); *University of Leiden, The Netherlands* (burton@strw. leidenuniv.nl)

F. BERTOLA, *University of Padua, Italy*
J. P. CASSINELLI, *University of Wisconsin, Madison, U.S.A.*
C. J. CESARSKY, *European Southern Observatory, Garching bei München, Germany*
P. EHRENFREUND, *Leiden University, The Netherlands*
O. ENGVOLD, *University of Oslo, Norway*
A. HECK, *Strasbourg Astronomical Observatory, France*
E. P. J. VAN DEN HEUVEL, *University of Amsterdam, The Netherlands*
V. M. KASPI, *McGill University, Montreal, Canada*
J. M. E. KUIJPERS, *University of Nijmegen, The Netherlands*
H. VAN DER LAAN, *University of Utrecht, The Netherlands*
P. G. MURDIN, *Institute of Astronomy, Cambridge, UK*
F. PACINI, *Istituto Astronomia Arcetri, Firenze, Italy*
V. RADHAKRISHNAN, *Raman Research Institute, Bangalore, India*
B. V. SOMOV, *Astronomical Institute, Moscow State University, Russia*
R. A. SUNYAEV, *Space Research Institute, Moscow, Russia*

Recently Published in the ASSL series

Volume 351: *High Time Resolution Astrophysics*, edited by Don Phelan, Oliver Ryan, Andrew Shearer. Hardbound ISBN: 978-1-4020-6517-0

Volume 350: *Hipparcos, the New Reduction of the Raw Data*, by Floor van Leeuwen. Hardbound ISBN: 978-1-4020-6341-1

Volume 349: *Lasers, Clocks and Drag-Free Control, Exploration of Relativistic Gravity in Space*, edited by Dittus, Hansjörg; Lämmerzahl, Claus; Turyshev, Slava. Hardbound ISBN: 978-3-540-34376-9

Volume 348: *The Paraboloidal Reflector Antenna in Radio Astronomy and Communication, Theory and Practice*, edited by Baars, Jacob W.M. Hardbound ISBN: 978-0-387-69733-8

Volume 347: *The Sun and Space Weather*, edited by Hanslmeier, Arnold. Hardbound ISBN: 978-1-4020-5603-1

Volume 346: *Exploring the Secrets of the Aurora*, edited by Akasofu, Syun-Ichi. Softcover ISBN: 978-0-387-45094-0

Volume 345: *Canonical Perturbation Theories, Degenerate Systems and Resonance*, edited by Ferraz-Mello, Sylvio. Hardbound ISBN: 978-0-387-38900-4

Volume 344: *Space Weather: Research Toward Applications in Europe*, edited by Jean Lilensten. Hardbound 1-4020-5445-9

Volume 343: *Organizations and Strategies in Astronomy: volume 7*, edited by A. Heck. Hardbound 1-4020-5300-2

Volume 342: *The Astrophysics of Emission Line Stars*, edited by Tomokazu Kogure, Kam-Ching Leung. Hardbound ISBN: 0-387-34500-0

Volume 341: *Plasma Astrophysics, Part II: Reconnection and Flares*, edited by Boris V. Somov. Hardbound ISBN: 0-387-34948-0

Volume 340: *Plasma Astrophysics, Part I: Fundamentals and Practice*, by Boris V. Somov. Hardbound ISBN 0-387-34916-9, September 2006

Volume 339: *Cosmic Ray Interactions, Propagation, and Acceleration in Space Plasmas*, by Lev Dorman. Hardbound ISBN 1-4020-5100-X, August 2006

Volume 338: *Solar Journey: The Significance of Our Galactic Environment for the Heliosphere and the Earth*, edited by Priscilla C. Frisch. Hardbound ISBN 1-4020-4397-0, September 2006

Volume 337: *Astrophysical Disks*, edited by A. M. Fridman, M. Y. Marov, I. G. Kovalenko. Hardbound ISBN 1-4020-4347-3, June 2006

Volume 336: *Scientific Detectors for Astronomy 2005*, edited by J. E. Beletic, J. W. Beletic, P. Amico. Hardbound ISBN 1-4020-4329-5, December 2005

For other titles see www.springer.com/astronomy

High Time Resolution Astrophysics

Don Phelan
NUI, Galway, Ireland

Oliver Ryan
NUI, Galway, Ireland

Andrew Shearer
NUI, Galway, Ireland

 Springer

Editors

Don Phelan
NUI, Galway,
Ireland

Oliver Ryan
NUI, Galway,
Ireland

Andrew Shearer
NUI, Galway,
Ireland

ISBN: 978-1-4020-6517-0   e-ISBN: 978-1-4020-6518-7

Library of Congress Control Number: 2007936158

© 2008 Springer Science+Business Media B.V.
No part of this work may be reproduced, stored in a retrieval system, or transmitted
in any form or by any means, electronic, mechanical, photocopying, microfilming, recording
or otherwise, without written permission from the Publisher, with the exception
of any material supplied specifically for the purpose of being entered
and executed on a computer system, for exclusive use by the purchaser of the work.

Printed on acid-free paper.

9 8 7 6 5 4 3 2 1

springer.com

Preface

High Time Resolution Astrophysics (HTRA) is an important new window to the universe and a vital tool in understanding a range of phenomena from diverse objects and radiative processes. This importance is demonstrated in this volume with the description of a number of topics in astrophysics, including quantum optics, cataclysmic variables, pulsars, X-ray binaries and stellar pulsations to name a few. Underlining this science foundation, technological developments in both instrumentation and detectors are described. These instruments and detectors combined cover a wide range of timescales and can measure fluxes, spectra and polarisation.

There are a number of new technological developments that are coming together at the present time to make it possible in the next decade for HTRA to make a big contribution to our understanding of the Universe. In the near future we will have detector systems providing high quantum efficiency detection with temporal resolution to nanoseconds or better. This will occur at a time when computer systems will be able to handle the vast quantities of data such detectors will deliver. Several extremely large telescopes (ELTs), with collection areas in excess of 1000 m^2, will become operational, and will deliver the fluxes to allow studies at high time resolution of many new objects to become feasible for the first time. A number of authors in this volume have pointed to the importance of the ELT to HTRA and particular attention has been given this topic in numerous chapters.

This volume was developed from the 2nd Galway workshop in High Time Resolution Astrophysics, held under the auspices of the OPTICON HTRA Network in June 2006. During the workshop, invited presentations were delivered that covered a wide range of topics, encapsulating the field of optical HTRA. The participants agreed to contribute a chapter on their own area of expertise to this volume. The Galway workshop brought together observers, theorists and instrumentalists who all could see the importance of opening up a new window in astronomy. The workshop concluded with a plan to further develop HTRA into the ELT era through collaborative programmes over the next number of years. This will include a regular programme of workshops, joint research activities and further development of the HTRA network.

We would like to thank the authors for their contributions and dedication throughout the process, and to the referees for their valuable comments on the manuscripts. The feedback and invaluable insights were greatly appreciated. We would like to thank OPTICON, which is funded from the European Community's Sixth Framework Programme under contract number RII3-CT-001566. It is this support that has enabled us to bring together the expertise from different teams to prepare this volume. We would also like to personally thank John Davies and Gerry Gilmore for their continued support.

NUI, Galway,
May 2007

Don Phelan
Oliver Ryan
Andrew Shearer

Contents

High Time Resolution Astrophysics and Pulsars
Andrew Shearer 1

High Time Resolution Observations of Cataclysmic Variables
S. P. Littlefair 21

High-Speed Optical Observations of X-ray Binaries
Tariq Shahbaz 37

Stellar Pulsation, Subdwarf B Stars and High Time Resolution Astrophysics
C. Simon Jeffery 53

High-Speed Optical Spectroscopy
T. R. Marsh 75

Photonic Astronomy and Quantum Optics
Dainis Dravins 95

ULTRACAM: An Ultra-Fast, Triple-Beam CCD Camera for High-Speed Astrophysics
V. S. Dhillon 133

OPTIMA: A High Time Resolution Optical Photo-Polarimeter
G. Kanbach, A. Stefanescu, S. Duscha, M. Mühlegger, F. Schrey, H. Steinle,
A. Slowikowska and H. Spruit 153

From QuantEYE to AquEYE—Instrumentation for Astrophysics on its Shortest Timescales
C. Barbieri, G. Naletto, F. Tamburini, T. Occhipinti, E. Giro
and M. D'Onofrio 171

Fast Spectroscopy and Imaging with the FORS2 HIT Mode
Kieran O'Brien 187

An Ultra-High-Speed Stokes Polarimeter for Astronomy
R. Michael Redfern and Patrick P. Collins 205

Use of an Extremely Large Telescope for HTRA
Oliver Ryan and Mike Redfern 229

EMCCD Technology in High Precision Photometry on Short Timescales
Niall Smith, Alan Giltinan, Aidan O'Connor, Stephen O'Driscoll,
Adrian Collins, Dylan Loughnan and Andreas Papageorgiou 257

The Development of Avalanche Amplifying pnCCDs: A Status Report
L. Strüder, G. Kanbach, N. Meidinger, F. Schopper, R. Hartmann, P. Holl,
H. Soltau, R. Richter and G. Lutz 281

Geiger-mode Avalanche Photodiodes for High Time Resolution Astrophysics
Don Phelan and Alan P. Morrison 291

Transition Edge Cameras for Fast Optical Spectrophotometry
Roger W. Romani, Thomas J. Bay, Jennifer Burney and Blas Cabrera 311

Imaging Photon Counting Detectors for High Time Resolution Astronomy
O. H. W. Siegmund, J. V. Vallerga, B. Welsh, A. S. Tremsin
and J. B. McPhate 327

Index 345

High Time Resolution Astrophysics and Pulsars

Andrew Shearer

Abstract The discovery of pulsars in 1968 heralded an era where the temporal characteristics of optical detectors had to be reassessed. Up to this point detector integration times would normally be measured in minutes rather seconds and definitely not on sub-second time scales. At the start of the 21st century pulsar observations are still pushing the limits of detector/telescope capabilities. Flux variations on times scales less than 1 nsec have been observed during giant radio pulses. Pulsar studies over the next 10–20 years will require instruments with time resolutions below a microsecond, high-quantum quantum efficiency, reasonable energy resolution and sensitive to circular possible and linear polarisation of stochastic signals. This chapter is a review of temporally resolved optical observations of pulsars. It concludes with estimates of the observability of pulsars with both existing telescopes and into the ELT era.

1 Introduction

Traditionally astrophysics has concerned itself with minimum time-scales measured in hours rather than seconds. This was understandable as the available recording media were slow; e.g. chart recorders and photographic plates. In the 1950s, 60s and 70s wavebands away from the optical were developed; from ground based radio studies to space and balloon borne high energy work. In contrast to optical wavelengths instrumentation in these (high and low energy) regimes was capable of time resolutions of less than a second. Vacuum tube and CCD technologies in the 1970s and 80s extended electronic detectors to the optical band pass thereby allowing for studies, in the optical regime, at all time scales. More recently superconducting detectors and avalanche photodiodes (APD) have given us the possibility of nanosecond observations with high quantum efficiency. To date high time resolution observations, in optical astronomy, have been driven predominantly by detector developments and possibilities—and not by the underlying science. This situation is now changing where studies of stochastic phenomena such as giant radio

---

Andrew Shearer
National University of Ireland, Galway
e-mail: andy.shearer@nuigalway.ie

Table 1 Small Timescale Variability of Astronomical Objects. Pulsars studies require observations on the shortest time-scales–variability in the radio region has been shown down to nanosecond scales and in the optical variability at the microsecond scale has been observed

		Time-scale	
		(now)	(ELT era)
Stellar flares pulsations		Seconds / Minutes	10–100 ms
Stellar surface oscillations	White Dwarfs	1–1000 μs	1–1000 μs
	Neutron Stars		0.1 μs
Close Binary Systems (accretion and turbulence)	Tomography	100 ms++	10ms+
	Eclipse in/egress	10 ms+	<1ms
	Disk flickering	10 ms	<1ms
	Correlations (e.g. X & Optical)	50 ms	<1ms
Pulsars	Magnetospheric	1 μs–100 ms	ns(?)
	Thermal	10 ms	<ms
AGN		Minutes	Seconds

pulses (GRP) and rotating radio transients (RRAT) require specific types of detector and instrumentation. Table 1 shows the characteristic time-scale for different astronomical objects. Clearly studies of pulsars require the ability to observe down to sub-millisecond and in some cases sub-microsecond times scales. Consequently, optical observations of pulsars probably represent the biggest instrumental challenge to high time resolution astrophysics. Radio observations [1] have shown flux variations from the Crab during GRPs on time scales of nanoseconds. Although over 1600 radio pulsars have been observed, only five normal pulsars and one anomalous X-ray pulsar (AXP) have been observed to pulse at optical wavelengths. Optical observations of pulsars are also limited by their intrinsic faintness in the optical regime see Table 2.

Table 2 Optical Pulsars—Basic Data. Pulsars names marked in bold have been observed to pulsate. The spectral index covers the region from 3500–7000 Å and is of the form $F_\nu \propto \nu^\alpha$. The spectral index are for time-averaged fluxes. The M31 pulsar flux is based upon extrapolating the Crab pulsar to the distance of M31 and increasing the noise background proportionately

Pulsar	m_B	Period (ms)	Spectral Index	VLT B photons /rotation	ELT
Crab	16.8	33	−0.11	3,300	120,000
PSR B0540−69	23	50	1.6	17	610
Vela	24	89	0.12	12	440
PSR B0656+14	25.5	385	0.45	13	470
Geminga	26	237	1.9	25.5	200
M31 (Crab)	30-31	33	–	0.02	1
PSR B0950+08	27.1(V)	253	–	–	–
PSR B1929+10	25.6(V)	227	–	–	–
PSR B1055−52	24.9(V)	197	–	–	–
PSR B1509−58	25.7(V)	151	–	–	–

In this chapter we present an analysis of optical observations of pulsars that have shown variability on time-scales from few nanoseconds to secular changes over years. In this regard they represent the most challenging target for optical High Time Resolution Astrophysics. Pulsars, neutron stars with an active magnetosphere, are created through one of two main processes—the compact core after a type II supernova explosion or as a result of accretion induced spin up in a binary system. Pulsars have strong fields up to 10^{13} G and rotation periods down to less than 2 milliseconds, which in combination of can produce exceptionally strong electric fields. The plasma accelerated by these fields radiates at frequencies ranging from the radio to TeV γ-rays. There is however a fundamental difference between the radio and higher energy emission. The former is probably a coherent process and the latter some form of synchrotron or curvature radiation. Despite nearly forty years of theory and observation a number of fundamental parameters are not known :

- What is the expected distribution of the plasma within the pulsar's magnetosphere?
- Where is the location and mechanism for accelerating the plasma?
- Where in the magnetosphere is radiation emitted?
- What is the emission mechanism in the radio and at higher photon energies?

Optical observations of pulsars are limited by their intrinsic faintness in the optical regime—early estimates indicated that the optical luminosity should scale as Period^{-10} [2] making many pulsars unobservable with current technology. However it is in the optical regime that we have two distinct benefits with regard to other high-energy wavelengths. In the optical regime we can readily measure all Stokes' parameters including polarisation and we are probably seeing a flux which scales linearly with local power density in the observer's line of sight. With the advent of increasingly large telescopes and more sensitive detectors we should soon be in a position where the number of observed optical pulsars will have increased from five to over hundred.

2 Normal Pulsars

2.1 Observations—The Phenomenology

Understanding the properties and behaviour of neutron stars has been one of the longest unsolved stories in modern astrophysics. They were first proposed as an end point in stellar evolution by Baade & Zwicky [3] in 1934. Possible emission mechanisms from rapidly rotating magnetised neutron stars were published *before* [4] their unexpected detection by Bell and Hewish in 1968 [5].

Figure 1 shows the P–$\dot{P}$ diagram for pulsars which gives a gross view of a pulsar's evolutionary position—normal pulsars are born towards the top of the diagram and move roughly towards the bottom right—i.e., they slow down and it is this spin-down energy which powers the pulsar. The optically and higher photon energy emitting pulsars can be seen to be younger and tend to have higher magnetic fields and higher $\dot{E}$ values. Millisecond pulsars, which have been spun up through

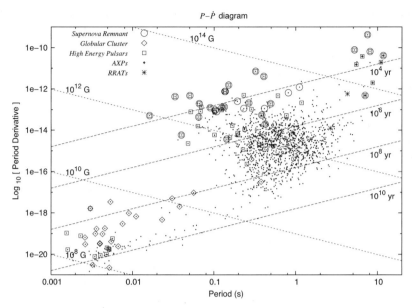

Fig. 1 The standard P–$\dot{P}$ diagram showing the location of optical pulsars, anomalous X-ray pulsars, rotating radio transients and pulsars associated with supernova remnants. Also marked are pulsars associated with globular clusters. The classifications are based upon data from ATNF Catalog [22]. On the diagram we have also marked the predicted surface magnetic field and age of each pulsar. These have been determined from the normal assumption that the pulsar's spin down energy is due to magnetic dipole braking. It can be shown the the surface magnetic field, $B_S \propto (P\dot{P})^{1/2}$ and a pulsar's characteristic age as $\tau \equiv P\dot{P}^{-1}$. It should be noted that the characteristic age is normally greater than the actual age determined from historical supernovae. In this diagram normal pulsar are expected to be born towards the top left of the diagram and will move towards the bottom right. At some point they cross the *death line* where there is insufficient energy for e^+e^- pair creation within the magnetosphere—this begins to explain the lack of pulsars with long periods towards the lower right of the diagram

accretion are found in the bottom left of diagram illustrating their relatively weak magnetic fields. The binary nature of their progenitors also increases the chance that they are found in globular clusters.

Recently two new members of the neutron star zoo have been identified—AXPs, now thought to be magnetars and RRATs. The former are characterised by extremely high magnetic fields ($>10^{14}$ G) and are probably powered by the decay of these fields. RRATs are the most recent neutron star observation characterised by their transient radio signal, which occurs for a few milliseconds, at random intervals ranging from a few to several hundred minutes. These different classes of pulsar are described below.

Since the first optical observations of the Crab pulsar in the late 1960s [6] only four more pulsars have been seen to pulsate optically (Vela [7]; PSR B0540−69 [8]; PSR B0656+14 [9]; PSR B0633+17 [10]). Four of these five pulsars are probably too distant to have any detectable optical thermal emission using currently available technologies. For the nearest and faintest pulsar, PSR B0633+17, spectroscopic studies have shown the emission to be predominantly non-thermal [11] with a flat featureless spectrum. For these objects we are seeing non-thermal emission, presumably from interactions between the pulsar's magnetic field and a stream of charged particles from the neutron star's surface. Four other pulsars have been observed to emit optical radiation, but so far without any detected pulsations [12]. Optical pulsars seem to be less efficient than their higher energy counterparts with an average efficiency[1] of about 10^{-9} compared to $\approx 10^{-2}$ at γ-ray energies [13]. Optical efficiency also decreases with age in contrast to γ-ray emission [14]. These factors combined indicate that pulsars are exceptionally dim optically, although it is in the optical where it is potentially easier to extract important parameters from the radiation—namely spectral distribution, flux and polarisation.

Detailed time-resolved spectral observations have only been made of the Crab pulsar, but through broad-band photometry the spectral index of the other pulsars has been determined. Table 2 shows the spectral index for pulsars with observed optical emission. If the emission mechanism is synchrotron and from a single location, you would expect the spectral index to be negative for high frequencies changing to a value of 5/2 below the critical frequency in the optically thick region where synchrotron self-absorption becomes dominant. One pulsar, PSR J0537−6910, which has not been observed optically despite showing many of the characteristics of the Crab pulsar, might be expected to be more luminous. It is likely that synchrotron-self absorption will play a significant part in reducing its optical flux [2], [15].

The other, optically fainter, pulsars have either had spectra taken with low signal-noise [11] or using broad band photometry had their gross spectral features determined. There have been suggestions of broadband features in the spectrum possibly associated with cyclotron absorption or emission features; Fig. 2 shows the IR-NUV spectrum for PSR B0656+14 and Geminga [16]. Until spectra with reasonable resolution have been taken we can not, with any degree of certainty, characterise the spectra in term of absorption or emission features. If a cyclotron origin for these features can be determined then we have the possibility of determining the local magnetic field strength independently of the characteristic surface magnetic field ($\propto (P\dot{P})^{-0.5}$). The overall spectral shape is consistent with the UV dominated by thermal radiation from the surface and the optical-infrared region dominated by magnetospheric non-thermal emission. Neither spectra show any evidence of synchrotron self absorption towards longer wavelengths.

Only one pulsar, the Crab, has had detailed polarisation measurements made [17], [18]. The polarisation profile shows emission aligned with the two main peaks and is consistent with synchrotron emission. One unusual feature is the location of maximum polarisation which for the Crab pulsar precedes the main pulse at a phase when the optical emission is at a minimum. The polarisation of one other, PSR B0656+14 [19] has been observed albeit at low significance. This pulsar has maximum polarisation coincident with the maximum luminosity. Three other pulsars have only had their time average polarisation measured in the optical regime [20].

---

[1] The optical efficiency is defined as the optical luminosity divided by the spindown energy—($\eta \equiv L_v/\dot{E}$).

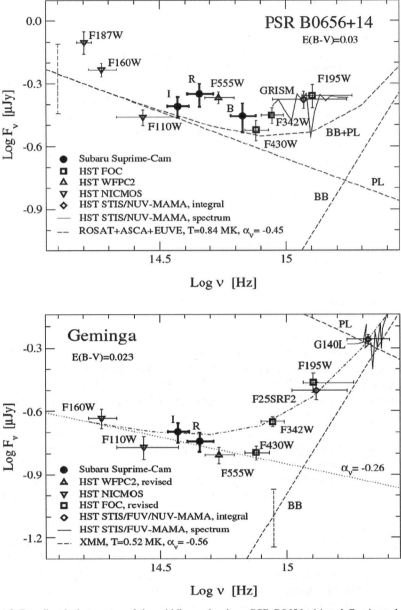

Fig. 2 Broadband photometry of the middle-aged pulsars PSR B0656+14 and Geminga showing some similarities in spectral shape. In particular both spectra show a increase in flux towards infra-red and a possible emission feature around the V band [16]. At the resolution of these measurements, it is not possible to determine whether this is a single feature or a number of emission lines from a pulsar wind nebula. Furthermore as these observations were time integrated it was not possible to determine how the spectral shape varied with pulse phase

2.2 Emission Theory

Many suggestions have been made concerning the optical emission process for these young and middle-aged pulsars. Despite many years of detailed theoretical studies and more recently limited numerical simulations, no convincing models have been derived which explain all of the high energy properties. There are similar problems in the radio, but as the emission mechanism is radically different (being coherent) only the high energy emission is considered here. In essence despite nearly forty years of observations, we still do not understand the mechanisms behind pulsar emission. The various competing theories have all failed to provide a comprehensive description of pulsar emission. *Probably the only point of agreement between all these theories is the association of pulsars with magnetized, rotating neutron stars* [21]. In the high energy regime there is more consensus as to the emission mechanism—either incoherent synchrotron or curvature radiation for outer gap models or inverse Compton scattering for polar gap models. However, like radio emission there is not a consensus as to the location of the emission region or the acceleration mechanisms for the plasma.

2.3 Crab Pulsar

The Crab pulsar has been observed for nearly forty years. The earliest photometric observations [25], [26] gave a visual magnitude of m_V 16.5 $\pm$ 0.1 compared to more recent observations [27] of m_V 16.74 $\pm$ 0.05 in the same period the pulsar has slowed by $\approx$1.5%. If we assume a $L \propto P^{-10}$ scalling law then in the same period the luminosity should have reduced by about 0.15 magnitudes—in reasonable agreement with the observed luminosity decrease.

As can be seen from Table 2 the Crab Pulsar is uniquely bright amongst the small population of optical pulsars. It is reasonably close ($\approx$2 kpc) and less than one thousand years old. It is the only pulsar which is bright enough for individual pulses to be observed with any significant flux—allowing for pulse-pulse spectral and polarimetric changes to be observed. As a pulsar it has unusual characteristics—for example it glitches and the radio emission is dominated by giant radio pulses (GRPs). The latter occur sporadically with a mean repetition rate of about 1 Hz—the highest of the GRP emitting pulsars—see below. The Crab pulsar is seen to emit at all energies—from radio to high-energy γ-rays with E > 1 TeV. In the optical (UBV) regime the emission is characterised by a flat power law $F_\nu \propto \nu^{-0.07\pm0.18}$[23]. Although no direct evidence of a roll-over in the spectrum has been observed there are hints that in the near infra-red that the slope steepens ([15] and references therein). Recent Spitzer observations [24] possibly show the beginning of a rollover at $\approx$2 μm. More detailed observations, particularly time-resolved from 1–50 μm will be required to confirm this.

The Crab pulsar's light curve, Fig. 3, shows four distinct regions
- Main pulse containing about 50% of the optical emission, the pulse has a FWHM of 4.5% of the pulsar's period
- Interpulse with about 30% of the emission

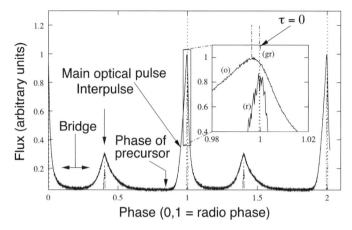

Fig. 3 Crab pulsar's light curve showing the main emission regions and the relationship between the optical light curve and its radio counterpart [30]. The optical data was taken using an APD based photometer on the William Herschel Telescope (WHT) each photon's arrival time has been folded in phase with the pulsar radio ephemeris to produce the time averaged light curve. The radio data has been treated in a similar fashion. Also shown is the location of a single giant radio pulse

- Bridge with about 20% of the emission
- Off-pulse just before the main pulse where the polarisation percentage is at a maximum

Spectral observations, of the Crab pulsar—both pulsed and time-integrated— have been made by a number of groups. All authors report spectra that is generally flat and featureless—consistent with a synchrotron origin for the radiation with an integrated spectral index of 0.11. A possible cyclotron absorption feature has been reported [27] although not seen by other observers. Of note are time resolved observations which indicate a change in the spectral index on the leading and falling edge of the main and secondary peaks [18], [28], [29]. As yet there is no clear explanation for either spectral shape changes, pulse width variations or polarimetry.

2.4 Giant Radio Pulse Pulsars—Optical Considerations

Six pulsars, see Table 3, exhibit giant radio pulse phenomena. Here, as well as the normal radio emission, we see enhanced radio emission up to several times the average occurring for a few percent of the total number of pulses. The Crab pulsar was itself discovered through its giant pulse phenomena [31]. Phenomenologically these pulsars are a very diverse group with the only possible common factor being their magnetic field strength at the light cylinder—the pulsars in Table 3 are all in the top six ranked by magnetic field strength at the light cylinder. The sixth pulsar in the list of pulsars with high magnetic field strength at the light cylinder, PSR J0537−6919 has only been observed at X-ray energies and hence no giant radio emission. Two of the pulsars, the Crab and PSR J0540−6919, have very similar properties—young [age < 1000 years] pulsars embedded in a plerion. The other three are all millisecond pulsars.

Observations of GRP pulsars at other energies has been difficult—primarily due to the low rate of GRP events (ranging from 10^{-6} to 2 Hz). γ-ray observations [32]

Table 3 Giant radio pulsars

Name	Period (ms)	Surface Field (10^9 G)	Light Cylinder Field (10^6 G)	GRP Rate (Hz)	Ref.
Crab	33.1	3800	9.8	1–2	
PSR J0540−6919	50.4	5000	3.7	0.001	[44]
PSR J1824−2452	3.1	2.2	0.7	0.0003	[45]
PSR J1959+2048	1.6	0.2	1.1		
PSR J1939+2134	1.6	0.4	1.0	0.0001	[46]
PSR J0218+4231				0.03	

showed an upper limit of 2.5 times the average γ-ray flux based upon 20 hours of simultaneous radio-γ-ray observations with CRGO/OSSE and the Greenbank 43m telescope. Optical observations [30] indicated a small ($\approx$3%) increase in the optical flux during the same spin cycle as a GRP. Although the percentage increase was small the energy increase in the optical pulse is comparable to the energy in the GRP. These results linked for the first time flux changes in the radio—where the emission is highly variable—to higher energy emission which is seen to be stable. Furthermore, given the different emission mechanisms [coherent versus incoherent] such a correlation was unexpected. Although small, the energy in the optical and radio GRPs are similar; if this were extrapolated to γ-ray energies we would expect an enhancement of <0.01%.

In two of these, PSR J1824−2452 and PSR J1939+2134, the GRP emission aligns in phase with the X-ray pulse rather than the normal radio pulse [33][34]. This possibly implies a different emission zone and/or mechanism. Furthermore the spectrum of GRP fluxes shows a different slope to the normal emission and cannot be regarded simply as its high flux extension. What characterises a GRP event compared to the high-energy tail of the normal distribution is very subjective. An acceptable definition of what is and what is not a GRP is that the pulses show a power-law energy distribution (cf. lognormal for typical pulsar emission) and have very short time-scales (typically $\approx$ of nanosecond duration) [33]. Most studies have used a 10 or 20 $\langle E \rangle$ criteria for GRPs.

From an optical observational perspective GRPs are significantly more difficult than normal pulsars to observe—the pulses arrive randomly albeit in phase with normal emission and at a rate significantly less than the normal pulsar rate. The random nature of GRP arrival times make synchronised systems, such as clocked CCDs, inappropriate. Polarisation studies of these type of event also restrict the type of polarimeter which can be used in the GRP studies, and makes the simultaneous measurements of all Stokes' parameters essential.

3 AXPs and RRATs

3.1 Anomalous X-Ray Pulsars

Anomalous X-Ray pulsars are a special class of neutron star characterised by very high inferred magnetic fields and variable X-ray emission. Soft Gamma-ray repeaters (SGRs), first observed as transient γ-ray sources and now known to be persistent sources of pulsed X-rays, show similar properties. Both AXPs and SGRs are thought to have similar physical properties and are likely to be magnetars—where the emission comes from the decay of the strong magnetic field rather than being rotation powered. The review by Woods and Thompson [35] is a recent review of general AXP and SGR properties.

In Fig. 1 AXPs are shown to be towards the top right of the P–$\dot{P}$ diagram. Models for their emission mechanism historically concentrated upon the interaction between the neutron star and an accretion disk. However the lack of a secondary star precluded all but a fossil disk around the star. The latter explanation was precluded by infrared observations [36]. The faintness of the infrared counterpart precluded a disk model and strengthened the case for the magnetar model where the high-energy emission comes from the decay of the strong magnetic field or from other magnetospheric phenomena.

Optical pulsed AXP emission was first observed from 4U 0142+61, using a phase-clocked CCD [37] and confirmed by UltraCam observations [38]. From Table 4 it can be seen that this pulsar is significantly brighter than other AXPs in the infra-red and all suffer from significant reddening. All of which combines to make future AXP optical observations very difficult and dependent on system such as Ultracam, but with better infra-red sensitivity.

3.2 Rotating Radio Transients

Rotating Radio Transients (RRATs) are a new class of pulsar that exhibit sporadic radio emission lasting for few milliseconds (2–30 ms) at random intervals ranging from a few hundred seconds to several hours. Table 5, based upon [39] details the basic RRAT parameters. For a few RRATs a periodicity and period derivative has been established by using the largest common divisor to estimate the period and in three cases the period derivative. It is not known where these lie in the now expanding pulsar menagerie, however one RRAT lies towards the magnetar region of the P–$\dot{P}$ diagram—see Fig. 1.

Table 4 Optical and Infra-red properties AXPs and SGRs

Source	P	$\dot{P}$	B	Magnitude		
	(s)	$10^{-11} s s^{-1}$	10^{14} G	V	I	J
SGR 0526−66	8.0	6.6	7.4	> 27.1	>25	–
SGR 1627−41	6.4					>21.5
SGR 1806−20	7.5	8.3–47	7.8			>21
SGR 1900+14	5.2	6.1–20	5.7			> 22.8
4U 0142+61	8.69	0.196	1.3	25.6	23.8	–
1E 10485937	6.45	3.9	3.9		26.2	21.7
RXS 17084009	11.00	1.86	4.7	–	–	20.9
1E 1841045	11.77	4.16	7.1	> 23(R)	–	–
1E 2259+586	6.98	0.0483	0.60	>26.4(R)	>25.6	>23.8
AX J1845.00258	6.97	–	–	–	–	
CXOU J0110043.1	8.02	–	–	–	–	
XTE J1810197	5.54	1.15	2.9	–	>24.3	

Table 5 Observational Properties of Rotating Radio Transients

Name	Period (s)	$\dot{P}$ $10^{-15} s s^{-1}$	Distance (kpc)	Rate (hr^{-1})	On Time (10^{-5})
J0848−43	5.97748	−	5.5	1.4	1.2
J1317−5759	2.6421979742	12.6	3.2	4.5	1.3
J1443−60	4.758565	−	5.5	0.8	0.4
J1754−30	0.422617		2.2	0.6	0.3
J1819−1458	4.263159894	50.16	3.6	17.6	1.5
J1826−14	0.7706187		3.3	1.1	0.06
J1839−01	0.93190		6.5	0.6	0.3
J1846−02	4.476739		5.2	1.1	0.5
J1848−12	6.7953		2.4	1.3	0.07
J1911+00		−	3.3	0.3	0.04
J1913+1333	0.9233885242	7.87	5.7	4.7	0.3

Todate no RRAT has had any optical counterpart observed, however UltraCam observations of PSRJ1819−1458 [40] produced an upper limit of 3.3, 0.4 and 0.8 mJy at 3560, 4820 and 7610 Å, respectively in 1800 seconds on the WHT. The characteristics of the emission is difficult to determine at this stage although the bursts do not seem to have the same form as Giant Radio Pulses, for example they do not show a power law size distribution. One suggestion is that we are seeing a selection effect ([41])—PSR B0656+14 would appear as an RRAT if it was located at a distance of greater than 3kpc. The spread of derived periods is from 0.4 to 8 seconds which when combined with the inferred surface magnetic field for one object being as high as 5 10^{13}G we have a possible link to AXPs and magnetars.

4 Future Observing Campaigns

Despite forty years of of observation there are still a number of fundamental unanswered question in pulsar astrophysics. In the optical regime we can be begin to answer these question if we have a larger and more comprehensive survey of normal pulsars. Most pulsed observations were made on medium sized (4–6 m) telescopes using detectors with modest or low quantum efficiencies. By moving to larger 8–10 m class telescopes we gain about a magnitude in our upper limits. By using more efficient detectors another two magnitude improvement is possible making pulsed observations of 27th magnitude objects plausible in one night. Fig. 4 shows a measure of the observability (ranked according to flux) ($\dot{E}/d^2$) of known pulsars. If we take, an admittedly arbitrary, level of 10^{35} ergs sec^{-1} kpc^{-2} we have 20–30 pulsars which could be observed using current technology. From Fig. 5 there is a rough divide around pulsar periods around 50–100 ms below which non-CCD technologies (Superconducting Tunnel Junction (STJ) devices [42], Transition Edge Sensors (TES) [43], APDs) are appropriate. In all cases polarisation measurements will be important as they give information of the local magnetic field strength and geometry. Spectral information, to look for synchrotron self absorption and cyclotron features, will give important clues to the strength

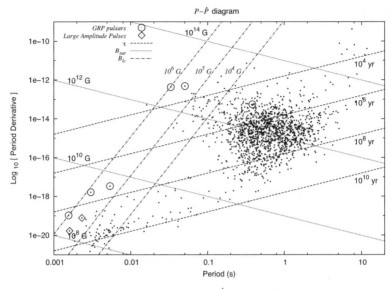

Fig. 4 The location of giant pulse emitters on the $P-\dot{P}$ diagram. We have also shown the inferred maximum magnetic field at the light cylinder based upon a 10 km radius neutron star

of the magnetic field in the emission zone and thus its height above the neutron star surface. In the future coordinated campaigns, particularly simultaneous radio-optical observations, can be used to look for specific phenomena such as GRPs and RRATs.

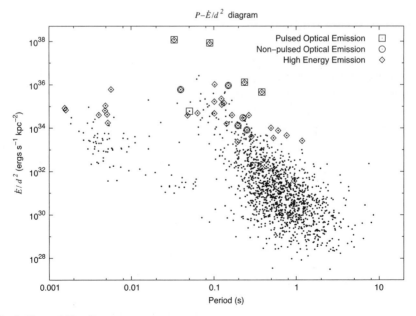

Fig. 5 Observability diagram

4.1 Normal Pulsars

Only five normal pulsars have had observed pulsed optical emission. From such a small sample it is not possible, particularly with the lack of an acceptable theory for pulsar emission, to make any strong conclusions. However with high quantum efficiency detectors and larger telescopes we can reasonably expect an increased sample of optical pulsars over the next 5–10 years. Specifically if we take the optical luminosity to scale with the spin down energy ($L_O \propto \dot{E}^{1.6}$) [14] then we can estimate the number of pulsars which are potentially observable. This is a very approximate relationship designed to indicate which pulsars are more likely to observed with current telescopes and instrumentation. From Table 6 the following pulsars can be potentially observed with 4–8m class telescopes with current optical detectors—PSR J0205+6449, PSR J1709−4429, PSR J1513−5908, PSR J1952+3252, PSR J1524−5625, PSR J0537−6910 and PSR J2229+6114.

Optical GRP observations are currently limited to the Crab pulsar and likely to remain so for the foreseeable future. The frequency of GRP events is very low with all other GRP emitters and only one of which (PSR J0540−6919) has been observed optically. If PSR J0540−6919 behaves in a similar manner to the Crab pulsar then the observed GRP rate of 0.001 Hz, about 1000 times lower than the Crab, will have a corresponding increase in observing time making optical observations unlikely in the near future. If PSR J0540−6919 shows a 3% increase in optical flux during GRP events then detecting this at the 3 σ level would require over 100 hours of observation using the 8m VLT and an APD based detector.

4.2 Millisecond Pulsars

With the exception of UV observations of the PSR J0437−4715 [83] there are no optical counterparts of any millisecond pulsar and in the case of PSR J0437−4715 the optical emission is probably thermal in origin. There have been a few upper limits and possible, but unconfirmed, counterparts. Until a number of optical millisecond pulsars has been observed, it will be difficult to even estimate the number of millisecond pulsars we can expect to observe optically. Table 7 shows those millisecond pulsars with the highest $\dot{E}/d^2$, which which we regard as a starting point for selecting those pulsars suitable for optical study.

4.3 AXPs

AXPs have typical periods in the range 8–12 seconds and consequently can be observed with conventional CCDs or Low-Light-Level CCDs (L3CCD). They also are characterised by high magnetic fields indicating that polarisation will be an

Table 6 Pulsar observability table, based upon the ATFN Pulsar Catalog[22], showing the top twenty normal, non-binary pulsars ranked according to $\dot{E}/d^2$. Only those pulsars with $\dot{E}/d^2 > 10^{35}$ and those with observed X and γ-ray emission have been included. In this we have not made any assumptions about the efficiency of optical emission with age, field, period or any of the normal phenomenological parameters. Where not explicitly referenced the A_V values are based upon $N_H = 1.79 \pm 0.03 A_V \, 10^{21}$ cm[49]. The predicted magnitude is based upon the transverse light cylinder magnetic field, Luminosity $\propto E^{1.6}$ [14]. This relationship should be treated with caution and is purely phenomenological. The aim here is to give a rough scaling between the different pulsars. No millisecond pulsar have been included in this table. Table 7 shows the similar table for millisecond pulsars without any prediction of optical luminosity

Name	Period (ms)	Distance (kpc)	log(age) years	log($\dot{E}$) ergs/s	log($\dot{E}/d^2$) ergs/s/kpc^2	N_H 10^{22}cm^{-2}	A_V	Pred. V Mag.	Association	Ref.
J0534+2200	0.0331	2	3.09	38.66	38.06	0.3	1.6	**16.8**	M1	[47]
J0835−4510	0.0893	0.29	4.05	36.84	37.93	0.04	0.2	**24**	Vela	[47]
J0205+6449	0.0657	3.2	3.73	37.43	36.42	0.3	1.7	23	3C58	[50] [51]
J1833−1034	0.0619	4.3	3.69	37.53	36.26	2.2	12.3	34	G21.5−0.9	[85] [84]
J0633+1746	0.2371	0.16	5.53	34.51	36.12	0.01	0.07	**26**		[47]
J1709−4429	0.1025	1.82	4.24	36.53	36.01	0.54	3.0	26.5	G343.1−2.3	[47]
J1513−5908	0.1507	4.4	3.19	37.25	35.96	0.8	5.2	28	G320.4-1.2	[47]
J1952+3252	0.0395	2.5	5.03	36.57	35.78	0.34	1.9	26	CTB80	[47]
J1930+1852	0.1369	5	3.46	37.06	35.66	1.6	8.9	33	G54.1+0.3	[61]
J0659+1414	0.3849	0.29	5.05	34.58	35.66	0.01	0.09	**25.5**	Mon. Ring	[47]
J1124−5916	0.1353	5.4	3.46	37.08	35.61	0.31	1.7	25	G292.0+1.8	[62]
J1747−2958	0.0988	2.49	4.41	36.40	35.61	2–3	1.4	26	G359.23−0.82	[63] [48]
J1617−5055	0.0694	6.46	3.91	37.20	35.58	6.8	38	62	RCW 103	[64] [65]
J1048−5832	0.1237	2.98	4.31	36.30	35.35	0.5	2.8	28		[47] [66]
J1524−5625	0.0782	3.84	4.50	36.51	35.34	–	–	*		[52]
J0537−6910	0.0161	49.4	3.69	38.69	35.30	0.1–0.6	2	24	N157B	[53] [54]

(continued)

Table 6 (continued)

Name	Period (ms)	Distance (kpc)	log(age) years	log($\dot{E}$) ergs/s	log($\dot{E}/d^2$) ergs/s/kpc^2	N_H 10^{22}cm^{-2}	A_V	Pred. V Mag.	Association	Ref.
J1357−6429	0.1661	4.03	3.86	36.49	35.28	–	–	*	G309.8−2.6?	[55]
J1420−6048	0.0682	7.69	4.11	37.02	35.24	2.2	12.3	37	Kookaburra	[56]
J1826−1334	0.1015	4.12	4.33	36.45	35.22	8.2	46	71		[47]
J2229+6114	0.0516	12.03	4.02	37.35	35.19	0.6	3.4	28	G106.6+2.9	[57]
J1913+1011	0.0359	4.48	5.23	36.46	35.16	–	–	*		[52]
J1803−2137	0.1336	3.94	4.20	36.35	35.16	14.0	78	104	G8.7−0.1(?)	[47]
J1809−1917	0.0827	3.71	4.71	36.25	35.11	–	–	*		[52]
J1740+1000	0.1541	1.36	5.06	35.37	35.10	–	–	*		[58]
J1801−2451	0.1249	4.61	4.19	36.41	35.09	–	–	*	G5.4−1.2	[52]
J0940−5428	0.0875	4.27	4.63	36.29	35.03	–	–	*		[52]
J0540−6919	0.0504	49.4	3.22	38.17	34.78	0.46	0.62	**23**	SNR 0540-693	[68]
J1105−6107	0.0632	7.07	4.8	36.39	34.69	0.7	3.9	31		[69]
J1932+1059	0.2265	0.36	6.49	33.59	34.48	0.04	0.2	32		[67]
J0538+2817	0.1432	1.77	5.79	34.69	34.20	0.3	1.7	33	S147	[70]
J1057−5226	0.1971	1.53	5.73	34.48	34.11	0.12	0.7	32		[71]
J0953+0755	0.2531	0.26	7.24	32.75	33.91	0.03	0.2	35		[59]
J0631+1036	0.2878	6.55	4.64	35.24	33.61	0.2	1.1	33		[72]
J0826+2637	0.5307	0.36	6.69	32.66	33.55	0.08	0.5	36		[60]

Table 7 Millisecond pulsar observability table, based upon the ATFN Pulsar Catalog [22]

Name	Period (ms)	Distance (kpc)	log(age) years	log($\dot{E}$) ergs/s	log($\dot{E}/d^2$) ergs/s/kpc^2	Ref.
J2124−3358	4.93	0.25	9.6	33.8	35.0	[73]
J1824−2452	3.05	4.9	7.5	36.4	35.0	[47]
J1939+2134	1.56	3.6	8.4	36.0	34.9	[47]
J0030+0451	4.87	0.23	9.9	33.5	34.8	[73]
J1024−0719	5.16	0.35	9.6	33.7	34.6	[74]
J1744−1134	4.08	0.36	9.9	33.7	34.6	[74]
J1843−1113	1.85	1.97	9.5	34.8	34.2	[75]
J1823−3021A	5.44	7.9	7.4	35.9	34.1	[47]
J0024−7204F	2.62	4.8	8.8	35.1	33.8	[76]
J1730−2304	8.12	0.51	9.8	33.2	33.8	[77]
J2322+2057	4.81	0.78	9.9	33.5	33.8	[47]
J0711−6830	5.49	1.04	9.8	33.6	33.5	[78]
J1910−5959D	9.04	4	8.2	34.7	33.5	[79]
J1944+0907	5.19	1.28	9.7	33.7	33.5	[80]
J1721−2457	3.50	1.56	10.0	33.7	33.4	[81]
J1905+0400	3.78	1.34	10.1	33.5	33.3	[75]
J1801−1417	3.63	1.8	10.0	33.6	33.1	[82]

important component of any future instrument and programme of observations. However the number of known AXPs is small and most have upper limits which are significantly lower than the detected flux from 4U 0142+61 making significant optical observations unlikely in the short term. In the near-IR recent observations of 4U 0142+61 [86] are consistent with the existence of a cool (T ≈ 91 K) debris disk around the neutron star. We would expect, if this is confirmed, that the pulse fraction in the near-IR should be lower than in the optical wavebands. It is interesting to speculate whether similar features could be observed in young normal pulsars.

4.4 RRATs

Observing RRATs in the optical will be a serious observational challenge requiring coordinated optical/radio-observations. For example PSRJ1819−1458 was observed for ≈1800 seconds with UltraCam which produced upper magnitude limits of 15.1, 17.4 and 16.6 in u', g' and i'. If only those frames coincident with radio events were recorded then this would have produced 3σ limits of 16.7, 18.8 and 18.1 respectively. If the link to AXPs is real then we must obtain peak i' magnitudes of about 24 or unrealistic exposure times of ≈ 71 hours using UltraCam. Even if we use a detector with zero read noise (e.g. L3CCDs or APDs) then our exposure times drop down to 57 hours, for a 3 σ detection. Alternatively if, as with GRPs, the energy in the radio and optical transient are similar then we would expect observation times of roughly 10 hours with UltraCam to achieve a 3 σ detection. Another way of looking at the RRAT detection probability is to scale according to the fractional on-time of each RRAT. PSR J1819−1458 is radiating for approximately 0.01% of the time or 10 magnitudes down from a pulsar which is radiating continuously. In contrast PSR J1848−12 is a further factor of 100 weaker.

5 Conclusion

Optical/infra-red observations of pulsars of all types are still in their relative infancy with very few objects detected. The likelihood is that in the next 5 years the number of pulsars with detected optical emission will have doubled and more significantly, new observations will include polarisation measurements. With respect to high energy emission, it is only at optical energies where it is possible to measure, with reasonable accuracy, all electro-magnetic aspects of pulsar radiation. These later observations, combined with detailed numerical models should elucidate, through geometric arguments, the structure of the emission zone for normal pulsars and hence provide stringent observational tests for the various models of pulsar emission. In the future ELTs with adaptive optics will enable pulsar observations down to 32–33 magnitude and importantly will enable spectra-polarimetry down to 29–30 magnitude.

Beyond this time into the ELT era there is the possibility of extending the number of optically observed pulsars (and pulsar types) dramatically. This will require detectors with :

- High-time resolution—both frame based systems from 1 ms+ and pixel read-outs for $\tau < 1$ ms
- Medium size arrays—wide-field operations are not required but nearby reference stars are needed—array sizes $> 32 \times 32$
- Polarisation—sensitive to polarisation changes—both circular and linear at the 1% level
- Energy Resolution—Broad-medium band energy resolution—1–10%

References

1. Hankins, T. H., Kern, J. S., Weatherall, J. C., & Eilek, J. A., 2003, Nature, 422, 141
2. Pacini, F., 1971, ApJ, 163, L17
3. Baade, W. & Zwicky, F., 1934, Proc. Nat. Acad. Sci, 20:5, 259
4. Pacini, F, 1967, Nature, 216, 567
5. Hewish, A., Bell, S. J., Pilkington, J. D., Scott, P. F. & Collins, R. A., 1968, Nature, 217, 709
6. Cocke, W. J., Disney, M. J. & Taylor, D. J., 1969, Nature, 221, 525
7. Wallace, P. T. et al., 1977, Nature, 266, 692
8. Middleditch, J. & Pennypacker, C., 1985, Nature, 313, 659
9. Shearer, A., Redfern, R. M., Gorman, G., Butler, Golden, A., R., O'Kane, P., Golden, A., Beskin, G. M., Neizvestny, S. I., Neustroev, V. V., Plokhotnichenko, V. L. & Cullum, M., 1997, ApJ, 487, L181
10. Shearer, A., Harfst, S., Redfern, R. M., Butler, R., O'Kane, P., Beskin, G. M., Neizvestny, S. I., Neustroev, V. V., Plokhotnichenko, V. L. & Cullum, M., 1998, A & A, 335, L21
11. Martin, C, Halpern, J.P. & Schiminovich, D., 1998, ApJ, 494, L211

12. Mignani, R. P., 2005, astro-ph/0502160
13. Lorimer, D. & Kramer, M., 2005, Handbook of Pulsar Astronomy, Cambridge University Press, ISBN 0 521 82823 6
14. Shearer, A. & Golden, G., 2001, ApJ, 547, 967
15. O Connor, P Golden, A & Shearer, A, 2005, ApJ, 631, 471
16. Shibanov, Y. A, Zharikov, S. V., Komarova, V. N., Kawai, N. , Urata, Y., A. B. Koptsevich, A. B., Sokolov, V. V., Shibata, S. & N. Shibazaki, N, 2006, A&A, 448, 313
17. Smith, F. G., Jones, D. H. P., Dick, J. S. B. & Pike, C. D., 1988, MNRAS, 233, 305
18. Romani, R. W., Miller, A. J., Cabrera, B., Nam, S. W. & Martinis, J. M., 2001, ApJ, 563, 221
19. Kern, B., Martin, C., Mazin, B. & Halpern, J. P., 2003, ApJ, 597, 1049
20. Wagner, Stefan J. & Seifert, W., 2000, ASPC, 202, 315
21. Lyutikov, M., Blandford, R., & Machabeli, G., 1999, MNRAS, 305, 338
22. Manchester, R. N., Hobbs, G. B., Teoh, A. & Hobbs, M., 2005, AJ, 129, 1993
23. Golden, A., Shearer, A., Redfern, R. M., Beskin, G. M., Neizvestny, S. I., Neustroev, V. V., Plokhotnichenko, V. L. & Cullum, M., 2000, A&A, 363, 617
24. Temim, T., Gehrz, R. D., Woodward, C. E., Roellig, T. L., Smith, N., Rudnick, L., Polomski, E. F., Davidson, K., Yuen, L. & Onaka, T., 2006, AJ, 132, 1610
25. Neugebauer, G., Becklin, E. E., Kristian, J., Leighton, R. B., Snellen, G. & Westphal, J. A., 1969, ApJ, 156, 133
26. Kristian, J., Visvanathan, N., Westphal, J. A. & Snellen, G. H., 1970, ApJ, 162, 475
27. Nasuti, F. P., Mignani, R., Caraveo, P. A. & Bignami, G. F., 1996, A&A, 314, 849
28. Fordham, J. L. A., Vranesevic, N., Carramiñana, A., Michel, R., Much, R., Wehinger, P. & Wyckoff, S., 2002, ApJ, 581, 485
29. Eikenberry, S. S., Fazio, G. G., Ransom, S. M., Middleditch, J., Kristian, J., & Pennypacker, C. R. 1997, ApJ, 477, 465
30. Shearer, A., Stappers, B., O'Connor, P., Golden, A., Strom, R., Redfern, M., and Ryan, O. , 2003, Science, 301, 493
31. Staelin, D. H. & Reifenstein, E. C. III, 1968, Science, 162, 1481
32. Lundgren, S. C., Cordes, J. M., Ulmer, M., Matz, S. M., Lomatch, S., Foster, R. S. & Hankins, T., 1995, ApJ, 453, 433
33. Romani, R., & Johnston, S., 2001, ApJ, 557, L93
34. Cusumano, G., Hermsen, W., Kramer, M., Kuiper, L., Lšhmer, O., Massaro, E., Mineo, T., Nicastro, L., & Stappers, B. W., 2003, A&A, 410, L9
35. Woods, P. & Thompson, C., 2004, astro-ph/0406133
36. Hulleman, F., van Kerkwijk, M. H. & Kulkarni, S. R., 2000, Nature, 408, 689
37. Kern, B. & Martin, C., 2002, Nature, 417, 527
38. Dhillon, V. S., Marsh, T. R., Hulleman, F., van Kerkwijk, M. H., Shearer, A., Littlefair, S. P., Gavriil, F. P. & Kaspi, V. M., 2005, MNRAS, 363, 609
39. McLaughlin M. A., Lyne A. G., Lorimer D. R., Kramer M., Faulkner A. J., Manchester R. N., Cordes J. M., Camilo F., Possenti A., Stairs I. H., Hobbs G., DÕAmico N., Burgay M., OBrien J. T., 2006, Nature, 439, 817
40. Dhillon, V. S., Marsh, T. R.& Littlefair, S. P., 2006, MNRAS, 372, 209
41. Weltevrede, P., Stappers, B. W., Rankin, J. M. & Wright, G. A. E., 2006, ApJ, 645, L149
42. Perryman, M. A. C., Favata, F., Peacock, A., Rando, N. & Taylor, B. G., 1999, A&A, 346, L30
43. Romani, R. W., Burney, J., Brink, P., Cabrera, B., Castle, P., Kenny, T., Wang, E., Young, B., Miller, A. J. & Nam, S. W., 2003, ASP Conference Proceedings, 291, 399
44. Johnston, S & Romani, R. W., 2003, ApJ, 590, L95
45. Johnston, S & Romani, R. W., 2003, ApJ, 557, L93
46. Soglasnov, V. A., Popov, M. V., Bartel, N., Cannon, W., Novikov, A. Yu., Kondratiev, V. I. & Altunin, V. I., 2004, ApJ, 616, 439
47. Becker, W. & Trumper, J., 1997, A&A, 326, 682
48. Gaensler, B. M., van der Swaluw, E., Camilo, F., Kaspi, V. M., Baganoff, F. K., Yusef-Zadeh, F. & Manchester, R. N., 2004, ApJ, 616, 383
49. Predehl, P. & Schmitt, J. H. M. M., 1995, A & A, 293, 889
50. Helfand, D. J., Becker, R. H., & White, R. L. 1995, ApJ, 453, 741

51. Torii, K., Slane, P. O., Kinigasa, K., Hashimotodani, K., & Tsunemi, H., 2000, PASJ, 52, 875
52. Kramer, M., et al., 2003, MNRAS, 342, 1299
53. Marshall, F. E., Gotthelf, E. V., Zhang, W., Middleditch, J. & Wang, Q. D, 1998, Apj, 499, 179
54. Townsley, L. K., Broos, P. S., Feigelson, E. D., Brandl, B. R.; Chu, Y-H, Garmire, G. P. & Pavlov, G. G., 2006, AJ, 131, 2140
55. Camilo, F. et al, 2004, ApJ, 611, L25
56. Roberts, M. S. E., Romani, R. W. & Johnston, S., 2001, ApJ, 561, 187
57. Halpern, J. P., Camilo, F., Gotthelf, E. V., Helfand, D. J., Kramer, M., Lyne, A. G., Leighly, K. M. & Eracleous, M., 2001, ApJ, 552, L125
58. McLaughlin, M. A., Arzoumanian, Z., Cordes, J. M., Backer, D. C., Lommen, A. N., Lorimer, D. R. & Zepka, A. F., 2002, ApJ, 564, 333
59. Zavlin, V. & Pavlov, G., 2004, A&A, 616, 452
60. Becker, W., Weisskopf, M. C., Tennant, Allyn F., Jessner, A., Dyks, J., Harding, A. K., Zhang, S. N., 2004, ApJ, 615, 908
61. Lu, F. J., Wang, Q. D., Aschenbach, B., Durouchoux, P. & Song, L. M., 2002, ApJ, 568, L49
62. Hughes, J. P., Slane, P. O., Park, S., Roming, P. W. A. & Burrows, D. N., 2003, ApJ, 591, 139
63. Camilo, F., Manchester, R. N, Gaensler, B. M. & Lorimer, D. R., 2002, ApJ, 579, L25
64. Torii, K., Kinigasa, K., Toneri, T., Asanuma, T., Tsunemi, H., Dotani, T., Mitsuda, K., Gotthelf, E. V. & Petre, R., 1998, ApJ, 494, 207
65. Gotthelf, E. V., Petre, R., & Hwang, U., 1997, ApJ, 487, L175
66. Pivovaroff, M. J., Kaspi, V. M. & Gotthelf, E. V., 2000, ApJ, 528, 436
67. Becker, W., Kramer, M., Jessner, A., Taam, R. E., Jia, J. J., Cheng, K. S., Mignani, R., Pellizzoni, A., de Luca, A., lowikowska, A. S. & Caraveo, P.A., 2006, ApJ, 645, 1421
68. Serafimovich, N. I., Shibanov, Yu. A.. Lundqvist, P. & Sollerman, J., 2004, A&A, 425 1041
69. Gotthelf, E. V. & Kaspi, V. M., 1998, ApJ, 497, L29
70. Romani, R. W. & Ng, C.-Y., 2003, ApJ, 585, L41
71. Mignani, R., Caraveo, P. A. & Bignami, G. F., 1997, ApJ, 474, L51
72. Torii, K., Saito, Y., Nagase, F., Yamagami, T., Kamae, T., Hirayama, M., Kawai, N., Sakurai, I., Namiki, M., Shibata, S., Gunji, S. & Finley, J. P., 2000, ApJ, 551, L151
73. Becker, W., Trümper, J., Lommen, A. N. & Backer, D. C., 2000, ApJ, 561, 308
74. Sutaria, F. K., Ray, A., Reisenegger, A., Hertling, G., Quintana, H. & Minniti, D., 2003, A&A, 406, 245
75. Hobbs, G., Faulkner, A., Stairs, I. H., Camilo, F., Manchester, R. N., Lyne, A. G., Kramer, M., D'Amico, N., Kaspi, V. M., Possenti, A., McLaughlin, M. A., Lorimer, D. R., Burgay, M., Joshi, B. C.& Crawford, F., 2004, MNRAS, 352, 1439
76. Robinson, C. R., Lyne, A. G., Manchester, A. G., Bailes, M., D'Amico, N., & Johnston, S. 1995, MNRAS, 274, 547
77. Lorimer, D. R., Nicastro, L., Lyne, A. G., Bailes, M., Manchester, R. N., Johnston, S., Bell, J. F., D'Amico, N. & Harrison, P. A., 1995, ApJ, 439, 933
78. Bailes, M., Johnston, S., Bell, J. F., Lorimer, D. R., Stappers, B. W., Manchester, R. N., Lyne, A. G., Nicastro, L., D'Amico, N. & Gaensler, B. M., 1997, ApJ, 481, 386
79. D'Amico, N., Lyne, A. G., Manchester, R. N., Possenti, A., & Camilo, F. 2001, ApJ, 548, L171
80. Champion, D. J., Lorimer, D. R., McLaughlin, M. A., Xilouris, K. M., Arzoumanian, Z., Freire, P. C. C., Lommen, A. N., Cordes, J. M. & Camilo, F., 2005, MNRAS, 363, 929
81. Edwards, R. T. & Bailes, M., 2001, ApJ, 553, 801
82. A. J. Faulkner, I. H. Stairs, M. Kramer, A. G. Lyne, G. Hobbs, A. Possenti, D. R. Lorimer, R. N. Manchester, M. A. McLaughlin, N. D'Amico, F. Camilo & M. Burgay, 2004, MNRAS, 355, 147
83. Kargaltsev, O., Pavlov, G. & Romani, R. W., 2004, ApJ, 602, 372
84. Camilo, F.,Ransom, S. M.,Gaensler, B. M., Slane, P. O., Lorimer, D. R., Reynolds, J., Manchester, R. N. & Murray, S. S, 2006, ApJ, 637, 456
85. Safi-Harb, S., Harrus, I. M., Petre, R., Pavlov, G. G., Koptsevich, A. B. & Sanwal, D. 2001,
86. Wang, Z., Chakrabarty, D. & Kaplan, D. L., 2006, Nature, 440, 772

High Time Resolution Observations of Cataclysmic Variables

S. P. Littlefair

Abstract Cataclysmic Variables (CVs) show a wide range of time-dependent phenomena which make them ideal objects of study with high time-resolution instruments. The most rapid timescales observed in CVs are of the order of a few seconds, and thus CVs are relatively slow objects compared to some phenomena studied by the high time-resolution community. However, the rich and varied processes which occur in CVs make high time-resolution observations powerful tests of cutting-edge astrophysics. Here we present a review of high time-resolution observations of CVs to date, and underline their relevance to modern astrophysics.

1 Introduction

The term cataclysmic variable (CV) covers a range of systems which are far from homogeneous. CVs are defined as semi-detached binary systems in which a white dwarf primary star (mass M_1) accretes material via Roche Lobe overflow from a secondary star (mass M_2), which is typically a red dwarf star. They derive their name from their violent but non-destructive outbursts, which can occur with a wide range of amplitudes and frequencies. Indeed, no cataclysm need be observed for a system to be categorised as a CV, provided it meets the definition above (e.g the nova-like variables). Most CVs accrete matter onto the primary via an accretion disc. Where the gas stream from the secondary impacts upon the disc, a shock-heated area may arise—producing an area of enhanced emission known as the *bright spot*. The *boundary layer* is the region over which gas moving at Keplerian velocities in the accretion disc is decelerated to match the surface velocity of the primary star. A schematic of such a disc CV is shown in Fig. 1. In some CVs the white dwarf's magnetic field is sufficient to disrupt the accretion flow. In these cases the white dwarf accretes material which threads down the field lines, either directly from the gas stream or via a truncated disc. These CVs are known as Polars and Intermediate Polars respectively. For an excellent and thorough review of CVs, see [36].

S. P. Littlefair
Dept of Physics and Astronomy, University of Sheffield, S3 7RH, UK
e-mail: sl@sheffield.ac.uk

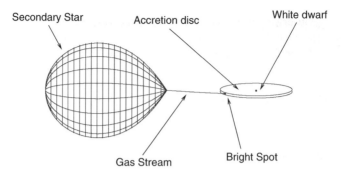

Fig. 1 A schematic of a disc CV

Cataclysmic Variables have been very popular objects of study for more than 30 years now. Their enduring popularity lies in the fact that CVs are excellent laboratories in which to study astrophysical processes which are ubiquitous amongst more "fashionable" objects of study. Below we outline two important science areas in which CVs can make a contribution, and in particular we highlight the potential uses of high time-resolution (HTR) observations.

- **Binary Star Evolution**
  Binary star evolution theory invokes, and can thus test, core astrophysics including: stellar structure and evolutionary models; angular momentum loss from gravitational radiation, magnetic wind braking and circumbinary discs ; and common envelope evolution. The parameters of binary evolution models have been tuned to reproduce the orbital period distribution of CVs, but make predictions of the secondary star mass and radius as a function of orbital period. Photometry of eclipsing CVs with ∼second time resolution allows determination of the mass and radius of the secondary star [40] (see also Section 5), and thus provides an independent test of the evolutionary models. Crucially, this technique allows measurement of the secondary star mass even when it is not directly detectable, allowing study of the low-mass secondary stars in short period systems. Such studies can therefore also contribute to our knowledge of the structure of low-mass objects like brown dwarfs and giant planets.
- **Magneto-Hydrodynamics (MHD)**
  A crucial property of CVs is the ability to resolve the accretion flow spatially [16], and in velocity space [23] with micro-arcsecond precision. Coupled with the fact that the accretion flow in CVs varies on observable timescales, this means that CVs are an excellent environment in which to study MHD phenomena. Photometry and spectroscopy of CVs with time resolutions of seconds could determine the dynamics of gas stream/magnetic field interactions, measure the precession and shape of warped accretion discs and determine the location and energetics of the flickering process. This ability means that HTR observations of CVs can provide benchmarks for top-end MHD codes.

This paper aims to review the current state of HTR observations of CVs, and to outline possible future research. In Section 2 we present a brief history of early high time-resolved observations of CVs. We then go on to discuss the various phenomena and resulting science that is accessed by high time-resolution observation of CVs in Sections 3, 4, 5 and 6. In Section 5 we draw our conclusions.

2 Early Time-Resolved Observations of CVs

The pioneering work on HTR observations of CVs occurred in the late 1940s, with the introduction of the 1P21 photomultiplier [21, 22]. These observations discovered the intrisic and stochastic variability known as flickering, which was later recognised to be characteristic of CVs. Another important discovery from this time was the observation of periodic variability with a period of 71.1 secs in DQ Her [35]. The application of pulse counting photometry in the early 1970s [26] led to many advances, notably the development of the canonical model of CVs [38], and the discovery of oscillations in dwarf novae in outburst [39]. In terms of time resolution, these early observations achieved resolutions of 2–3 seconds. The main limitation of these observations was not speed, but photometric sensitivity and accuracy. This limitation remained in place until the development of the frame-transfer CCD cameras [27, 9], which provided the required time resolution alongside the high quantum efficiency and imaging capabilities of CCDs.

3 Flickering and Oscillations

3.1 DNOs

Dwarf Nova oscillations (DNOs) are rapid (8–40 s) modulations with rich and varied phenomenology. They are seen in the light of cataclysmic variables with high mass transfer rates (dwarf novae in outburst and nova-like variables). Typically they are low-Q modulations—their periods can change by seconds on timescales of hours, or be stable to milliseconds for many hours, giving $10^3 < Q < 10^7$. They show a monotonic relationship between the period of modulation and the EUV flux, which is a good measure of accretion rate in these systems [37]. Furthermore, there is reason to believe that DNOs are the same phenomenon as the kiloherz QPOs of low-mass X-ray binaries, allowing one to study that phenomenon without the use of expensive X-ray telescope time [25]. However, there is still no accepted explanation for how DNOs are excited or sustained for long periods.

Whilst the DNOs still remain a puzzling phenomena, interest in them has been revitalised by the new detector technology (see [37] for a review of recent observations), and there are prospects for a deeper understanding of DNOs from HTR spectroscopy [31]. An example of a DNO observed with ULTRACAM is presented in Fig. 2.

Quite separate to the DNOs observed in dwarf novae are the QPOs which are seen with periods from 1–3 s in the magnetic catalcysmic variables known as

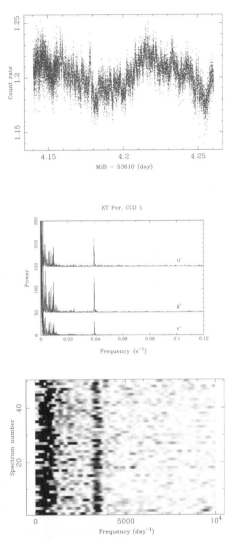

Fig. 2 ULTRACAM observations of DNOs in KT Per (taken from Marsh et al *in prep*). Top: ULTRACAM lightcurve of KT Per taken a few days after outburst. Lots of low frequency noise is visible due to flickering. Middle: The Lomb-Scargle periodogram of the lightcurve. The DNO is clearly visible. Bottom: a trail of Lomb-Scargle periodograms of the above data. The DNO makes small shifts in period, and at times disappears completely

Polars. These QPOS are believed to arise either from inhomogeneities (blobs) in the accretion flow, or from standing Alfvén waves in the magnetic flux tubes which thread the accretion stream. There are few recent optical observations of this phenomenon, but photomultiplier tube observations on the 3.6m telescope at ESO show that the QPOs are associated with the cyclotron emitting region [19].

3.2 Flickering

One of the long-standing unsolved problems in accretion physics is related to the cause of flickering. Flickering is seen in many accreting systems from T Tauri stars [15], close binaries [2] and active galactic nuclei [12]. CVs offer two ways of discovering the physical mechanisms behind flickering. Eclipse mapping (see Section 4) can be used to determine the location of the flickering sources, whilst the shape of the power spectrum can constrain the size of the flickering regions.

The eclipse mapping method has been applied to several systems, e.g. [8], with the result that the location of flickering varies from system to system, but is usually confined to either the bright spot or inner disc/white dwarf. The application of this method with frame transfer CCDs [4] has allowed better spatial resolution, and the study of higher frequency flickering. Whilst the low-frequency flickering in V2051 Oph arises from the gas stream, the high-frequency flickering source is located in the entire accretion disc, and may be caused by fluctuations in MHD turbulence [13].

Analysis of the power spectrum of flickering has been more challenging. Figure 3 highlights the problem. This figure shows the $u'g'r'$ lightcurves and power spectra of SS Cyg, obtained with ULTRACAM+WHT. The high frequency break is not a real feature of the power spectrum, but the point at which the lightcurves become dominated by scintillation noise. Clearly, observations with larger telescopes, or sophisticated pupil imaging techniques [10] will be required to reduce the scintillation noise, and thus find high-frequency breaks in the flickering of CVs.

4 Eclipse Mapping

Eclipse mapping [16] is an inversion technique: it makes use of the information contained within the lightcurves of eclipsing cataclysmic variables to deduce the structure of the accretion disc. The three basic assumptions of the method are that the secondary star fills its Roche Lobe, that the brightness distribution is constrained to the orbital plane, and that the emitted radiation is independent of the orbital phase. Whilst the first assumption is robust, the other two will not hold in general. For a discussion of the consequences of this, and a review of eclipse mapping in general, see [3]. Briefly, the *eclipse map* is a grid of intensities centred on the white dwarf, which is iteratively adjusted to find the brightness distribution which fits the eclipse light curve. Because the lightcurve alone is not sufficient to fully constrain a two-dimensional map it is usual to find the eclipse map by maximising the image entropy with respect to a smooth default map, subject to a constraint on the quality of the fit, usually estimated by χ^2. Eclipse mapping requires high quality lightcurves of the eclipse itself. A typical eclipse lasts of the order of twenty minutes, so time resolutions of a few seconds are ample.

Early eclipse mapping results showed that the radial temperature profiles in outbursting dwarf novae [17], and novalike variables [28] closely follow the $T \propto R^{-3/4}$ law expected for a steady state disc, whilst the temperature profile is essentially flat in quiescent dwarf novae [42]. This important observational result lent support to the

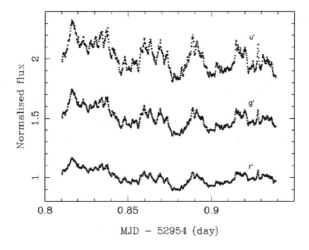

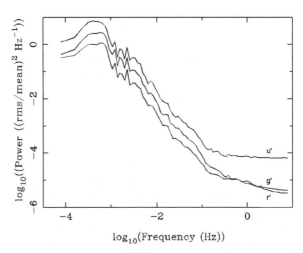

Fig. 3 Top: Normalised ULTRACAM + WHT $u'g'r'$ lightcurves of SS Cyg (taken from Marsh et al *in prep*). Flickering is clearly visible, with relative amplitude strongly increasing towards the blue. The lightcurves have been offset for clarity. Bottom: $u'g'r'$ power spectra, computed from the lightcurves above. The break in the power spectra around 0.2 Hz is due to the lightcurves becoming dominated by scintillation noise at higher frequencies

disc instability model of dwarf nova outbursts [20]. As might be expected, eclipse mapping has also proved useful in spatial studies of accretion discs. Tidally induced spiral shocks are expected to appear in dwarf novae discs during outbursts, and may help with angular momentum transport within the disc [24]. Comparing eclipse [6] and doppler [30] maps of the spiral shocks in IP Peg show that the gas in the shocks has sub-Keplerian velocities.

4.1 Accretion Disc "Movies"

By calculating eclipse maps as a function of time, eclipse maps have been used to assess changes in disc structure, for example to follow the evolution of surface brightness distributions during a dwarf nova outburst cycle. The difficulties in obtaining and scheduling large numbers of time-resolved observations has meant that this technique has been limited to obtaining a few "frames" of the movie, with large amounts of time elapsing between each frame. The coupling of modern queue-scheduled observing with a facility instrument capable of high-time resolution observations should overcome this limitation.

Perhaps the best examples to date are the observations of OY Car [28], and EX Dra [5]. In OY Car, eclipse maps show that the outburst starts in the outer disc regions with the development of a bright ring, while the inner disc regions remain at constant brightness during the rise. The flat radial temperature profile of quiescence and early rise changes, within one day, into a steep distribution that matches a steady state disc model at outburst maximum. Their results suggest that an uneclipsed component develops during the rise and contributes up to 15 per cent of the total light at outburst maximum. This may indicate the development of a vertically-extended disc wind, or a flared disc. For EX Dra, eclipse maps covering the full outburst cycle (see Fig. 4) reveal how the disc expands during the rise until it fills most of the primary Roche lobe at maximum light. During the decline phase, the disc becomes progressively fainter until only a small bright region around the white dwarf is left

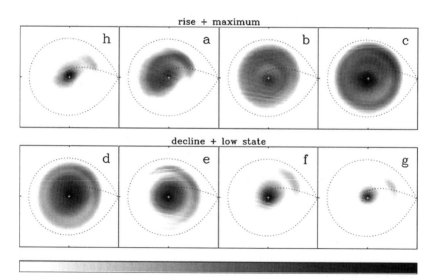

Fig. 4 Sequence of eclipse maps of EX Dra through the outburst cycle in a logarithmic grey-scale. Brighter regions are indicated in black; fainter regions in white. A cross marks the centre of the disc; dotted lines show the Roche lobe and the gas stream trajectory; the secondary is to the right of each map and the stars rotate counterclockwise. The grey-scale bar corresponds to a linear scale in log of intensity from -5.8 to -2.7. The labels correspond to disc states; (a,b) on the rise, (c) at maximum, (d–f) during the decline, (g) in the low state, and (h) in quiescence. Taken from [5]

at minimum light. The evolution of the radial brightness distribution suggests the presence of an inward and an outward-moving heating wave during the rise and an inward-moving cooling wave in the decline.

4.2 Spectral Studies

Eclipse mapping is not limited to single-channel photometric data. Time-resolved spectroscopy can be used to produce spectra of the accretion disc, spatially resolved on micro-arcsecond scales. The application of this technique has been limited so

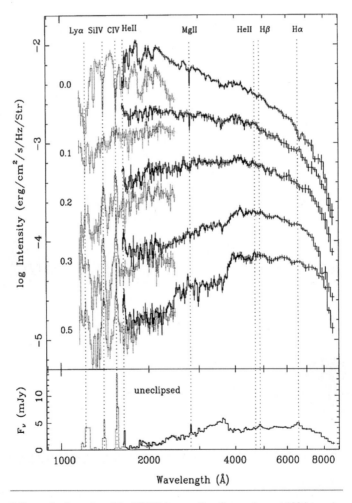

Fig. 5 Spatially resolved spectra of the UX UMa accretion disc on August 1994 (gray) and November 1994 (black). The spectra were computed for a set of concentric annular rings (mean radius shown on the left in units of the distance from disc centre to the inner lagrangian point). The time-resolved spectroscopy was obtained using HST/FOS. Taken from [7]

far, in large part because of the prohibitive dead-time of spectroscopic CCD observations. The advent of fast, L3CCD spectrographs on large telescopes should see a resurgence of interest in this technique. The detailed, spatially resolved accretion disc spectra which result will allow powerful tests of accretion disc models, and may allow us to determine the vertical structure of accretion discs with some accuracy. The current state of the art is shown in Fig. 5. The spectral mapping shows that steady-state novalike discs are hot and optically thick in their inner part and cool and optically thin in their outer part, whilst the uneclipsed component shows emission lines and optically thin continuum, presumably from a disc chromosphere and wind.

5 Evolution: Mass Determinations

High time resolution photometry of eclipsing catalcysmic variables has an important role to play in studies of binary evolution. The parameters of binary star evolution models are tuned to reproduce the observed orbital period distribution of CVs, but make predictions of the donor star mass as a function of orbital period. The donor star mass can thus be used as an independent test of the evolutionary models (see Fig. 6). As such determinations of masses in CV donor stars can test important physics such as stellar structure, magnetic braking and gravitational radiation. Unfortunately accurate masses are hard to derive, and to make matters worse the donor star is often not visible above the contaminating background from the white dwarf and accretion disc. This situation is most marked at short orbital periods, where the very low-mass donor is hardly ever visible. As a result the evolutionary models are practically unconstrained by current measured donor star masses in CVs. In particular the existence of *post period-minimum* systems is still unconfirmed. As the donor star transfers mass to the white dwarf, the orbital period of the system drops until, at a certain point, the donor's mass drops below the hydrogen burning limit. Changes in the donor's mass-radius relationship at this point mean that further mass loss results in an increase in the orbital period of the system [18]. Thus, cataclysmic variables are expected to show a minimum orbital period, with post-period minimum CVs possessing brown dwarf donor stars. Population models predict that around 70% of the current CV population should have evolved past the orbital period minimum. However, to date there are no donor star masses measured below the hydrogen burning limit.

High time resolution photometry of eclipsing CVs allows these problems to be overcome, because the shape of the eclipse contains the information necessary to deduce the radius of the white dwarf, and the mass ratio of the system (assuming the bright spot lies on the ballistic trajectory of the gas stream). The donor star mass can then be deduced, assuming the white dwarf follows a theoretical mass-radius relationship [41]. To apply the technique it is necessary to obtain simultaneous multicolour photometry with time resolutions of a few seconds, which allows the white dwarf eclipse to be well-resolved yielding a radius and approximate temperature for the white dwarf. The design of ULTRACAM has allowed this technique to be applied to faint CVs, allowing a meaningful sample of CVs to be observed. Importantly, it is not necessary to be able to detect the secondary star in order to deduce the

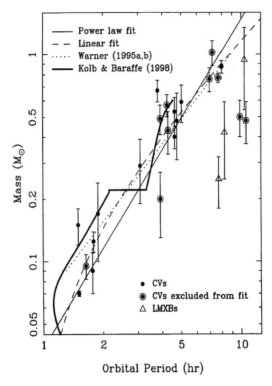

Fig. 6 The masses and radii of the secondary stars in CVs versus their orbital periods. Power law and linear fits are shown, whilst the dotted line is a semi-empirical fit [36]. The thick solid line is a current evolutionary model, accounting for the effects of mass loss from the donor star and angular momentum loss from the system [18]. The ringed points have been omitted from the fits. Taken from [29]

component masses. Applying this method can therefore give donor masses accurate to a few percent across the *entire* orbital period distribution.

An example of the application of this technique to ULTRACAM data is given in Fig. 7. In this case the derived donor mass is 0.0842 ± 0.0024 $M_\odot$, placing the donor close to the upper limit on the mass of a brown dwarf. This system is thus a very old cataclysmic variable which has almost reached the period minimum. The ULTRACAM team is currently undertaking a survey of all known eclipsing short period CVs, which should finally yield confirmation of the existence of post period-minimum systems, and will provide important tests of stellar models for low-mass objects like brown dwarfs and giant planets.

6 Asteroseismology

The temperatures of the white dwarfs in CVs are much hotter than expected from the age of the binary. The high effective temperatures are a result of compressional heating from the ongoing accretion [34]. For the shortest period CVs, the

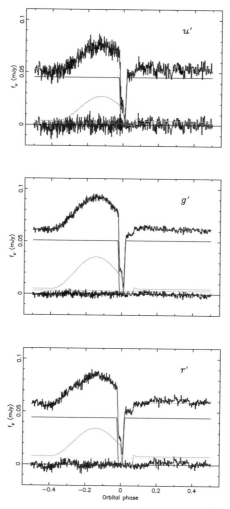

Fig. 7 The phased folded $u'g'r'$ (from top to bottom) light curves of XZ Eri, fitted separately using the model described in ref [11]. The data are shown with the fit (solid line) overlaid and the residuals plotted below. Below are the separate light curves of the white dwarf (thick dashed line), bright spot (dotted line), accretion disc (dash-dotted line) and the secondary star (light dashed line). Taken from [11]

accretion rate (and thus the heating) is sufficiently low that the WD cools into a region unstable to pulsation [1]. These pulsations are exciting because they offer the possibility of determining masses, not just for the white dwarf itself, but also for the accreted envelope (e.g. [33]). There are possible implications for the progenitors of Type Ia supernovae; the progenitors could be closely linked to the CV population, but the exact relationship is unclear. Determining how the accreted envelope mass changes with time would go a long way towards resolving this uncertainty. There are also important implications for how much white dwarf material enriches the interstellar medium in classical nova outbursts [14]. Comparing white dwarf

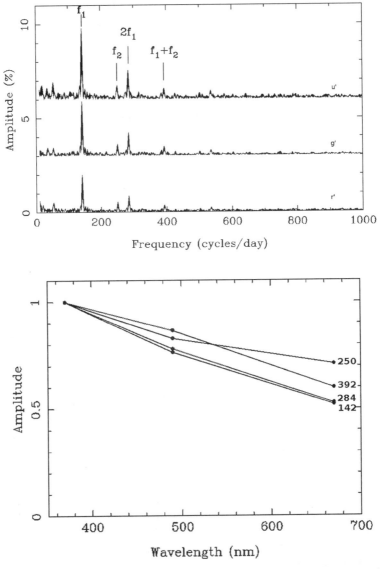

Fig. 8 ULTRACAM and VLT power spectra (top) and mode amplitude (bottom) diagrams of SDSS 1610-0102. The power spectra show the two previously known modes, f_1 and f_2 [43] as well as signs of up to 15 possible new modes of pulsation. The mode amplitude diagram shows that the $f_1 = 142$ cycles/day and $2f_1 = 284$ cycles/day modes appear bluer than the $f_2 = 250$ cycles/day and $f_1 + f_2 = 392$ cycles/day modes. Taken from Marsh et. al. (in prep)

masses from seismology with those found from eclipse analysis can also allow us to determine if the white dwarfs in CVs follow theoretical mass radius relationships.

Despite the promise held by asteroseismology of the white dwarfs in CVs progress has so far been slow. To date there are 10 pulsating white dwarfs in CVs

[1], but seismology has only been performed on one, GW Lib [33], and that analysis relied on parameters for the binary derived via other methods. The reason for this limited success has been the problem of mode identification, which is hampered by the small number of known frequencies. It is possible that multi-colour photometry could provide important information on the modes of known frequencies. Figure 8 shows the results of observing the known pulsator SDSS1602-0102 with ULTRACAM and VLT. The power spectra show the advantage that modern instrumentation on large telescopes can give, with the identification of up to 15 new frequencies. In addition, the colours of the two previously known frequencies appear to be different, providing possible evidence for different l values for these modes. However, the fact that the spectra of two modes cross in the bottom panel of Fig. 8 shows that the errors are comparable with the separations of these curves, so this early result should be treated with caution.

Time-resolved *spectroscopy* of pulsating white dwarfs [32] is potentially a very powerful method of mode indentification in pulsating white dwarfs. However, obtaining time-resolved spectroscopy with high signal/noise ratios on $g' \sim 19$ objects will require the combination of efficient L3CCD spectrographs giving low read noise and minimal dead time, with the largest telescopes. Finally, we note here that the white dwarfs in CVs are likely to be rotating rapidly, and that seismology of rotating stars is an area of active theoretical interest. Hence, accreting white dwarfs may provide the proving ground for asteroseismological techniques for rapidly rotating stars.

7 Conclusions

Compared to some phenomena studied by the HTR community, CVs are slowly varying objects. The dynamical timescale of white dwarfs are of the order of a few seconds. The *most* rapid observable phenomena in CVs are likely to be eclipses of substructure on the white dwarf surface. Thus, observing CVs with time-resolution of greater than ~ 0.1 seconds is not justified. However, CVs do present an environment in which to test cutting-edge astrophysics. The combination of high speed observations with tomographic techniques allow accretion disc spectra to be obtained with spatial resolutions of micro-arcseconds. With appropriate modelling, such data represent the best currently available measure of the vertical structure of accretion discs. Eclipse mapping can measure the shape and evolution of accretion discs— results that can be directly compared with the results of cutting edge MHD models. The donor mass-period relation can be determined through high speed photometry of eclipsing CVs, and tests theories of angular momentum loss and the structure of low-mass objects. Finally the results from asteroseismology of the white dwarfs in CVs have implications for the progenitors of type-Ia supernova and the enrichment of the interstellar medium, as well as providing the best test of the mass-radius relationship for accreting white dwarfs. It is the author's hope that continued support for high time-resolution instrumentation on the largest telescopes allows this important work to continue.

References

1. P. Arras, D. M. Townsley, and L. Bildsten. Pulsational Instabilities in Accreting White Dwarfs. ApJ, 643:L119–L122, June 2006.
2. T. Augusteijn, K. Karatasos, M. Papadakis, G. Paterakis, S. Kikuchi, N. Brosch, E. Leibowitz, P. Hertz, K. Mitsuda, T. Dotani, W. H. G. Lewin, M. van del Klis, and J. van Paradijs. Co-ordinated X-ray and optical observations of Scorpius X-1. A&A, 265:177–182, November 1992.
3. R. Baptista. What can we learn from accretion disc eclipse mapping experiments? *Astronomische Nachrichten*, 325:181–184, March 2004.
4. R. Baptista and A. Bortoletto. Eclipse Mapping of the Flickering Sources in the Dwarf Nova V2051 Ophiuchi. AJ, 128:411–425, July 2004.
5. R. Baptista and M. S. Catalán. Changes in the structure of the accretion disc of EX Draconis through the outburst cycle. MNRAS, 324:599–611, June 2001.
6. R. Baptista, E. T. Harlaftis, and D. Steeghs. Eclipse maps of spiral shocks in the accretion disc of IP Pegasi in outburst. MNRAS, 314:727–732, June 2000.
7. R. Baptista, K. Horne, R. A. Wade, I. Hubeny, K. S. Long, and R. G. M. Rutten. HST spatially resolved spectra of the accretion disc and gas stream of the nova-like variable UX Ursae Majoris. MNRAS, 298:1079–1091, August 1998.
8. A. Bruch. Studies of the flickering in cataclysmic variables. VI. The location of the flickering light source in HT Cassiopeiae, V2051 Ophiuchi, IP Pegasi and UX Ursae Majoris. A&A, 359:998–1010, July 2000.
9. V. Dhillon and T. Marsh. ULTRACAM - studying astrophysics on the fastest timescales. *New Astronomy Review*, 45:91–95, January 2001.
10. D. Dravins, L. Lindegren, E. Mezey, and A. T. Young. Atmospheric Intensity Scintillation of Stars. III. Effects for Different Telescope Apertures. PASP, 110:610–633, May 1998.
11. W. J. Feline, V. S. Dhillon, T. R. Marsh, and C. S. Brinkworth. ULTRACAM photometry of the eclipsing cataclysmic variables XZ Eri and DV UMa. *MNRAS*, 355:1–10, November 2004.
12. A. Garcia, L. Sodré, F. J. Jablonski, and R. J. Terlevich. Optical monitoring of quasars - I. Variability. MNRAS, 309:803–816, November 1999.
13. G. T. Geertsema and A. Achterberg. Turbulence in differentially rotating thin disks - A multicomponent cascade model. A&A, 255:427–442, February 1992.
14. R. D. Gehrz, J. W. Truran, R. E. Williams, and S. Starrfield. Nucleosynthesis in Classical Novae and Its Contribution to the Interstellar Medium. PASP, 110:3–26, January 1998.
15. W. Herbst and V. S. Shevchenko. A Photometric Catalog of Herbig AE/BE Stars and Discussion of the Nature and Cause of the Variations of UX Orionis Stars. AJ, 118:1043–1060, August 1999.
16. K. Horne. Images of accretion discs. I - The eclipse mapping method. MNRAS, 213:129–141, March 1985.
17. K. Horne and M. C. Cook. UBV images of the Z Cha accretion disc in outburst. MNRAS, 214:307–317, May 1985.
18. U. Kolb and I. Baraffe. Brown dwarfs and the cataclysmic variable period minimum. *MNRAS*, 309:1034–1042, November 1999.
19. S. Larsson. Discovery of 1 second optical QPO in VV Puppis. *Advances in Space Research*, 8:305–308, 1988.
20. J-P. Lasota. *New Astronomy Review*, 45:449, 2001.
21. A. P. Linnell. UX Ursae Majoris. S&T, 8:166–+, May 1949.
22. A. P. Linnell. A Study of UX Ursae Majoris. *Harvard College Observatory Circular*, 455:1–13, 1950.
23. T. R. Marsh and K. Horne. *MNRAS*, 235:269, 1988.
24. T. Matsuda, M. Makita, H. Fujiwara, T. Nagae, K. Haraguchi, E. Hayashi, and H. M. J. Boffin. Numerical Simulation of Accretion Discs in Close Binary Systems and Discovery of Spiral Shocks. Ap&SS, 274:259–273, 2000.

25. C. W. Mauche. Correlation of the Quasi-Periodic Oscillation Frequencies of White Dwarf, Neutron Star, and Black Hole Binaries. ApJ, 580:423–428, November 2002.
26. R. E. Nather and B. Warner. Observations of rapid blue variables. I. Techniques. MNRAS, 152:209, 1971.
27. D. O'Donoghue. High Speed CCD Photometry. *Baltic Astronomy*, 4:519–526, 1995.
28. R. G. M. Rutten, J. van Paradijs, and J. Tinbergen. *A&A*, 260:213, 1992.
29. D. A. Smith and V. S. Dhillon. *MNRAS*, 301:767, December 1998.
30. D. Steeghs, E. T. Harlaftis, and K. Horne. Spiral structure in the accretion disc of the binary IP Pegasi. MNRAS, 290:L28–L32, September 1997.
31. D. Steeghs, K. O'Brien, K. Horne, R. Gomer, and J. B. Oke. Emission-line oscillations in the dwarf nova V2051 Ophiuchi. MNRAS, 323:484–496, May 2001.
32. S. E. Thompson. Mode Identification of DAVs with Time Series Spectroscopy. *Bulletin of the American Astronomical Society*, 37:1158–+, December 2005.
33. D. M. Townsley, P. Arras, and L. Bildsten. Seismology of the Accreting White Dwarf in GW Librae. ApJ, 608:L105–L108, June 2004.
34. D. M. Townsley and L. Bildsten. Measuring White Dwarf Accretion Rates via Their Effective Temperatures. ApJ, 596:L227–L230, October 2003.
35. M. F. Walker. A Photometric Investigation of the Short-Period Eclipsing Binary, Nova DQ Herculis (1934). ApJ, 123:68–+, January 1956.
36. B. Warner. *Cataclysmic Variable Stars*. Cambridge University Press, Cambridge, 1995.
37. B. Warner. Rapid Oscillations in Cataclysmic Variables. PASP, 116:115–132, February 2004.
38. B. Warner and R. E. Nather. Observations of rapid blues variables. II. U Gem. MNRAS, 152:219, 1971.
39. B. Warner and E. L. Robinson. White dwarfs-More rapid variables. Nature, 239:2–7, September 1972.
40. J. H. Wood and C. S. Crawford. *MNRAS*, 222:645, 1986.
41. J. H. Wood, K. Horne, G. Berriman, R. Wade, D. O'Donoghue, and B. Warner. *MNRAS*, 219:629, 1986.
42. J. H. Wood, K. Horne, G. Berriman, and R. A. Wade. *MNRAS*, 341:974, 1989.
43. P. A. Woudt and B. Warner. SDSS J161033.64-010223.3: a second cataclysmic variable with a non-radially pulsating primary. MNRAS, 348:599–602, February 2004.

High-Speed Optical Observations of X-ray Binaries

Tariq Shahbaz

1 Introduction

Low-mass X-ray binaries are interacting binaries in which a low-mass star transfers material to a neutron star or black hole via an accretion disc. The orbital periods range from 42 min for the exotic degenerate pulsar 4U 1627–67, up to 33.5 days for GRS 1915+105. The X-ray emission is dominated by thermal emission from the hot inner accretion disc (there is also a component due to a jet [8]), and the UV/optical emission is thought to be mainly produced by reprocessing of X-rays, implying that some or all of the UV/optical variability is actually reprocessed X-ray variability [39]. The X-ray variability is usually dominated by instabilities in the inner accretion flow. X-rays irradiate the outer regions of the accretion disk and the inner face of the companion star, resulting in reprocessed optical/UV radiation which is an echo of the X-ray variability. From the time delay between the X-ray and the reprocessed optical/UV variability it is possible to infer information about the geometry and scale of the reprocessing region [29]. Multi-wavelength studies can identify causal connections between different energy bands and provide measurements of their relative energy output. Timescales can be identified; e.g. a quasi-periodic oscillation (QPO) or a break in the frequency power density spectrum (PDS) may indicate the characteristic timescale at some transition radius in the disk, and lags between energy bands constrain possible mechanisms for causal links between different regions of the inner and outer accretion flow. The PDS of the X-ray lightcurves observed in the low- and high-states of the X-ray transients (XRTs) and many X-ray binaries often show features such as band-limited noise and quasi-periodic oscillations (QPOs) [40]. Lower frequency timing signatures (breaks and QPOs) are also expected in the quiescent states. From the variability properties of quiescent XRTs one can study low-luminosity accretion flows. In this review I will focus on a few key past and present fast CCD optical observations of X-ray binaries.

Tariq Shahbaz
Instituto de Astrofísica de Canarias, C/ Via Lactea s/n, 38200 La Laguna, Tenerife, Spain
e-mail: tsh@iac.es

2 Early Fast CCD Observations

In the 1990's, probably the most popular fast CCDs with frame transfer were the Stover & Allen CCD [35] and the UCT CCD (University of Cape Town; [5]), which were primarily used to study Cataclysmic Variables. However, it was only in the late 1990's that these instruments were starting to be used to study bright X-ray binaries.

One of the first high speed CCD observations of an X-ray binary was of the black hole XRT V404 Cyg during decline from its outburst in 1989 (see Fig. 1 [9]). Observations with the 1 m Lick telescope and the Stover & Allen CCD in frame transfer mode [35] were obtained with an exposure time of 15 sec. The PDS of the time-resolved lightcurves revealed broad 3–10 min QPO features superimposed on a power-law spectrum. A few years later, CCD images of the accretion powered X-ray pulsar GX 1+4, with exposure times as short as 1 sec were obtained with GPS time stamping [17]. The optical lightcurves clearly showed a 124 sec pulse period, which is the echo of the X-ray pulse period rotation of the neutron star, similar to what is seen in Her X–1 [23] A key observation in the late 1990's was of the X-ray binary pulsar 4U 1626–67. SAAO 1.9 m + UCT CCD white light observations, with exposure times as short as 2 sec were taken simultaneous to *HST* UV observations and *RXTE*. X-ray observations [2]. A 1 mHz QPO was detected in the optical but

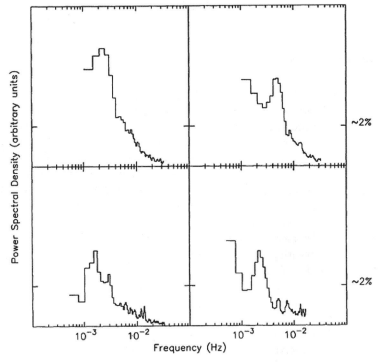

Fig. 1 The power density spectrum of the V404 Cyg optical lightcurve during its outburst in 1989. A 210 sec QPO is clearly seen [9]

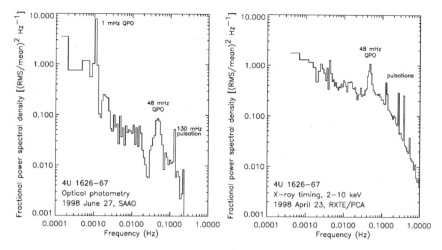

Fig. 2 Left: The power density spectrum of the white lightcurve taken with the UCT CCD. The 1 mHz QPO is visible in the data. Right: The power density spectrum of the X-ray data. The 1 mHz QPO is not detected [2]

not in the X-rays and was also distinct from the 130 mHz pulsar spin frequency and the previously known 48 mHz QPO (see Fig. 2). The 1 mHz QPO is thought to be related to the warping of the inner accretion disc.

3 Optical/X-ray Bursts

Together with simultaneous optical observations, X-ray bursts (thermonuclear explosions on the surface of a neutron star [36]) have been used for echo mapping experiments. Reprocessed optical bursts were first discovered in the 1970's using photomultiplier tubes in 4U 1735–444 and Ser X–1 [10],[19],[11], where the optical burst was found to lag a few seconds behind the X-rays. Given that the optical flux was too high to be due to direct emission from the neutron star surface, it was soon appreciated that the brightening is due to reprocessing of X-ray flux by regions in the accretion disc and/or companion star. Observations of simultaneous optical/X-ray bursts in 4U 1636–536 revealed lags of up to 4 sec [22]. One burst in particular showed an optical burst rise-time longer than the X-ray burst, suggesting significant light-travel time smearing of a few seconds [38].

Since these studies, there have been significant steps forward in the development of X-ray and optical detectors. *RXTE* has made immense contributions in timing studies of X-ray binaries and the replacement of photomultiplier tubes with frame transfer and higher quantum efficiencies has provided significant improvements in the quality and time resolution of the optical data. An excellent example of this is the simultaneous observations of type 1 bursts from GS 1826–24 [18]. SAAO 1.9 m + UCT CCD observations allowed the optical burst to be resolved and despite the source being fainter than 4U 1636–536 at both X-ray and optical wavelengths, a lag of ∼3 sec with respect to the X-ray burst was found. The large dispersion in the lag of ∼3 sec was determined [38],[18], and the rapid optical response showed without a doubt that the accretion disk plays a major part in the reprocessing of X-ray bursts.

Currently, the best dataset of an X-ray burst is the multi-wavelength observations obtained for EXO 0748–676 using *RXTE*, a fast CCD camera attached to the Gemini-South telescope and *HST/STIS* [16]. One burst was recorded simultaneously in X-rays, optical and UV and so provided a unique observation (see Fig. 3). The optical burst was fitted with a model in which the X-ray burst was converted to V and far-UV fluxes assuming that the reprocessor is a single-temperature blackbody. This yielded a mean lag of about 4 sec, a smearing width of 2.5 sec, a rise in temperature during the burst from 18,000 to 35,000 K and an emitting area comparable to that expected for the disk and/or irradiated companion star. However, the fits require a renormalization between the optical and UV bandpasses, suggesting that a single-temperature blackbody model is not sufficient. The X-rays are absorbed at relatively high optical depths, thermalized and re-emitted with a quasi-blackbody spectrum, and the shape of the UV continuum is consistent with blackbody emission.

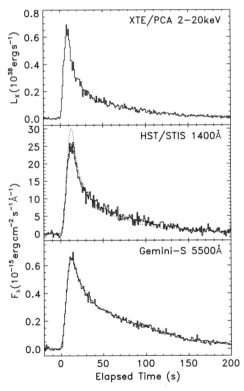

Fig. 3 Simultaneous burst profiles observed in EXO 0748–676. Fits to the UV/optical data are shown, obtained by using the X-ray burst to determine the bolometric reprocessed luminosity and then converting to optical fluxes assuming that the reprocessor is a single blackbody [16]

4 Short-Term Variability in Quiescent XRTs

X-ray transients (XRTs) are a subset of the low-mass X-ray binaries that display episodic, dramatic X-ray and optical outbursts, usually lasting for several months. More than 70 percent of XRTs are thought to contain black holes [3]. The black hole XRT, V404 Cyg, not only exhibits short-term optical variability in outburst, but also in quiescence [28]. The variability superposed on the secondary star's ellipsoidal modulation has been observed since the onset of the quiescent phase [41] but its origin is still unknown. To study whether this feature was peculiar to V404 Cyg alone, or a common feature of all quiescent XRTs, a campaign of fast optical photometry of quiescent XRTs was initiated [43].

Fast optical variations were observed superposed on the secondary star's double-humped ellipsoidal modulation. The variability resembles typical flare activity and has amplitudes ranging from 6 to 60 percent occurring on timescales of minutes to a few hours, with no dependency on orbital phase (see Fig. 4). The observed level of flaring activity seems to be veiled by the light of the companion star, and therefore systems with cool companions (e.g. GRO J0422+32) exhibit stronger variability.

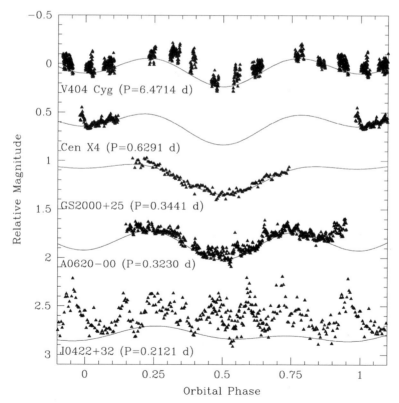

Fig. 4 Optical lightcurves for the five quiescent XRT. Short-term flaring activity is clearly seen and have amplitudes ranging from a few percent to tens of percent [43]

The optical luminosities are too large to be due to chromospheric activity in the rapidly rotating companions. QPOs ranging in the range 30 and 90 min are observed for the short-period XRTs and longer than 1 hr for V404 Cyg. The associated dynamical (Keplerian) timescales suggest that flares are produced at regions that lie between 30 and 70 percent of the disc's radius. The possible formation mechanisms are magnetic loop reconnection events in the disk or optical reprocessing of X-ray flares.

4.1 The Spectrum of the Flares

In order to determine the location of the flares a time-resolved spectrophotometric study of the optical variability in the black hole XRTs V404 Cyg and A0620–00 were obtained [12], [32]. Dramatic Hα variability was found in V404 Cyg with similar correlated behaviour seen in the continuum. Low-level flickering involving changes in the line flux, without large changes in the profile shape, were seen. Furthermore, large flares, seen approximately once per night and possibly associated with the 6 hr QPO, were observed and are thought to involve the development of a pronounced asymmetry in the line profile with the red wing strongest. Given that the line profile were observed to change during the flares, the most likely origin for the variability is variable photoionization by the central source [12].

A0620–00 was observed with the Very Large Telescope (VLT) using FORS+600V grism and with exposure times of 120 sec. Superimposed on the double-humped continuum lightcurve are the well known flare events which last tens of minutes. Some of the flare events that appear in the continuum lightcurve were also present in the emission line lightcurves. From the Balmer line flux and variations, the persistent emission is optically thin and during the flare events there is a significant increase in temperature. The data suggests that there are two HI emitting regions, the accretion disc and the accretion stream/disc region, with different Balmer decrements. The orbital modulation of Hα with the continuum suggests that the steeper Balmer decrement is most likely associated with the stream/disc impact region. The spectrum of the flares has a frequency power-law index of -1.40 and can be described by an optically thin gas with a temperature in the range 10000–14000 K that covers 0.05–0.08 percent of the accretion disc's surface. Given these parameters the possibility that the flares arise from the bright-spot (regions where the gas stream interacts with the edge of the disc) cannot be ruled out [32].

5 Correlated Optical X-ray Variability

Of the quiescent black hole XRTs, the most accessible for detailed studies is V404 Cyg, due to its brightness, which varies at X-ray, optical, IR, and radio. The optical emission line variations were observed to be correlated with the optical continuum [12], and the emission line flares were found to exhibit a double-peaked line profile, suggestive of emission distributed across the accretion disk rather than arising in localized regions. This was attributed to irradiation of the outer disk by the variable X-ray source, and hence it was predicted that the X-ray variations should be correlated with the optical. Such correlated variability is commonly seen in X-ray bright states in both neutron star systems and black holes (e.g. [30]), but had not been directly observed in quiescent systems.

Simultaneous X-ray and optical observations of V404 Cyg in quiescence have shown that the X-ray variability is indeed well correlated with the Hα emission line variability (see Fig. 5.). The Hα profile peak separation implies that the outer edge of the emitting region is at or outside the circularization radius and the prompt response in the entire Hα emission line confirms that the variability is powered by X-ray (and/or EUV) irradiation. Therefore it seems that optical observations can be used to perform an indirect study of the X-ray variability, at least for V404 Cyg [15].

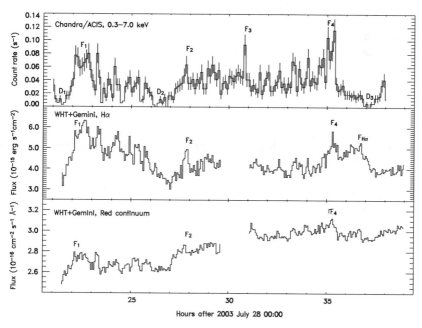

Fig. 5 Simultaneous X-ray and optical lightcurves of V404 Cyg in quiescence. The WHT data corresponds to individual 200 sec exposures, and the Gemini shows 40 s exposures binned by a factor of four to match the WHT resolution [15]

6 ULTRACAM Observations of Quiescent X-ray Transients

The XRTs are key laboratories where we can study the accretion onto compact objects primarily because they go through a large range of mass accretion rate during their outbursts, therefore allowing us to study how the accretion properties change with accretion rate. To date the timing properties of the X-ray states, in particularly the low and high luminosity states, have been studied most extensively at X-ray energies [40].

Black hole X-ray transients are known to exhibit five distinct X-ray spectral states, distinguished by the presence or absence of a soft blackbody component at 1 keV and the luminosity and spectral slope of emission at harder energies; the quiescent, low, intermediate, high and very high state [40]. Four of them are successfully explained with the advection dominated accretion flow (ADAF) model [25],[6]. In the context of the ADAF model, properties similar to the low/hard state are expected for the quiescent state, as there is no distinction between the two except that the mass accretion rate is much higher and the size of the ADAF region is smaller for the former state.

Unlike the transition between the low/hard and high/soft (thermal-dominant) state, where there is a reconfiguration of the accretion flow [6], observational evidence for a transition between the low/hard and quiescent states is limited. In both these states, the ADAF model predicts that the inner edge of the disc is truncated at some large radius, with the interior region filled by an ADAF. Strong evidence for such a truncated disc is provided by observations of J1118+480 in the low/hard state during outburst (e.g. [20],[4]) and a hot optically-thin plasma in the inner regions. Similar properties to the low-state are expected for the quiescent state, as the accretion flow is believed to be similar. ADAF models predict that interactions between the hot inner ADAF and the cool outer disk at or near the transition radius (r_{tr}) can produce QPO variability. The ADAF model predicts that the inner disc edge will move outward to larger radii [6] compared to the low-state. Hydrodynamical models for the PDS of X-ray binaries predict the observed 1/f-like fluctuations as well as a low-frequency break, which is related to the size of the ADAF region [37].

Given the low X-ray luminosities ($< 10^{32}$ erg s^{-1}; [27]), the quiescent state has been difficult to explore at X-ray wavelengths. However, the variability can be effectively studied in the optical, where reasonable count rates are now possible for quiescent studies with efficient and high-time resolution detectors attached to large aperture telescopes. It is then possible to observe at higher time resolutions than in previous studies. Indeed, the variability extends to the shortest timescales observable, with pronounced changes sometimes seen in a few tens of seconds or less. This is also indicated by the extension of the red noise component in the PDS to 0.05 Hz or even higher. Below we give the result of studies of XRTs in quiescence, where there is evidence for QPOs and breaks in the PDS which are related to the transition radius.

6.1 V404 Cyg: Evidence for a QPO

High time-resolution multicolour observations of the black hole XRT V404 Cyg were obtained with the 4.2 m WHT (William Herschel Telescope) and ULTRACAM [31]. Superimposed on the secondary star's ellipsoidal modulation are large flares on timescales of a few hours, as well as several distinct rapid flares on

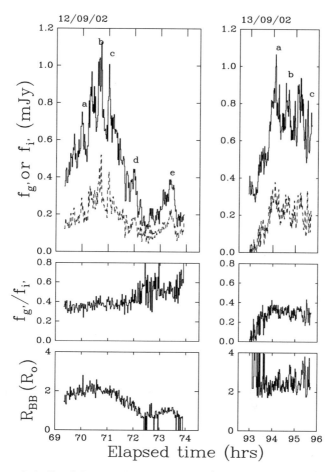

Fig. 6 Top panel: the flare lightcurve of V404 Cyg in the g' (dashed line) and i' (solid line) bands obtained by subtracting the secondary star's ellipsoidal modulation. The exposure time was 5 sec. Middle panel: the flux density ratio $f_{g'}/f_{i'}$. Bottom panel: the projected blackbody radius of the region producing the flares [31]

timescales of tens of minutes (see Fig. 6). The rapid flares, most of which show further variability and unresolved peaks, cover shorter timescales than those reported in previous observations. The Sloan i' band flux of the large flare events is larger than the the Sloan g' band flux, which is is consistent with optically thin gas with a temperature of ~ 8000 K arising from a region with an equivalent blackbody radius of at least 2 $R_\odot$, which covers 3 percent of the accretion disc's surface. The PDS of the Sloan g' and Sloan i' lightcurves show a QPO feature at 0.78 mHz ($= 21.5$ min) (see Fig. 7).

The outer regions of the accretion disc in quiescent XRTs and dwarf novae (DNe) are assumed to be very similar. Therefore if the optical variability observed in the quiescent XRTs originates from these outer regions, then they should have similar properties to those observed in quiescent DNe. A study of the rapid variability in

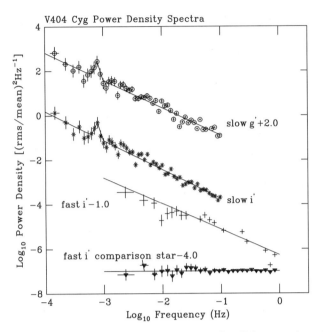

Fig. 7 The power density spectrum of V404 Cyg quiescent flare lightcurve taken with ULTRA-CAM on the WHT. The 0.78 mHz QPO is clearly seen [31]

cataclysmic variables [1] reveals the slope of the PDS in quiescent DNe to be in the range −1.6 to −2.6, which seems to be in general steeper than those observed in quiescent XRTs (−1.0 to −1.52; [43], [13]). This suggests that there is either more low-frequency variability in the lightcurves of quiescent DNe compared to quiescent XRTs, or that there is more high-frequency variability in the XRTs. Note that the former could be attributed to intrinsic variability associated with the stream/disc impact region or the white dwarf (e.g. [42]). On average the slope of the Sloan i' PDS in V404 Cyg is steeper than the Sloan g' slope, suggesting that there is more low-frequency disc variability in Sloan i' compared to Sloan g'. This difference could be attributed to more variability from outer regions of the disc, such as the stream/disc impact region or more high frequency variability in Sloan g' from the inner disk. Thus one expects the slope of the PDS to be dependent on the observed waveband.

An ADAF has turbulent gas at all radii, with a variety of timescales, ranging from a slow timescale at the transition radius down to nearly the free-fall time close to the black hole. In principle, interactions between the hot inner ADAF and the cool, outer thin disk, at or near the transition radius, can be a source of optical variability, due to synchrotron emission by the hot electrons in the ADAF. For an ADAF the variability could be quasi-periodic and would have a characteristic timescale given by a multiple of the Keplerian rotation period at r_{tr}. One also expects slower variations, since the mass supply to the ADAF originates at r_{tr} [26]. It should be noted that it is difficult to produce such rapid variations using the thin disc models. If we assume that the 0.78 mHz QPO feature observed in the PDS of V404 Cyg is the dynamical timescale at the transition radius, then r_{tr} lies at 10^4 Schwarzschild radii, which is consistent with that estimated by [26] using the maximum velocity observed in the Hα emission line. Thus the timing and spectral analysis results support the idea that the rapid flares (i.e. the QPO feature) most likely arise from regions near the transition radius.

6.2 Change in the Size of the ADAF Region

A0620–00 was observed with the acquisition camera attached to Gemini-South. Superimposed on the ellipsoidal modulation are several prominent flares together with weaker continual variability. The PDS of the flare lightcurve exhibits band-limited noise closely resembling the X-ray PDS of black hole candidates in their low-states, but with the low-frequency break at a lower frequency (see Fig. 8), most likely related to the change in size of the ADAF region. The most convincing example for a change in the size of the transition region between the low/hard and quiescent state is for J1118+480. Observations of J1118+480 in the low/hard state reveal the presence of band-limited noise at X-ray and optical/UV wavelengths (see Fig. 9). WHT+ULTRACAM observations of J1118+480 show rapid flare events which typically last a few minutes, superimposed on the double-humped ellipsoidal lightcurves [33]. The PDS of the lightcurves is described by either a broken power—law model with a break frequency at $\sim$2 mHz or a power—law model plus a broad quasi-periodic oscillation (QPO) at $\sim$2 mHz (see Fig. 9). The low/hard state X-ray PDS shows a low-frequency break at 23 mHz and a QPO at $\sim$80 mHz, whereas the optical quiescent PDS shows either a break at a much lower frequency of $\sim$2 mHz or a broad QPO at $\sim$2 mHz. It is interesting to compare the optical and X-ray PDS assuming that the optical PDS can be described by either a break-frequency or a broad QPO model. The position of the quiescent optical broad QPO is a factor of $\sim$40 lower than the low/hard state QPO, and the quiescent optical break-frequency is $\sim$12 lower than the low/hard state break-frequency. The presence of the pos-

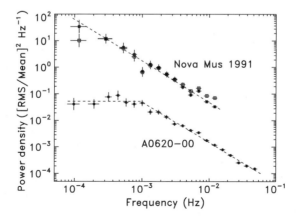

Fig. 8 The power density spectrum of A0620–00 and Nova Mus 1991 (GU Mus). The break frequency in A0620–00 is at 9.5×10^{-4} Hz, corresponding to a timescale of $\sim$20 min [13]

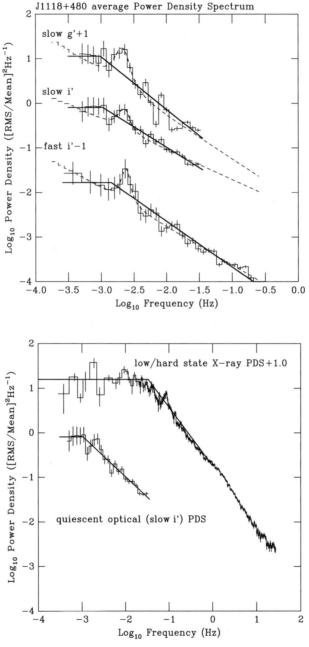

Fig. 9 Top: The power density spectrum (PDS) of the Sloan g' and i' band flare lightcurves for J1118+480 in quiescence taken with ULTRACAM. There is a vertical offset between the spectra. The solid line is a fit with a broken power–law and the dashed line is a fit with a power–law+QPO model. A QPO at ∼2 mHz is noticeable in all the PDS. Bottom: The quiescent optical ULTRACAM Sloan (i') and low/hard state X-ray PDS of J1118+480 (taken from [14]). Note the shift in the break-frequency and the broad QPO

sible break-frequency or a broad QPO (possibly with a multiple frequency ratio) provides evidence that we are seeing the same phenomenon in outburst. The similarity between the low/hard and quiescent state PDS suggest that the optical variability could have a similar origin to the X-ray PDS and might be associated with the size of the direct emission from the self-absorbed synchrotron emission arising from an advective-dominated flow (see [24] and references therein) or from optically thin synchrotron emission directly from a jet [21],[7].

The size of the quiescent ADAF region is estimated to be $\sim 10^4$ Schwarzschild radii, similar to that observed in other quiescent black hole XRTs such as V404 Cyg [31] and A0620–00 [13], suggesting the same underlying physics. The similarities between the low/hard and quiescent state PDS suggest a similar origin for the optical and X-ray variability, most likely from regions at/near the ADAF.

6.3 GU Mus

Perhaps the best dataset to date of the fast optical variability of an XRT is of GU Mus observed with the VLT+ULTRACAM [34]. Previous V-band observations with the Acquisition Camera on Gemini-South of GU Mus showed the strongest and most dramatic flaring behavior of any XRT [13]. Figure 10 shows the Sloan u', g' and

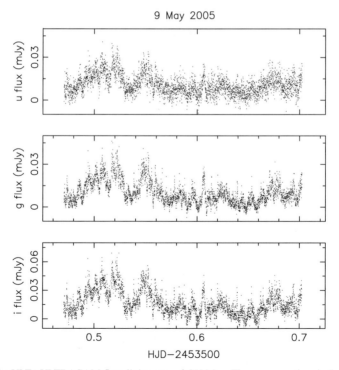

Fig. 10 The VLT+ULTRACAM flare lightcurve of GU Mus. The exposure time is 5 sec for this $V = 20.5$ quiescent black hole XRT. From bottom to top are the Sloan $u'g'$ and i'-band lightcurves. Dramatic flaring activity with amplitudes of up to 40 percent are clearly seen [34]

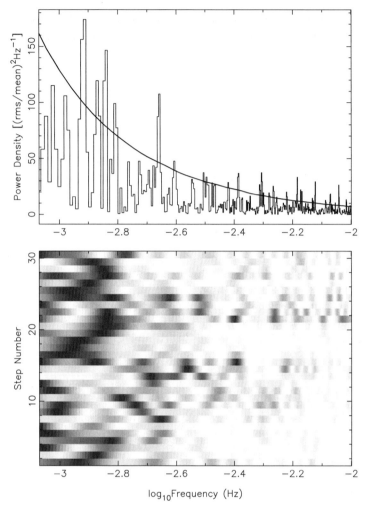

Fig. 11 The dynamical power density spectrum (PDS) of the GU Mus Sloan i'-band flare lightcurve. Each grey-scale is the PDS of a 10 min section of the lightcurve. The top plot shows the PDS, where feature above the dashed line are significant at the 95 percent level. One can see that some of these features are coherent but only last a few tens of minutes [34]

i'-band ULTRACAM lightcurves. Flaring activity with amplitudes of up to 40 percent are clearly seen in all bands. Even with an exposure time of 5 sec, some of the flares events are still unresolved. What makes these observations unique is that there is simultaneous three-band colour information for the flares.

The dynamical PDS of the lightcurves clearly shows many QPOs in the range 1–3 mHz with quality factors ($Q = \nu/FWHM$) >10 (see Fig. 11), some of which are highly coherent, e.g. folding the data on the 7.6 min QPO, one clearly sees a pulse-like profile in all bands. As in V404 Cyg and J1118+480 the origin of these QPOs is most likely from regions at/near the ADAF. However, GU Mus is the first XRT in which more than one QPO has been detected in three colors. Fast spectroscopic observations would shed light on the origin and location of these flare events.

7 Conclusions

The replacement of photomultiplier tubes with CCDs in the 1980's was initially a drawback as high time-resolution optical photometry was not possible. However, the advent of CCDs with frame transfer and higher quantum efficiencies provided significant improvements in the quality of the optical data. Indeed, simultaneous multiwavelength studies of X-ray bursts and bright X-ray binaries have provided significant advances in our understanding of the location and size of the reprocessed optical bursts.

Over the last few years we have seen a similar trend. High time-resolution instruments with sub-second time-resolution capabilities attached to large aperture telescopes has allowed us to probe not only bright X-ray binaries at higher time resolution than in previous studies, but also the quiescent state of the XRTs. The PDS of the optical lightcurves observed in the low- and high-states of the XRTs have shown features such as band-limited noise and QPOs which indicate a change in the size of the ADAF region in the inner disk.

Acknowledgments I would like to thank my collaborators in this field; Phil Charles, Jorge Casares, Cristina Zurita, Rob Hynes and Carole Haswell. This work has made extensive use of the NASA ADS Abstract Service. I acknowledge support from the Spanish Ministry of Science and Technology under the project AYA2004-02646.

References

1. A. Bruch: A&A, **266**, 237, (1992)
2. D. Chakrabarty, L. Homer, P.A. Charles & D. O'Donoghue, ApJ, **562**, 985, (2001)
3. P.A. Charles, M. Coe: Optical, ultraviolet and infrared observations of X-ray binaries. In: *Compact Stellar X–ray Sources*, vol 39, ed W.H.G. Lewin and M. van der Klis (Cambridge University Press, 2006)
4. S. Chaty, C.A. Haswell, J. Malzac, R.I. Hynes, C.R., Shrader, W. Cui: MNRAS, **346**, 689, (2003)
5. D. O'Donoghue: Baltic Astron, **320**, 485, (1995)
6. A.A. Esin, J.E. McClintock & R. Narayan: ApJ, **489**, 865, (1997)
7. R.P. Fender, E. Gallo & P.G. Jonker: MNRAS, **343**, L99, (2003)
8. R.P. Fender, Jets from X-ray binaries. In: *Compact Stellar X–ray Sources*, vol 39, ed W.H.G. Lewin and M. van der Klis (Cambridge University Press, 2006)
9. E. Gotthelf, J. Patterson & R.J. Stover: ApJ, **374**, 340, (1991)
10. J.E. Grindlay, J.E. McClintock, C.R. Canizares, L., Cominsky, F.K. Li, W.H.G. Lewin & J. van Paradijs: Nature, **274**, 567, (1978)
11. J.A. Hackwell, G.L. Grasdalen, R.D. Gehrz, L. Cominsky, W.H.G. Lewin & J. van Paradijs: ApJL, **233**, L115, (1997)

12. R.I. Hynes, C. Zurita, C.A. Haswell, J., Casares, P.A. Charles, E.P. Pavlenko, S.Yu. Shugarov & D.A. Lott: MNRAS, **330**, 1009, (2002)
13. R.I. Hynes, P.A. Charles, J. Casares, C.A. Haswell, C. Zurita & T. Shahbaz: MNRAS, **340**, 447, (2003)
14. R.I: Hynes, C.A. Haswell, W. Cui, C.R. Shrader, K. O'Brien, S. Chaty, D.R. Skillman, J. Patterson & K. Horne: MNRAS, **345**, 292, (2003)
15. R.I. Hynes, P.A. Charles, M.R., Garcia, E.L. Robinson, J. Casares, C.A. Haswell, A.K.H. Kong, M. Rupen, R.P. Fender, R. Wagner, E. Gallo, B.A.C. Eves, T. Shahbaz & C. Zurita: ApJ, **611**, L125, (2004)
16. R.I. Hynes, K. Horne, K. O'Brien, C.A. Haswell, E.L. Robinson, A.R. King, P.A. Charles, K.J. Pearson: ApJ, in press, (2006)
17. F.J. Jablonski, M.G. Pereira, J. Braga & C.D. Gneiding: ApJ, **482**, L171, (1997)
18. A.K.H. Kong, L. Homer, E. Kuulkers, P.A. Charles, A.P. Smale: MNRAS, **311**, 405, (2000)
19. J.E. McClintock, C.R. Canizares, L. Cominsky, F.K., Li, W.H.G. Lewin & J. van Paradijs, J.E. Grindlay: Nature, **279**, 47, (1979)
20. J.E. McClintock, R. Narayan, M.R. Garcia, J.E. Orosz, R.A., Remillard & S.S. Murray: ApJ, **593**, 435, (2003)
21. S. Markoff, H. Falcke & R.P. Fender, A&A, **372**, L25, (2001)
22. M. Matsuoka, K. Mitsuda, T Ohashi, H. Inoue, K. Koyama, F. Makino, K. Makishima, T. Murakami, M. Oda, Y. Ogawara, N. Shibazaki, Y. Tanaka, K. Tsuno, S. Miyamoto, H. Tsunemi, K. Yamashita, S. Hayakawa, H. Kunieda, K. Masai, F., Nagase, Y. Tawara, I. Kondo, J. Cominsky, J.G. Jernigan, A. Lawrence, W.H.G. Lewin, H. Pedersen, C. Motch & J. van Paradijs: ApJ, **283**, 774, (1984)
23. J. Middleditch, K.O. Mason, J.E. Nelson & N.E. White: ApJ, **244**, 1001, (1981)
24. R. Narayan, I. Yi: ApJ, **428**, L13, (1994)
25. R. Narayan, J.E. McClintock & I. Yi: ApJ, **457**, 821, (1996)
26. R. Narayan, D. Barret & J.E. McClintock: ApJ, **482**, 448, (1997)
27. R. Narayan, in Two Years of Science with Chandra, Abstracts from the Symposium held in Washington, (2001)
28. E.P. Pavlenko, A.C. Martin, J. Casares, P.A, Charles & N.A. Ketsaris : MNRAS, **281**, 1094, (1996)
29. K. O'Brien, K. Horne, R.I. Hynes, W. Chen, C.A. Haswell & M.D. Still: MNRAS, **334**, 426, (2002)
30. L.D. Petro, H.V. Bradt, R.L. Kelley, K. Horne & R. Gomer: ApJL, **251**, L7, (1981)
31. T. Shahbaz, V.S. Dhillon, T.R. Marsh: MNRAS, **346**, 1116, (2003)
32. T. Shahbaz, R.I. Hynes, P.A. Charles, C. Zurita, J. Casares, C.A. Haswell, S. Araujo-Betancor & C. Powell: MNRAS, **354**, 31, (2004)
33. T. Shahbaz, V.S. Dhillon & T.R. Marsh: MNRAS, **362**, 975, (2005)
34. T. Shahbaz, V.S. Dhillon, T.R. Marsh, R.I. Hynes, C.A. Haswell, J. Casares, C. Zurita & P.A. Charles: MNRAS, in preparation (2007)
35. R.J. Stover, S.L. Allen: ApJ, **99**, 877, (1987)
36. T. Strohmayer, L. Bildsten: New views of thermonuclear bursts. In: *Compact Stellar X–ray Sources*, vol 39, ed b W.H.G. Lewin and M. van der Klis (Cambridge University Press, 2006)
37. M. Takeuchi S. Mineshige: ApJ, **486**, 160, (1997)
38. J. Truemper, M.W: Sztajno, w. Pietsch, J. van Paradijs & W.H.G. Lewin: Space Science Reviews, **40**, 255, (1985)
39. J. van Paradijs & J.E. McClintock: Optical and Ultraviolet Observations of X-ray Binaries In: *X-ray Binaries*, ed W.H.G. Lewin, J. van Paradijs, E.P.J. van den Heuvel, (Cambridge University Press, 1995), pp 58
40. M. van der Klis, Rapid X-ray variability. In: *Compact Stellar X–ray Sources*, vol 39, ed b W.H.G. Lewin and M. van der Klis (Cambridge University Press, 2006)
41. R.M. Wagner, T.J. Kreidl & S.B. Howell, S.g. Starrfield: ApJ, **401**, L97, (1992)
42. B., Warner, in Cataclysmic Variable Stars, CUP, (1995)
43. C. Zurita, J. Casares & T. Shahbaz: ApJ, **582**, 369, (2003)

Stellar Pulsation, Subdwarf B Stars and High Time Resolution Astrophysics

C. Simon Jeffery

Abstract This chapter asks what high time resolution can bring to the study of stellar pulsation, with particular focus on the pulsations of subdwarf B stars. We give a brief introduction to the different types of pulsation found in stars, and identify which short-period oscillations might benefit from high-speed observations. We describe the basic properties of radial and non-radial oscillations, their effect on the stellar surface and emitted flux, and the consequences for the observer. The particular case of non-radial oscillations in subdwarf B stars and their potential for asteroseismology is outlined. Various campaigns to obtain high time resolution spectroscopy and photometry, and hence to identify pulsation modes in subdwarf B stars, are described in more detail. We close with a brief summary of problems in stellar pulsation likely to benefit from advances in high time resolution technology.

Keywords: Stars: oscillations · Stars: subdwarfs

1 Introduction

Pulsation is a manifestation of a star's normal modes of oscillation. It can occur in stars right across the Hertzsprung-Russell diagram, from neutron stars to red supergiants, from M dwarfs to OB supergiants. The phenomenon can cause spectacular variability in the luminosity and radius of a star, or it can cause almost imperceptible waves running through its atmosphere. Its importance lies in the fact that normal modes are sensitive both to the global properties of the star and to its internal structure—especially the distribution of temperature, density and opacity. Consequently, pulsations have enabled us to use stars as standard candles for measuring the size of the Universe, as well as seismic probes to see inside the Sun, and as tests of theoretical atomic physics.

C. Simon Jeffery
Armagh Observatory, College Hill, Armagh BT61 9DG, N.Ireland
e-mail: csj@arm.ac.uk

The prevalence of pulsation across the HR diagram is associated with a vast range in the time-scale of variability. Classically, pulsations are associated with the fundamental radial mode of oscillation—in which a star pulsates as a sphere, with no stationary points.

Linear theory shows that the period of the fundamental radial mode P_F may be written in terms of the mean density ρ

$$P_F \sqrt{\rho} = Q \tag{1}$$

where Q is a constant ~ 0.3 if P_F is expressed in days and ρ in solar units [23]. Since $\rho \propto M/R^3$ we have, for a given mass M,

$$P_F \propto R^{3/2} \tag{2}$$

where R is the stellar radius. Thus it is evident that short-period oscillations will be found in the smallest stars. It is also evident, since stars can be found covering eight decades in radius—from neutron stars at $10^{-5}\,R_\odot$ to red supergiants at $10^3\,R_\odot$, that stellar oscillations may theoretically cover some 12 decades in period—from milliseconds to gigaseconds.

This review addresses the following: when are high time resolution techniques required for the study of stellar pulsations? Which objects pulsate so fast that conventional equipment is inadequate? What techniques are required to resolve their behaviour? What science can we learn by pushing the technology?

These questions are introduced by a discussion of the types of pulsating variable for which high time resolution techniques are necessary, the types of observations which can be made and the uses to which they can be put (Sect. 2). As illustration, some of the scientific challenges presented by subdwarf B stars are outlined (Sect. 3). Our attempts to address these using high time resolution techniques are presented in Sect. 4. Some conclusions to the question "does pulsation need high time resolution?" are drawn in Sect. 5.

2 Stellar Pulsation and High Time Resolution

2.1 Radial and Non-Radial oscillations

The largest amplitude stellar pulsations are generally the radial modes, where the displacement is purely radial and spherically symmetric. With no stationary surfaces between the centre and surface of a star, an oscillation may be called the fundamental (F) radial mode. With one or more stationary surfaces, or nodes, it is called the first or higher-order harmonic or overtone. Every star possesses such *normal modes*. However, for the star to actually pulsate, these normal modes must be excited by resonance with a suitable driving mechanism; this only occurs for a relatively small fraction of all stars.

Being three-dimensional objects, stars also possess a large number of *non-radial* normal modes. In non- or slowly-rotating stars, the displacement becomes manifest as a spherical harmonic

$$s(t, r, \theta, \phi) = s_k(r) Y_{\ell m}(\theta, \phi) e^{i\sigma t}, \tag{3}$$

where r, θ, ϕ describe position (radius, longitude, latitude) within the star, σ represents the frequency of the oscillation, and t represents time. The integer eigenvalues k, ℓ, m characterize the normal mode:

- k: order—related to the number of nodes in the radial direction (in the linear adiabatic approximation).
- ℓ: degree—the number of lines of nodes on the surface.
- m: azimuthal number—the number of lines of nodes passing through the polar axis.

Note that a radial oscillation is a special case where $\ell = m = 0$, and that the fundamental radial mode has $k = \ell = m = 0$. Many radial and non-radial modes (nro's) may be simultaneously excited and superposed.

As in any oscillation, a driving mechanism is coupled with a restoring force. The nature of this restoring force dictates the type and period of oscillation within the star. Examples of restoring force include pressure, gravity, the Coriolis force, and electrostatic forces. Each of these is responsible for different types of oscillation, and propagates under different conditions.

- p-modes: the restoring force is pressure. Pure radial modes are normally a special case of a p-mode. Periods are generally less than or equal to that of the F-mode, and the oscillations propagate near the stellar surface. A common example of a p-mode would be the surface oscillations of a soap bubble or balloon.
- g-modes: the restoring force is gravity. There are no radial g-modes. Periods are always greater than that of the F-mode. Oscillations generally propagate in the deep interior. Terrestrial examples of g-modes include water waves in the ocean.
- r-modes: (or quasi-toroidal modes): the restoring force is the Coriolis force. With negligible radial motion, they are rarely observed in stars, but have been recognized as Rossby waves in planets.
- t-modes: the restoring force is the electrostatic force between atomic particles. Possibly occurring as disruptions of crystal structure in neutron stars, they are more commonly recognized on earth as earthquakes.

By far the most common form of driving is caused by a peak in the radiative opacity in the stellar interior (the κ-mechanism). Two conditions must be satisfied. First, the opacity must occur in a zone where the thermal timescale $(L/c_V T \Delta m)^{-1}$ is close to the star's dynamical timescale $(G\bar{\rho})^{-1/2}$. $L, c_V, T, \Delta m, \bar{\rho}$ are, respectively, luminosity, specific heat per unit mass at constant volume, temperature, zone mass and mean density. Second, the opacity gradients, $\kappa_T = (\partial \ln \kappa / \partial \ln T)_\rho$, $\kappa_\rho = (\partial \ln \kappa / \partial \ln \rho)_T$ must be sufficiently high that work done in the driving region exceeds that lost in the damping layers above and below.

2.2 Short-Period Pulsations

With potential periods ranging from milliseconds to gigaseconds, we arbitrarily restrict our discussion to periods $\lesssim 10^5$ s.

Pulsating stars with periods in the range $\sim 10^3$–10^5 s include the well-known RR Lyrae, β Cepheid, and δ Scuti variables [73, 26, 37]. Less well known are the V652 Her and PG 1716+426 type variables. V652 Her and BX Cir are the only known members of a class of short-period radially pulsating extreme helium stars [48, 78]. They have effective temperatures $\sim 22\,000$ K and pulsation periods $\sim 10^4$ s. The PG 1716+426 variables are subdwarf B stars (Sect. 3) with effective temperatures $\sim 22\,000$–$29\,000$ K and periods $\sim 10^4$ s [32]. They appear to pulsate non-radially [29, 47].

For most purposes, conventional techniques seem adequate to observe these stars, either photometrically or spectroscopically. By "conventional techniques", we assume the current use of charge coupled devices (CCDs) configured so that the required detector area is read out in a single step, which may take anything from 30 s upwards. If each exposure takes 10 s, then a cycle time of, say, 40 s, limits serious investigation of time-dependent phenomena to those with timescales longer than about 400 s. Below this range, we need better methods, which we here take to imply *high time resolution*

In the ~ 10–10^3 s period range, the number and type of pulsations observed become quite diverse. The most well-known are probably the five-minute oscillations seen in the Sun [74]. Substantial efforts are being made to identify and exploit similar oscillations in solar analogues [6]. These oscillations are of very low-amplitude and are generally multi-periodic. In the Sun, many thousands of modes are excited simultaneously. Ultra-high signal-to-noise and frequency resolution are more important issues than time resolution.

Staying on the main-sequence, rapidly-oscillating Ap (ro-Ap) stars exhibit multiple oscillations with periods of $\sim 10^2$–10^3 s [59]. Originally detected photometrically, these have provided much material for asteroseismology [24]. Recent advances include the spectroscopic measurement of oscillations as a function of height (depth) in the atmosphere and their interaction with the strong magnetic fields [60]. These oscillations are primarily high-order non-radial p-modes.

Stellar oscillations in the same period range are found in numerous variables below the main-sequence. For the most part, these may be found amongst low-mass stars contracting from the asymptotic-giant branch to the white dwarf sequence. The sequence includes the GW Virginis (or PG1159) variables [22], the DB [36] and ZZ Ceti (or DA) variables [31], the latter being pulsating white dwarfs with spectra dominated by helium and hydrogen respectively. These are primarily multi-periodic high-order g-mode oscillators.

The final group in this period range are the V361 Hya variables, subdwarf B stars (sdB stars: Sect 3) with effective temperatures $\sim$29 000–35 000 K and periods $\sim 10^2$ s [53, 52]. These are primarily multi-periodic low-order p-mode oscillators [12, 13, 14].

Normal modes with periods shorter than these would include high-order p-modes for sdB stars, or any p-modes for white dwarfs. These have not been observed so far. Neutron stars are much smaller, but their structures are so stiff that conventional oscillations are not expected. There have however been reports of oscillations in the X-ray emission from soft gamma repeaters on timescales of 50–200 $\times 10^{-3}$ s that may be related to oscillations of the underlying neutron star [28].

2.3 Asteroseismology

Oscillations are probably excited at some level in a very large fraction of all stars. Many hundreds of modes have been identified in the Sun.

Since different modes propagate in different layers in the interior of a star, the study of stellar oscillations provides a means to explore both global and local properties. Low-order p-modes penetrate deep into the stellar envelope and provide information about mean density and overall radius. High-order p-modes are reflected more strongly toward the surface and indicate conditions in the outer layers of the star. Rotation causes the frequency degeneracy of modes with different azimuthal wavenumber to be broken by an amount proportional to the rotation frequency, and so on.

Except in rare cases where many hundreds of oscillation frequencies can be measured, the general procedure for using asteroseismology to discover the structure of the stellar interior uses the "forward" method. In this, a family of stellar models is constructed covering a significant volume of parameter space. Parameters may include total mass, core mass (for horizontal-branch stars), composition (i.e. helium abundance, metallicity), and age. Rotation and other physical processes may also be included.

Each model is analyzed using a suitable linear pulsation code to determine frequencies and growth-rates for all of the normal modes. Other fundamental properties of the models such as effective temperature T_{eff}, radius R and surface gravity g are also recorded. These theoretical properties are then compared with the observed frequency spectrum and other observables (T_{eff}, $\log g$) in order to identify the best-fit model according to an appropriate statistic. In principle, this best fit then constitutes the asteroseismic solution.

Much detailed asteroseismology relies upon the accurate measurement of frequencies and frequency splittings in stars which exhibit many pulsation modes simultaneously. In such cases, it may be possible to identify the modes unambiguously from their frequencies alone. However, relatively few modes have been detected in the majority of non-radially oscillating stars. In such cases, the number of predicted normal modes may be much greater than the number observed. It then becomes difficult to demonstrate that any particular model matches the observation sufficiently well. Since the model predicts both the frequency and the mode of oscillation, it is extremely valuable if the mode of the matching observed oscillation can also be identified.

2.4 Observational Diagnostics

Most low-degree oscillations are discovered from periodic variations in brightness—photometry is the most common tool for the accurate measurement of their frequencies. However, single-channel photometry alone gives very little information about the mode of oscillation.

Since pulsations affect the stellar surface locally, each surface element is associated with a local effective temperature, radius, tilt and velocity. Generally, such local effects cannot be resolved, but are integrated over the stellar disk in the line of sight. Because of superposition and cancellation between hot and cool, or inward and outward moving elements, non-radial oscillations become visible only when they are both of sufficiently low degree *and* of significant amplitude. They may then be observable as variations in total light, colour, radial velocity or absorption line profile. In the case of rapidly rotating stars, higher-degree modes may also be visible because the broadened absorption line profiles help to resolve the stellar surface longitudinally.

The effects of low-degree oscillations on a stellar spectrum are illustrated in Figs. 1 and 2. Figure 1 shows the flux as a function of wavelength and time through one cycle of a radial pulsation with a simulated velocity amplitude of 10 km s^{-1}. The top panel in Fig. 2 shows the the average spectrum of a star in the region of

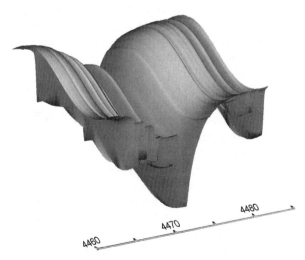

Fig. 1 Spectral variation near HeI4471 due to one cycle of a radial mode ($\ell, m = 0, 0$) in a pulsating sdB star. The marked represents wavelength (in Ångström), flux is displayed vertically, and time runs into the diagram

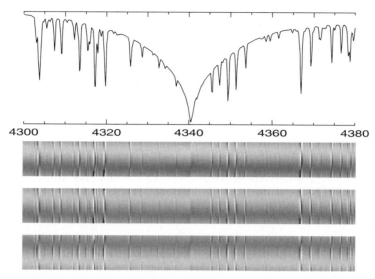

Fig. 2 Mean and trailed residual spectra for modes $l, m = (0, 0), (2, 0)$ and $(2, 2)$ (*top to bottom*). The mean spectrum shows flux averaged over a pulsation cycle as a function of wavelength (in Ångström) in the region of Hγ. The residual spectra represent time series similar to that shown in Fig. 1 from which the mean spectrum has been subtracted. Thus wavelength runs horizontally, time runs vertically, and the greyscale represents residual flux, light shades being positive and dark shades being negative

Hγ. Below are shown grey-scale representations of residual time series (i.e. the average spectrum subtracted from the time-varying spectrum) for three non-radial modes (amplitudes = 10 km s^{-1}). The horizontal shading shows how the total flux varies through the pulsation cycle. S-shaped waves may be seen running through the vertical bands. These demonstrate how the disk-averaged radial motion of the stellar surface modifies the profiles of the absorption lines. Note the subtly different character of the s-waves associated with each mode.

The spectrum also varies on a larger scale. The most straightforward of these is *colour*. As the non-radial oscillation distorts the stellar surface, inward moving regions heat whilst outward moving regions cool. Hotter regions are systematically bluer than cool regions, and so the amplitude of the light variation is strongly wavelength dependent. The total apparent amplitude depends on the actual aspect of the pulsation and, in particular, on the inclination of the pulsation axis i and the azimuthal wavenumber m. However, the relative amplitude at one wavelength to that at another turns out to be independent of i and m, but is sensitive to ℓ [41] (Figs. 3 and 4). By measuring the amplitude of a particular mode at several wavelengths, it should be possible to identify its spherical degree. Since this can be carried out using broad-band photometry it is, in principle, the most economical means of mode identification. Figure 4 demonstrates how this method can work for sdB stars. Modes with $\ell = 1$ or 2, are difficult to distinguish from $\ell = 0$ photometrically, although indiviudal absorption lines are more sensitive to the degree. For $\ell > 2$, the amplitude ratios carry a quite distinctive signature.

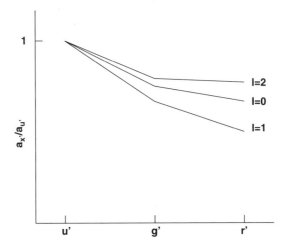

Fig. 3 Schematic illustration of the amplitude ratio method. For a mode of given spherical degree ℓ, the amplitude of the flux variation is a function of wavelength and is independent of i and m. By measuring the amplitude in three wavebands, for example, and normalizing, the spherical degree may be identified by comparing with a predicted amplitude distribution (Fig. 4). In practice, modes of different ℓ are not always so clearly separated

Other observables which have been used to identify modes include the ratio of light variation to radial velocity and the behaviour of the absorption line profile as a function of pulsation phase (method of moments [1]).

2.5 Resolution: Frequency Versus Time

We have argued that, at the present, high time resolution techniques are necessary for stars with pulsation periods shorter than about 400 s (Sect. 2.2). This does not necessarily mean that all phenomena with periods shorter must be studied with purpose-built instruments. Where studies of periodic and multi-periodic phenomena require very high *frequency* resolution (small $\Delta \nu$), it is the timebase of the observations (T) rather than the interval between the observations, that is crucial. Recall that the lower and upper frequencies measurable from a time series of N observations in an interval T are given by

$$\nu_l = 2/T \quad \text{and} \quad \nu_u = 2N/T \qquad (4)$$

respectively [10], while the nominal frequency resolution in a continuous data series is

$$\Delta \nu \sim 1.5/T \qquad (5)$$

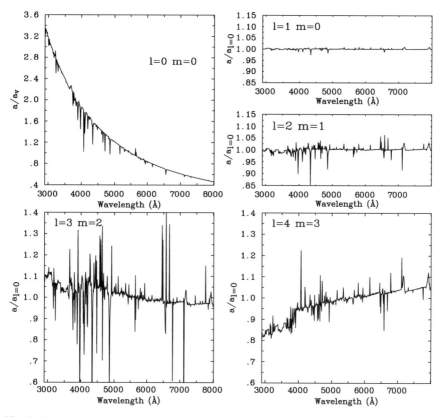

Fig. 4 Theoretical amplitudes as a function of wavelength and spherical degree ℓ. The first panel *top left* shows the amplitude of the radial mode ($\ell = 0$) normalized at V. The other panels show the amplitude relative to that of the radial mode for modes of higher spherical degree (*as labelled*). These calculations are for a cool V361 Hya variable and were made in the adiabatic approximation (courtesy C. Pereira)

(these equations need to be modified where the data series contains significant breaks [25]). To first order, the signal-to-noise ratio in the signal is

$$S \propto \sqrt{ATf\Delta\lambda} \qquad (6)$$

where A is the aperture (collecting area) of the telescope, f is the on-target efficiency and $\Delta\lambda$ is the pass-band in wavelength of collected light.

Thus, the measurement of accurate frequencies for use in asteroseismology need not depend on high time resolution, providing a sufficiently long time series can be established. The amplitudes of signals with different frequencies are accessible because they are assumed to persist over many cycles, so that relatively few observations per cycle are permissible. However, some information may be lost because the observation is smeared out over too large a fraction of the pulsation cycle. Moreover, for ground-based observations, the rotation of the earth and the intervention of the atmosphere impose limitations. Only by observing from space, or with a large network of identical instruments, can one maintain the integrity of several data properties over the intervals necessary; observations obtained from single sites and inhomogeneous networks suffer severe aliasing.

High time resolution offers a partial solution to this problem. By increasing N in a given time interval, ν_u is increased, although $\Delta \nu$ remains unchanged. The greatest contribution to increasing S comes from higher observing efficiency f. In the above example of conventional CCD observations, f may need to be 0.25 or less for short-period phenomena. The first task of high time resolution must be to increase $f \sim 1$. Then, a substantially larger S can be achieved with a shorter T by increasing A and $\Delta \lambda$. Thus it becomes realistic to explore some phenomena with large telescopes over one or a few nights, particularly when precise frequencies are less important or are independently well established.

One area in which high time resolution can and has played an important contribution to stellar pulsations is in the area of mode identification. This has been attempted using both high-speed spectroscopy and high-speed colorimetry, in particular for subdwarf B stars.

3 Subdwarf B Stars

Subluminous B (or sdB) stars were first identified as an important spectral class by [33]. In their survey of faint blue stars in the Galactic halo, they demonstrated that, on the Hertzsprung-Russell diagram, the sdB stars lie between the upper main sequence and the white dwarf sequence, possibly on a blue extension of the horizontal branch observed in globular clusters. This was confirmed by subsequent fine analyses of sdB stars in the field and extremely blue horizontal-branch stars in globular clusters [39, 38, 40]. From these studies it was established that sdB stars are core-helium burning stars of $\sim 0.5\,M_\odot$ with a very thin surface layer of hydrogen—often referred to as "extreme horizontal-branch stars". With high surface gravities ($5 \lesssim \log g \lesssim 6$), diffusion affects chemical stratification in the atmosphere, leading many sdB stars to appear depleted in helium [62].

The sdB stars represent the most extreme case of the horizontal branch problem, namely: how can a star lose virtually all of its hydrogen-rich outer layers before it ignites helium reactions in its core? Originally thought to be mainly single stars, an increasingly large fraction are now known to be binaries; the companions being either G–F-type main-sequence stars [4, 5], white dwarfs [61] or M-dwarfs. This provides mass transfer by Roche lobe overflow or the ejection of a common-envelope while the star was on the giant-branch as a natural mechanism for removing the outer hydrogen [34, 35]. However a number of single sdB stars, and one system containing two helium-rich sdB stars are also known [3]. Some are likely to be single, formed as the result of a merger of two helium white dwarfs [43, 44, 42, 72]. Further information, such as that provided by asteroseismology, is essential for probing the structure and evolution of these stars.

3.1 V361 Hya and PG 1716 Variables

The discovery of pulsations in sdB stars was quite serendipitous. Kilkenny et al. [53] had been searching for variability in a number of white dwarf candidates including one EC 14026–2647 (subsequently named V361 Hya). Finding short-period low-amplitude variability they investigated further and discovered that the star was an sdB star with an F-type companion. Extended photometric monitoring demonstrated a light curve with several independent frequencies which could only be attributed to a non-radial oscillation in one or other but most likely the sdB star.

Subsequently, photometric monitoring of a much larger sample of sdB stars led to the discovery of several more V361 Hya variables, although the non-variables far outnumbered the variables. The second to be analyzed was PB 8783 [57]. Following the discovery paper, an extended multi-site photometric campaign was able to resolve a spectrum of closely-spaced frequencies between 7 and 10 mHz [63], providing powerful ammunition for asteroseismological studies. Independently of the SAAO group, Charpinet et al. [12, 13] had been making theoretical calculations which predicted that some sdB stars should pulsate. The Montreal group had been searching for sdB pulsators with little success until they observed PG 1047+003 [7] and KPD 2109+4401 [8]. Meanwhile the latter star had formed part of the SAAO survey, who published its power spectrum in the same year [58]. Both groups obtained nearly identical light curves, with a clutch of frequencies between 5 and 6 mHz.

The number of known V361 Hya variables stars continues to grow. Their frequency spectra and general properties are reviewed elsewhere [56]. The driving mechanism is interesting: sdB stars lie in the so-called Z-bump instability strip. The pulsations are believed to be driven by opacity due to M-shell electrons in iron-group elements. This same peak is also believed to be responsible for pulsations in β Cep and V652 Her variables. However it is not actually capable of driving the pulsations *unless* it is enhanced in some way [46]. The solution appears to be that in the high-gravity surface layers of sdB stars, diffusion allows the competition between specific radiative and gravitational forces acting on different ions to concentrate elements in layers of high specific opacity. Thus iron will accumulate in regions of high iron opacity, and amplify the local opacity peak even further [21, 13].

In the process of searching for new V361 Hya stars, [32] extended their survey to sdB stars cooler than 29 000 K and adapted their technique by monitoring stars for several hours. As a result, they discovered that a much larger fraction of these cooler sdB stars were variable on much longer timescales, and were possibly multi-periodic with periods in the range 1800–6000 s. In particular, the class prototype PG 1716+426 shows up to six independent modes [70]. Since the periods are substantially greater than the sound crossing time for these stars, the oscillations are attributed to g-modes. Initial theoretical studies showed that high-order g-modes could be excited in sdB stars, but *only* for modes of relatively high degree ($l \geq 3$) and *only* for stars cooler than about 25 000 K [29]. A more recent study has shown that this can be mostly solved by using more modern opacity calculations and including an enhancement of nickel, as well as iron, in the driving zone [47].

3.2 Asteroseismology

The potential of V361 Hya stars for asteroseismological study has been exploited extensively. The first detailed study provided a best-fit model to 23 distinct pulsation modes in PG 0014+067 [11, 19]. Significant results were the measurement of its mass, $M = 0.490 \pm 0.019$ $M_\odot$ and envelope mass $M = 0.25 \times 10^{-5}$ $M_\odot$, confirming it as an extreme horizontal-branch star. Studies of two additional stars (PG 1219+534, Feige 48) have led to similar conclusions [17, 18].

Attempts are being made to carry out asteroseismology of PG 1716 stars, most recently of PG 1627+017 and PG 1338+481 [66, 67]. These have met with limited success, possibly because issues regarding the driving mechanism have still to be resolved [47].

3.3 Problems to Solve

A feature of the V361 Hya stars is that the instability strip is not universal. Although all the known variables lie within the area where such stars are unstable against pulsation, providing iron is sufficiently abundant in the driving zone, there exist nine times as many "non-variables" as variables within this zone, and there exist pairs of spectroscopically identical sdB stars of which only one is variable.

A second problem is that, for most V361 Hya stars, relatively few modes are excited compared with the number predicted. This may be a question of detection threshold; some modes may simply be of such low amplitude as to evade detection without intensive observation using large telescopes. On the other hand, there are also stars in which the mode density appears to be too high. That is to say, while modes with $\ell \geq 3$ are difficult to observe because of cancellation, there are too many modes within a given frequency range for all to have $\ell \leq 3$. Both issues raise questions about the uniqueness of any asteroseismological solution thus derived.

4 Mode Identification in V361 Hya Variables

It was with questions like these in mind that the Armagh group set out to make high time resolution observations of V361 Hya variables shortly after their discovery. Initially, we had the idea to use high-speed spectroscopy, primarily using intermediate-dispersion spectroscopy with a CCD read out in "drift-mode" on the William Herschel (WHT) and Anglo-Australian telescopes (AAT). Our aim was to measure the radial-velocity amplitude of individual modes. Whilst partially successful, this technique was necessarily limited to bright stars with relatively few pulsation modes. The construction of ULTRACAM [27] offered the opportunity to identify modes photometrically using the amplitude ratio method.

4.1 High-Speed Spectroscopy

Observations with ISIS on the WHT yielded an instrumental resolution (2 pixel) $R = 5000$ in the wavelength interval 4020–4420 Å, which contains two strong Balmer lines and, potentially, a number of neutral helium and minor species lines normally observed in early-B stars. The CCD was read out in low-smear drift mode [71] in which only a small number j of CCD rows (parallel to the dispersion direction) are read out at one time. A dekker is used to limit the slit-length, thus only a fraction of the CCD window, $\simeq j$ rows, is exposed at one time. After exposing for a short interval, typically 10 s, the CCD contents are stepped down by j rows. Each set of j rows is accumulated into a data cube containing n individual 2D spectra, stacked adjacent to one another. The technique provided high time resolution and good wavelength stability with reasonable photon statistics.

To measure radial velocities, a template spectrum was constructed by co-adding all of the data in a time series. Individual velocities $v(t)$ were measured by cross-correlation with the template. The functions $v(t)$ were then investigated for periodic content by means of the discrete Fourier transform $\mathcal{F}$.

4.1.1 PB 8783 and KPD 2109+4401 [45]

Our first targets were PB 8783 (= EO Cet) and KPD 2109+4401 (= V2203 Cyg). The velocity data $v(t)$ and amplitude spectra $\mathcal{F}$ are shown in Figs. 5 and 6. Figure 6 also shows a representation of the amplitude spectrum measured photometrically ($\mathcal{P}$) [63, 58].

Over the entire range $v_l = 0.1 < v/\text{mHz} < v_u \sim 90$, $\mathcal{F}$ shows many peaks of comparable amplitude which on their own have little statistical significance. However the coincidence of the highest peaks in $\mathcal{F}$ with the highest peaks in $\mathcal{P}$ (Fig. 6) provides an estimate of the velocity amplitudes of these particular modes. The resolution $\Delta v \sim 0.1 \text{mHz}$ in $\mathcal{F}$ is insufficient to resolve the fine structure observed in much longer photometric time series.

4.1.2 PG 1605+072, PG 1336–018, KPD 1930+2752 [79, 65, 81, 80]

Three further targets were observed in subsequent seasons.

Amongst V361 Hya stars, PG 1605+072 (= V338 Ser) has both the largest number of observed frequencies (>50, [55]) and, until recently, the largest amplitudes (>0.025 mag), making it a potentially ideal target for asteroseismological study. A low-dispersion campaign had already identified velocity variations at the same frequencies as the largest photometric variations [64], but the cycle time for individual observations was ~60–75 s. We first executed a two-site campaign, obtaining some 16 hours of near-contiguous data on the AAT and WHT with a time-resolution of 13–22 s [79] and subsequently led a part of the Multi-Site Spectroscopic (MSST)

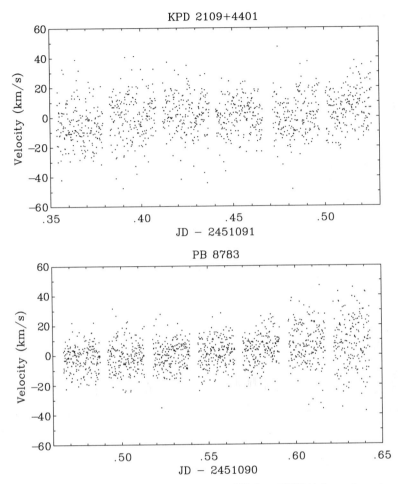

Fig. 5 Radial velocities of KPD 2109+4401 and PB 8783 from WHT high-speed spectroscopy. The horizontal axis is time given in Julian Date

campaign [51]. In both cases, data quality and weather problems compromised the results. Both the AAT/WHT and subsequent MSST observations showed further evolution in the power spectrum of PG 1605+072. Whether this is due to unresolved beating, to stochastic excitation of different modes, or to changes in internal structure, remains to be shown.

PG 1336–018 (= NY Vir) [54, 69] offers the spectroscopist a potentially discovery rich environment. It is both a V361 Hya variable, and an HW Virginis variable. That is to say, it is a short-period (0.1 d) eclipsing binary with an M-dwarf companion that is itself unseen, and yet which reflects the light of the hot primary. Our high-speed spectroscopy successfully resolved the orbit of the primary, but was unsuccessful in detecting underlying pulsations [81].

KPD 1930+2752 (= V2214 Cyg) is also remarkable, being both a V361 Hya variable and *and* a very short-period binary (P = 0.095113 d, [9]) in which the hot

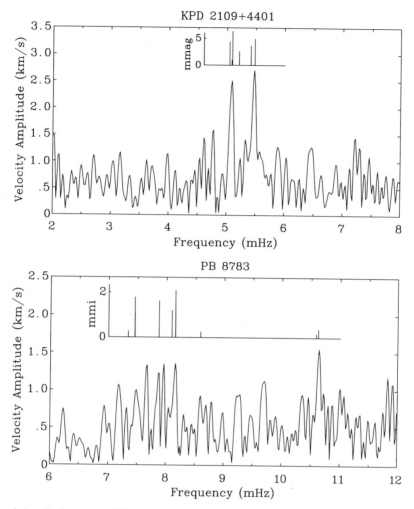

Fig. 6 Amplitude spectra of KPD 2109+4401 and PB 8783 from WHT high-speed spectroscopy. Representations of the power spectra measured from photometry ($\mathcal{P}$) are inset

subdwarf is tidally distorted by its nearby white dwarf companion. Its discovery prompted us to obtain high-speed spectroscopy with the WHT over two nights in June 2000. Meanwhile, Maxted et al. [61] announced their measurement of an orbital semi-amplitude of 349 km s^{-1}, prompting speculation that KPD 1930+2752 could be the progenitor of a Type Ia supernova. Our own high-speed observations confirmed the orbital semi-amplitude and provided a secure ephemeris and a much improved period, accurate to 1 part in 10^6 [80]. The predicted orbital spin-up due to gravitational radiation produces a phase shift of one minute after 46 years. However we found no significant variations >4 km s^{-1} commensurate with any pulsation period [80].

4.2 ULTRACAM Photometry

Mounted on either the WHT or Very Large Telescope (VLT), the combination of high throughput and high time resolution offered by ULTRACAM provides an outstanding tool with which to explore pulsating sdB stars. In our observations of V361 Hya variables, frame rates of 0.1–10 Hz are typical, primarily dictated by the pulsation frequencies ($\sim$5–8 mHz) and, for the brightest targets, the need to avoid CCD saturation.

4.2.1 Mode Identification for KPD 2109+4401 and HS 0039+4302 [49]

First observations were obtained in 2002 September for the targets KPD 2109 +4401 and HS 0039+4302 ($=$ V429 And). A portion of the light curves for each of these targets is shown in Fig. 8. Note how the amplitude is clearly modulated by beating between several closely-spaced frequencies. From amplitude spectra obtained with these light curves, the amplitudes of several modes were measured in each of the three ULTRACAM channels, being u', g' and r' of the SDSS system [30]. The ratios $a_{g'}/a_{u'}$ and $a_{r'}/a_{u'}$ were computed and are shown in Fig. 8. Meanwhile, theoretical models of the colour amplitude ratios expected for non-radial oscillations of different spherical degree in sdBVs had been computed [68] for comparison with these measurements.

From Fig. 7 it was evident that at least one oscillation in each target should be a relatively high-degree ($\ell = 4$) mode, although in the case of KPD 2109+4401 this has since been disputed [82]. Predicted ratios for low-degree modes ($\ell = 0,1,2$) lie close together, so it is not possible to identify the degree of the observed modes so easily. There were no modes identified with $\ell = 3$. Assuming the observed frequencies belong to modes with unique k, ℓ values (i.e. there is no rotational splitting), an additional constraint can be used. For a given degree ℓ, modes of successive radial order k must be well-spaced in frequency, so modes of similar frequency cannot have the same ℓ. By imposing such a constraint and by comparing the colour-amplitude ratios, it is possible to assign k and ℓ values with some confidence (cf. [75]). The next step is to compare the observed frequency spectrum with theoretical models for non-radial oscillations in extended horizontal branch stars [16], and hence to select the models which best represent the star observed in terms of total mass, envelope mass, age, and other characteristics.

4.2.2 PG 0014+067, SDSS J1717+5805, PG 1336–018 and Balloon 090100001

Subsequent ULTRACAM runs have been carried out to observe three primary targets, PG 0014+067 [50], PG 1336–018 and Balloon 090100001 and one secondary target SDSS J1717+5805 [2]. Combining WHT time from both Netherlands and UK allocation committees allowed six nights of data to be obtained for PG 0014+067. Although of magnitude $V = 16.5$, the extended data coverage allowed us to obtain

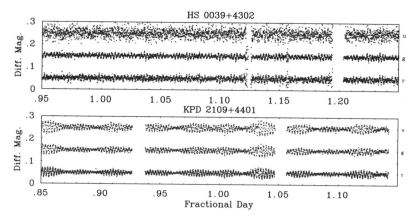

Fig. 7 Partial ULTRACAM light curves in u', g' and r' for sdBVs KPD 2109+4401 and HS 0039+3202 (from [49]). Gaps are due to clouds

an outstanding light curve with a frequency resolution comparable to that of previous asteroseismological studies [11, 19]. The star showed ∼20 independent frequencies in its light curve with frequencies between 5.7 and 12.9 mHz. Our aim was to check mode identifications deduced from the existing asteroseismological model,

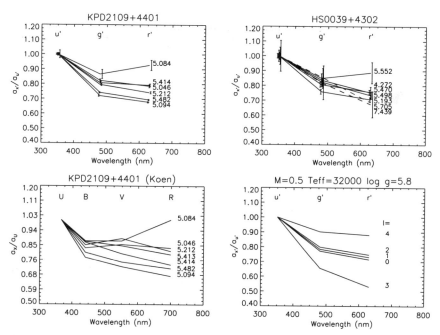

Fig. 8 Observed amplitude ratios for KPD 2109+4401 and HS 0039+4302 (*top*). Modes with $a_{u'}$ <1.4 mmag shown as dashed lines. The figures on the right of the curves represent the frequency of the oscillations in mHz. Lower panels show (*left* previous photometry of KPD 2109+4401 [58] and (*right*) a set of theoretical colour-amplitude ratios $\ell = 0, \ldots, 4$ (as labelled, [68]). From [49]

and also to support a subsequent Whole Earth Telescope campaign to improve the overall frequency resolution.

Although less conclusive than for the targets observed previously, it was possible to show that the two dominant modes must be $\ell = 0, 1$ or 2, and to find evidence that the rotational period should be nearer to 4 d than to the 1.35 d reported previously. The first result is significant because it contradicts the best seismic models [11, 19], in which the dominant mode has $\ell = 3$. Shortly after the ULTRACAM run, a Whole Earth Telescope campaign obtained very secure frequencies for ten of the frequencies identified by us, and reported further evolution of the relative amplitudes in the power spectrum [76].

Limited observations of a secondary target, the 17th mag. sdB SDSS J171722.08 +58055.8 enabled the identification of a second frequency [2].

The V361 Hya / HW Vir variable PG 1336–018 was the subject of a combined high time resolution spectroscopy (UVES) and photometry ULTRACAM campaign at the European Southern Observatory's VLT. Here the goal was to use the eclipse of the subdwarf by the secondary as a diagnostic of the pulsation modes. As the dwarf moves in front of the pulsating star, one sees only those parts of the line profile or light curve contributed by the visible portion of the disk. The fact that the oscillations are still visible at light minimum (Fig. 9) shows the eclipse is not completely total. So far, these data have been used in a precise analysis of the orbital light and orbital velocity curves [77]. The solution is degenerate, indicating possible masses of 0.389 and 0.466 $M_\odot$ for the sdB star—a situation that should be resolved by further analysis of the pulsations.

A three-site high time resolution study of the recently discovered V361 Hya / PG 1716 variable Balloon 090100001 was carried out using ULTRACAM at the WHT, the Montreal Lapoune photometer at the Canada-France-Hawaii telescope and a CCD photometer at the Himalaya Chandra Telescope over a period of 5 nights in 2005 August. Again, these results will be reported in detail in due course, but a preview of a part of the ULTRACAM light curve is shown in Fig. 10.

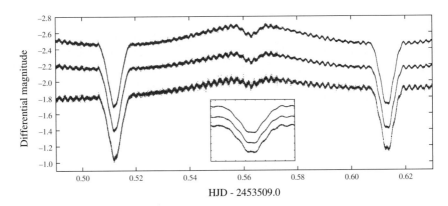

Fig. 9 Segment of the ULTRACAM light curve for sdBV PG1336–018. The second eclipse is magnified *inset* and the lightcurves for filters r', g' and u' are shown running *top to bottom* (adapted from [77], courtesy M. Vučković)

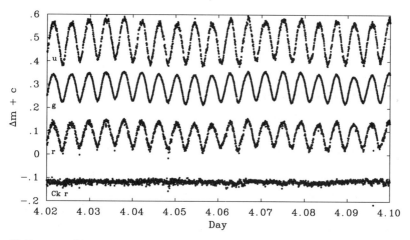

Fig. 10 Segment of the ULTRACAM light curve for sdBV Balloon 090100001

Important lessons have been learned from these campaigns. The goal is to obtain positive mode identifications in variables with complex power spectra. Even when the periods are well known from prior studies, it is necessary to observe for several nights to obtain adequate frequency resolution (i.e. to deconvolve the power in closely spaced frequencies) *and* to observe a star of sufficient brightness or with telescopes of sufficient aperture that amplitudes can be measured with a precision of less than 0.001 magnitudes in all wavebands. To do this for V361 Hya stars with periods of ∼80–500 s, the detectors will necessarily be count-rate limited, so that high time resolution and high efficiency are essential.

4.3 Goals

The ultimate goals of studying pulsation in any star, including the sdB stars, reach well beyond an immediate asteroseismic solution. The identification of pulsation frequencies and modes, either through multicolour photometry or through spectroscopy, are pre-requisite to the latter and require considerable observational effort. Evidence is accruing that sdB stars do have masses in the range 0.46–0.50 $M_\odot$, with errors ∼ ±0.01 $M_\odot$ (cf. [20]). Envelope masses are being measured with similar precision, and both are in reasonable agreement with models for the internal structure of these stars. The question is now whether it will be possible to distinguish between sdB stars produced from different evolutionary histories—do mergers produce sdB stars with a different mass distribution from those produced from common-envelope evolution or from Roche-lobe overflow? Results from many more painstaking analyses will be necessary to answer this. Similarly, it is still too early to know whether asteroseismology can tell us whether iron is significantly enhanced in the driving zone—as required by pulsation theory. If it can, we will be able to make a direct test of models for both diffusion and opacity in the interior of sdB stars. Meanwhile, there is still much work to be done.

5 Does Pulsation need HTRA?

The use of pulsations for measuring stellar dimensions and probing their interiors requires the precise measurement of frequencies and amplitudes, and the accurate identification of the modes of oscillation. For measuring frequencies, time resolution is less important than a good timebase T, since $\delta \nu \propto 1/T$.

The need for high time resolution depends partially on what is meant. The ability to make repeat observations with time-intervals in the range 0.1–5 s is vital for the analysis of pulsations with periods in the range $1-10^4$ s. Pulsating stars in this category include variables of type V361 Hya, ro-Ap stars, β Cep, GW Vir, DBV and ZZ Ceti. High-speed spectroscopy and multi-wavelength photometry is necessary for the direct measurement of radius using Baade's method. One or both of these are necessary to provide the mode identifications required for asteroseismology. In particular, photometric amplitudes have to be measured to sub milli-magnitude precision and are thus limited by photon statistics. High time resolution is necessary because charge-storage detectors have to be read out frequently in order to deal with the necessary count-rates. Radial-velocity measurements must be made with short exposure times in order to avoid smearing as the surface accelerates through the pulsation cycle. Similarly, line profile variations can only be accurately interpreted if they are not badly smeared out.

High time resolution is also important for the study of local phenomena within the atmosphere of a pulsating star. It has not been possible to discuss these here, but it is noted that shocks at minimum radius will be of very short duration ($<P/20$). The study of differential motion of separate layers within the stellar atmosphere demands very high spectral resolution and high signal-to-noise with a time resolution at least one order of magnitude shorter than the pulsation period.

In the study of stellar pulsation, the need for even higher time resolution is less well-established. Millisecond resolution would be necessary to study neutron star oscillations, but these remain to be discovered.

References

1. Aerts, C., de Pauw, M., Waelkens, C., 1992, A&A 266, 294
2. Aerts, C., Jeffery, C.S., Fontaine, G., Dhillon, V.S., Marsh T.R., Groot P., 2006, MNRAS 367, 1317
3. Ahmad, A., Jeffery, C.S., Fullerton A.W., 2004, A&A 418, 275
4. Allard, F., Wesemael, F., Fontaine, G., Bergeron, G., Lamontagne, R.,q 1994, AJ 197, 1565
5. Aznar Cuadrado, R., Jeffery, C.S., 2002, A&A 385, 131
6. Bedding, T.R., Kjeldsen, H., 2006, Mem. Soc. Astr. Ital. 77, 384 Ber96
7. Billères, M., Fontaine, G., Brassard, P., et al., 1997. ApJ 487, L81

8. Billères, M., Fontaine, G., Brassard, P., et al., 1998. ApJ 494, L75
9. Billères, M., Fontaine, G., Brassard, P., Charpinet, S., Liebert J., Saffer R.A., 2000, ApJ 530, 441
10. Bracewell, R.N., 1999, "The Fourier Transform and Its Applications", McGraw Hill
11. Brassard, P., Fontaine, G., Billères, M., et al., 2001, ApJ 563, 1013
12. Charpinet, S., Fontaine, G., Brassard, P., Dorman, B., 1996, ApJ 471, L103
13. Charpinet, S., Fontaine, G., Brassard, P., et al., 1997, ApJ 483, L123
14. Charpinet, S., Fontaine, G., Brassard, P., 2000, ApJS, 131, 223
15. Charpinet, S., Fontaine, G., Brassard, P., 2001, PASP 113, 775
16. Charpinet, S., Fontaine, G., Brassard, P., Dorman, B., 2002, ApJS 140, 469
17. Charpinet, S., Fontaine, G., Brassard, P., Billères, M., Green, E.M., Chayer, P., 2005, A&A 443, 251
18. Charpinet, S., Fontaine, G., Brassard, P., Green, E.M., Chayer, P., 2005, A&A 437, 575
19. Charpinet, S., Fontaine, G., Brassard, P., Green, E.M., Chayer, P., 2005, ASP Conf Ser., 334, 619
20. Charpinet, S., Silvotti, R., Bonanno, A., et al., 2006, A&A 459, 565
21. Chayer, P., Fontaine, G., Wesemael, F., 1995, ApJS 99, 189
22. Córsico, A.H., Althaus, L.G., 2006, A&A 454, 863
23. Cox, J.P., 1980. "Theory of Stellar Pulsation", Princeton University Press
24. Cunha, M., 2005, JA&A 26, 213
25. Cuypers, J., 1987, "The Period Analysis of Variable Stars", Academiae Analecta, Royal Academy of Sciences, Volume 49, Number 3, Belgium
26. de Cat, P., 2002, ASP Conf. Ser. 259, 196
27. Dhillon, V., Marsh, T.R., & the ULTRACAM Team, 2002, ING Newsl. No. 6, p. 25.
28. Duncan, R., 1998, ApJL 498, 45
29. Fontaine, G., Brassard, P., Charpinet, S., Green, E.M., Chayer, P., Billères, M., Randall, S.K., 2003, ApJ 597, 518
30. Fukugita, M., Ichikawa, T., Gunn, J.E., Doi, M., 1996, AJ 111, 1748
31. Gianninas, A., Bergeron, P., Fontaine, G., 2006, AJ 132, 831
32. Green, E.M., Fontaine, G., Reed, M., et al., 2003, ApJ 583, 31.
33. Greenstein, J.L., Sargent, A.I., 1974, ApJS 28, 157
34. Han, Z., Podziadlowski, Ph., Maxted, P.F.L., Marsh, T.R., Ivanova N., 2002, MNRAS 336, 449
35. Han, Z., Podziadlowski, Ph., Maxted, P.F.L., Marsh, T.R., 2003, MNRAS 341, 669
36. Handler, G., Wood, M.A., Nitta, A., et al., 2002, ASP Conf. Ser. 259, 608
37. Handler, G., 2005, JA&A 26, 241
38. Heber, U., 1986, A&A 155, 33
39. Heber, U., Hunger, K., Jonas, G., & Kudritzki, R.P., 1984, A&A 130, 119
40. Heber, U., Kudritzki, R.P., Caloi, V., Castellani, V., Danziger J., 1986, A&A 162, 171
41. Heynderickx, D., Waelkens, C., Smeyers, P., 1994, A&AS 105, 447
42. Iben, I.,Jr., 1990, ApJ 352, 215
43. Iben, I.,Jr., Tutukov, A., 1985, ApJS 58, 661
44. Iben, I.,Jr., Tutukov, A., 1986, ApJ 311, 753
45. Jeffery, C.S., Pollacco, D., 2000, MNRAS 318, 974
46. Jeffery, C.S., Saio, H., 2006, MNRAS 371, 659
47. Jeffery, C.S., Saio, H., 2006, MNRAS 372, L48
48. Jeffery, C.S., Woolf, V.M., Pollacco, D., 2001, A&A, 376, 497
49. Jeffery, C.S., Dhillon, V.S., Marsh, T.R., Ramachandran, B., 2004, MNRAS 352, 699
50. Jeffery, C.S., Aerts, C., Dhillon, V.S., Marsh, T.R., Gänsicke B., 2005, MNRAS 362, 66
51. Jeffery, C.S., Heber, U., Dreizler, S., et al., 2006, Balt. Ast. 15, 321
52. Kikenny, D., 2002, ASP Conf. Ser., 259, 356
53. Kilkenny, D., Koen, C., O'Donoghue, D., Stobie, R.S., 1997, MNRAS 285, 640
54. Kilkenny, D., O'Donoghue, D., Koen, C., Lynas-Gray, A.E., Van Wyk, F., 1998, MNRAS 296, 329
55. Kilkenny, D., Koen, C., O'Donoghue, D., et al., 1999, MNRAS 303, 525

56. Kilkenny, D., Billères, M., Stobie, R.S., et al., 2002, MNRAS 331, 399
57. Koen, C., Kilkenny, D., O'Donoghue, D., Van Wyk, F., Stobie R.S., 1997, MNRAS 285, 645
58. Koen, C., 1998. MNRAS 300, 567
59. Kurtz, D.W., 1990, Ann.Rev.A&A 28, 607
60. Kurtz, D.W., Elkin, V.G., Mathys, G. 2006, MNRAS 370, 1274
61. Maxted, P.F.L., Marsh, T.R., North, R.C., 2000, MNRAS 317, L41
62. Michaud, G., Bergeron, P., Wesemael, F., Heber, U., 1989, ApJ 338, 417
63. O'Donoghue, D., Koen, C., Solheim, J.-E., et al., 1998, MNRAS 296, 296
64. O'Toole, S.J., Bedding, T.R., Kjeldsen, H., et al., 2000, ApJL 537, L53
65. O'Toole, S.J., Heber, U., Jeffery, C.S., et al., 2005. A&A 440, 6670
66. Randall, S.K., Fontaine, G., Green, E.M., et al., 2006, ApJ 643, 1198
67. Randall, S.K., Green, E.M., Fontaine, G., et al., 2006, ApJ 645, 1464
68. Ramachandran, B., Jeffery, C.S., Townsend, R.H.D., 2004, A&A 428, 209
69. Reed, M., Kilkenny, D., Kawaler, S.D., et al., 2000, Baltic Astronomy 9, 183
70. Reed, M.D., Green, E.M., Callerame, K., et al., 2004, ApJ 607, 445
71. Rutten, R., Gribbin, F., Ives, D., Bennett, T., Dhillon, V., 1997. "Drift-Mode CCD readout - User's Manual", Isaac Newton Group, La Palma
72. Saio, H., Jeffery, C.S., 2000, MNRAS 313, 671
73. Smith, H.A., 2004, "RR Lyrae Stars". Cambridge Astrophysics Series Vol. 27, Cambridge University Press
74. Thompson, M.J., Christensen-Dalsgaard, J., Miesch, M.S., Toomre, J., 2003, Ann.Rev.A&A 41, 599
75. Tremblay, P.-E., Fontaine, G., Brassard, P., Bergeron, P., Randall, S.K., 2006, ApJS 165, 551
76. Vučković, M., Kawaler, S.D., O'Toole, S., et al., 2006, ApJ 646, 1230
77. Vučković, M., Aerts, C, Østensen, R., et al., 2007, A&A submitted
78. Woolf, V.M., Jeffery, C.S., 2000, A&A 358, 1001
79. Woolf, V.M., Jeffery, C.S., Pollacco, D.L., 2002, MNRAS 329, 497
80. Woolf, V.M., Jeffery, C.S., Pollacco, D.L., 2002, MNRAS 332, 34
81. Woolf, V.M., Jeffery, C.S., Pollacco, D.L., 2003, "13th European Workshop on White Dwarfs". NATO-ARW Workshop Ser. p.95
82. Zhou, A.-Y., Reed, M.D., Harms, S., et al., 2006, MNRAS 367, 179

High-Speed Optical Spectroscopy

T. R. Marsh

Abstract The large surveys and sensitive instruments of modern astronomy are turning ever more examples of variable objects, many of which are extending the parameter space to testing theories of stellar evolution and accretion. Future projects such as the Laser Interferometer Space Antenna (*LISA*) and the Large Synoptic Survey Telescope (*LSST*) will only add more challenging candidates to this list. Understanding such objects often requires fast spectroscopy, but the trend for ever larger detectors makes this difficult. In this contribution I outline the science made possible by high-speed spectroscopy, and consider how a combination of the well-known progress in computer technology combined with recent advances in CCD detectors may finally enable it to become a standard tool of astrophysics.

1 Introduction

High-speed photometry has a long history in optical astronomy. The late 1960s and early 1970s saw the combination of digital recording techniques and photomultiplier tubes to give photon counting high-speed photometry, e.g. [30]. Equivalent high-speed spectroscopy has never been as straightforward and is still not a standard technique. There are several reasons for this. First, it is technically more difficult and expensive to record spectra at high-speed as one needs fast 2D imagers rather than single pixel devices. An early example, which was developed for faint object spectroscopy but could also take fast spectra, was the Image Photon Counting System (IPCS) [2]. The IPCS illustrates a problem typical of high-speed spectroscopy, because although it could take spectra with exposure times well below 1 second, it was rarely used to do so because its design limited its maximum count rate in any given pixel to less than 1 photon/second. A second difficulty of high-speed spectroscopy in the era of the IPCS was simply that it could produce more data than the computer technology of the 1970s and early 1980s could easily deal with; the author

---

T. R. Marsh
Department of Physics, University of Warwick Coventry CV4 7AL, UK
e-mail: tom.marsh@warwick.ac.uk

of this chapter recalls an observing run on the Anglo-Australian telescope which produced enough magnetic tape to run to and from the local town Coonabarabran 40 km away twice over. Computer technology has since taken enormous strides of course, but the 1980s brought the biggest obstacle of all for high-speed spectroscopy, namely Charge-Coupled Devices (CCDs). Although wonderful detectors in many respects, the CCDs employed in most observatories have two significant disadvantages: they are slow, often taking several tens of seconds to read out, and significant noise is added to each pixel, burying the small signals characteristic of high-speed spectroscopy. The slow readouts of standard CCDs has lead to a small revival of the old photographic method of "trailed spectra" where the target is moved along the slit to give time-resolved spectra [13, 35], but the disadvantages of this method, such as seeing-dependent time-resolution, mean that it is really only a stop gap measure and I will not consider it further.

Despite the problems of CCDs, their other excellent characteristics, in particular their high quantum efficiency (QE), have lead to their complete dominance of optical astronomy, and so the focus of this chapter will be tilted towards the use of CCDs for high-speed spectroscopy. Moreover, CCDs are so dominant that they provide us with an empirical definition of what "high-speed" means when applied to spectroscopy, i.e. any spectroscopic observation that is difficult to carry out with normal CCDs is "high-speed". This does not always mean very fast: readout noise can make echelle spectroscopy of an 18th magnitude object difficult to carry out with a time resolution shorter than 10 minutes.

With the above broad definition in mind, an overview is given of the science that high-speed spectroscopy makes possible followed by the technical requirements and the possible application of electron-multiplying CCDs to this area.

2 Scientific Motivation

As the introduction hinted, for a variety of reasons, high-speed spectroscopy has yet to take off fully and many applications exist only in the imagination. There are nonetheless a reasonable number of published examples which give an idea of what to expect, and in addition we have the many applications of high-speed photometry to draw on as a resource when considering what can be learned from high-speed spectroscopy. In this section I will look at what high-speed spectra can tell us when applied to the following phenomena, sticking as far as possible to what is already known:

- pulsating white dwarfs and subdwarf B (sdB) stars
- accreting binary stars
- white dwarf binary stars

Before doing so, it is worth mentioning objects yet to be discovered because surveys are being planned to look for short-timescale variable phenomena which are bound to discover many objects that will need high-speed spectroscopy. Pre-eminent perhaps, although still far-off, is the Large Synoptic Survey Telescope (*LSST*), in which it is planned to use an 8.4 m telescope to survey the observable sky once every 3 nights in 15-second exposures.

2.1 Pulsating White Dwarfs and sdB Stars

As well demonstrated by helioseismology, pulsations in stars can give us unique insights into stellar structure. The area of asteroseismology tries to use pulsations in other stars in this manner. It is observationally demanding first and foremost because of the need for long, uninterrupted time series in order to obtain clean time series. In the case of white dwarfs and sdB stars, which have pulsation periods of order 100 to 1000 seconds, fairly fast observations are also a requirement. If all goes well, it is possible to pin down stellar parameters with great precision. For instance studies of the pulsating sdB stars PG0014+067, PG1219+534 and Feige 48 have measured their surface gravities to $\sim 1\%$, their masses to $\sim 2\%$ and also determined the masses of their hydrogen envelopes, for which no other method exists, to $\sim 10\%$ [4, 8, 7]. Despite this, even large datasets can leave stars unsolved, especially if they display few pulsation modes, the difficulty being the secure identification of a particular eigenmode with a given frequency.

It is unlikely that spectroscopic data can ever be taken with the same time coverage and uniformity as the photometric data. However spectra can add extra information missing from the photometric studies because the variation of limb darkening with wavelength in combination with the different patterns of different eigenmodes can lead to different variations of pulsation amplitude with wavelength [34]. Figure 1 shows predictions [9] of the amplitudes of different eigenmodes for the ZZ Ceti star G29-38. This shows that spectra may allow $l=3$ and $l=4$ eigenmodes to be distinguished from $l=1$ and $l=2$ eigenmodes. A signal-to-noise ratio of 1000 only gives a signal-to-noise of 10 in the amplitude spectrum when the amplitude itself is of order one percent as it is in ZZ Ceti stars, and therefore extremely high quality data are needed to separate $l=1$ from $l=2$. Thus this application requires moderately fast data acquisition to resolve the pulsations, but also a large aperture to give high signal-to-noise. Figure 2 shows the result of 4 hours of Keck II/LRIS time devoted to the brightest ZZ Ceti star, G29-38 ($V=13$) [41, 9]. The mean spectrum of these data appears almost noiseless [9], as might be expected for a bright star on a large telescope, but comparing Figs. 1 and 2 shows how necessary the large aperture is in this case.

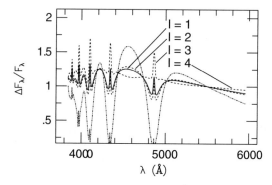

Fig. 1 The pulsation amplitudes versus wavelength for spherical harmonics Y_{lm} with different l-values. Each curve is normalised by its value at 5500Å. The figure is Fig. 3 taken from [9]

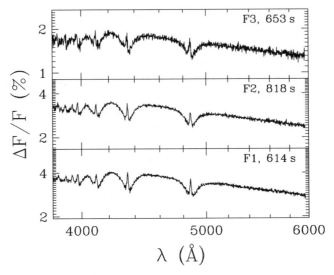

Fig. 2 Spectra of the first pulsation amplitudes of the ZZ Ceti star G29-38 observed on Keck-II/LRIS by [41]. The figure is adapted from Fig. 4 of [9]

This case illustrates another important feature of high-speed spectroscopy: it does not have to be especially "high-speed" by photometric standards to strain the capabilities of standard instrumentation: the data were taken as 700 × 12 second exposures, but for each exposure there were a further 12 seconds "deadtime" for readout, etc., so the observations were only 50% efficient. This is quite common; indeed 12 seconds is commendably short in this respect. Reducing deadtime is probably the single most effective way of enabling high-speed spectroscopy and would save countless hours of precious telescope time.

The end result of the work shown in Figs. 1 and 2 is that most of the modes are l = 1 modes. This shows the power of spectroscopy because in principle this method can be applied to stars showing just a few modes where it would certainly not be possible to unravel the modes purely photometrically.

2.2 Accreting Binary Stars

I now move on to systems which display variability on much shorter timescales than the white dwarf and sdB pulsators. The shortest dynamical timescales in accreting binaries with compact objects range from 1 to 10 seconds in the case of accreting white dwarfs down to about a millisecond in the case of the neutron stars and stellar mass black holes. Variability on seconds timescales is well established from the "dwarf nova oscillations" displayed by dwarf novae during their outbursts. To my knowledge, dwarf nova oscillations have only been observed spectroscopically with sufficient speed to detect them once [38], but this single observation was enough to show the great diagnostic potential spectra may hold for these poorly-understood but remarkable phenomena.

The variability timescales seen in black-hole and neutron star systems can depend upon the system brightness and instrumental limitations as much as the object. X-ray variability at kilohertz frequencies has been seen in X-ray binaries, while optical variability on timescales well below 0.1 seconds has been seen in bright systems [28]. Using ULTRACAM on the Very Large Telescope (VLT) we found significant flaring on timescales of a few seconds in a faint quiescent black-hole accretor (Shahbaz et al, in prep). One can only speculate upon the spectroscopic signatures of this variability at optical wavelengths as no such observations have been made, but it is very likely that there will be some, although in the quiescent black-hole case only an "Extremely Large Telescope" (ELT) would be up to the task. A case of particular interest is the fluorescent Bowen blend emission from the donor stars seen in some X-ray binaries and which in some cases has revealed the donor star for the very first time [36]. This emission is presumably X-ray driven, and can be expected to respond to X-ray variability; some evidence of this has been obtained using narrow-band photometry with ULTRACAM [6], but a much cleaner signature could come from high-speed spectra.

A dramatic example of spectral variability from an accreting binary is shown in Fig. 3. This shows the eclipse phases of the white dwarf accretor IP Peg in one of its temporary high states (outburst) [18]. The lines from this system come primarily

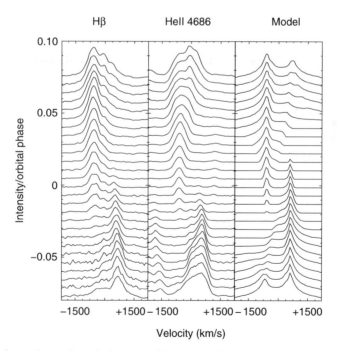

Fig. 3 The first and second panels show two of the emission lines of the dwarf nova IP Peg as they are eclipsed, with orbital phase increasing up the plot. The spectra were taken on Subaru during an outburst of IP Peg; the lines are dominated by light from the bright accretion disc during this state. The second panel shows a simple computation of the eclipse of a disc in prograde, Keplerian rotation. Figure adapted from Fig. 2 of [18]

from the accretion disc and have the well-known double-peaked profiles that come from Doppler shifting in accretion discs [17]. The spectra plotted cover about 30 minutes and show very large changes as first the approaching side of the disc is obscured, affecting the blue-shifted parts of the lines, followed soon after by their red-shifted counterparts. The right-most panel of Fig. 3 shows a simple model which captures much of the phenomenology of the data while at the same time showing significant discrepancies. The nature of these is not understood although it is possible that the disc during outburst is far from Keplerian in nature. This example is an even starker indication of the problems often faced with facility instruments: IP Peg reaches $V=12$ in outburst, yet despite using the 8 m Subaru, the time resolution here was a sluggish 80 seconds, each spectrum consisting of 30 seconds exposure followed by 50 seconds of deadtime. The telescope, instrument and object would have allowed *much* higher speed than this. Ideally one would want to sample fast enough to resolve structure comparable in size to the white dwarf, which would be about 5 seconds in this case. Unfortunately the detector/data acquisition system was not up to the job, a not-uncommon situation, as instruments are rarely built with high-speed applications in mind.

I finish off this section with a look at Doppler imaging as applied to accreting binary stars in the form of Doppler tomography [24, 22]. Doppler tomography uses the information in line profiles as a function of phase to image binaries. Notable successes have been the discovery of spiral structure in accretion discs during outburst [37, 14] and the unravelling of the complex accretion structures in the magnetic polar class of accreting white dwarfs [35]. Doppler tomography provides a quantitative illustration of the need for high-speed spectroscopy and is useful in defining requirements for such work. This is because the resolution in Doppler tomography is limited by *both* spectral *and* time resolution. If one is aiming for a spatial resolution Δx, then this imposes the following restrictions on the spectral resolution $R_\lambda = \lambda/\Delta\lambda$ and the time resolution Δt:

$$R_\lambda \sim (c/\Omega)(\Delta x)^{-1}, \tag{1}$$
$$\Delta t \sim V^{-1}\Delta x, \tag{2}$$

where the orbital angular frequency $\Omega = 2\pi/P$ where P is the orbital period, and V is the velocity of the feature being imaged. Small Δx requires large R_λ and small Δt. These relations can be combined into a single one relating spectral and time resolution:

$$\Delta t \sim \Omega^{-1} \frac{c}{V} R_\lambda^{-1}. \tag{3}$$

Taking typical values $P = 1.5$ hours and $V = 700$ km s^{-1}, and assuming that we are working on a spectrograph with $R_\lambda = 10{,}000$, then we require $\Delta t < 30$ seconds in order that smearing during the exposures does not degrade the resolution. Equations 1 and 2 imply that the number of counts per detector resolution element per exposure scales as $(\Delta x)^2$, and thus Doppler tomography can become challenging even on quite bright objects and large telescopes.

2.3 White Dwarf Binary Stars

For my final example of applications of high-speed spectroscopy I turn to binary stars with white dwarf components. This is another case where there are only a few examples, but where one can point to future applications which will prove tough for even the largest telescopes. Work over the past decade has established the existence of a huge population of detached, close double white dwarfs, with orbital periods of a few days or less [23, 29]. The shortest period of these is WD0957-666 with $P = 88$ minutes (Fig. 4) [3, 26, 27]. There are estimated to be of order 100 million such systems in our galaxy, and they should be steadily spiralling towards shorter periods under the action of gravitational wave radiation. The existence of systems such as WD0957-666, which will merge within 200 million years, proves that there must be a population of much shorter period systems that we have yet to detect. This was realised some time ago and it is thought that this population, along with their accreting counterparts, the AM CVn stars, will be the dominant gravitational wave emission sources at frequencies of order 10^{-3} Hz, right in the waveband of the Laser Interferometer Space Antenna (*LISA*) [15, 31, 32] which unlike ground-based detectors is sensitive to gravitational waves of relatively long period. *LISA* is predicted to be able to detect several thousand of these sources and will be able to narrow down the location of many of them to less than one

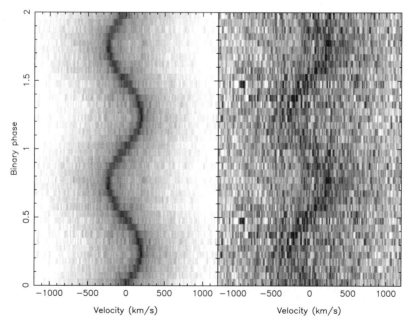

Fig. 4 Trailed spectra around Hα of the shortest period double white dwarf, WD0957-666, based upon the data of [26]. In the left panel, the brightest of the two white dwarfs is clearly seen; the right-hand panel shows the result of subtracting a fit to the brightest star which reveals its companion. The two stars will merge under gravitational radiation in about 200 million years

degree [10, 39], and it should be possible to optically identify some of them. Optical follow-up can help in the determination of parameters from the *LISA* data, but will not be easy: the majority of sources will be fainter than $V = 22$, and yet have periods of order 10 minutes. Obtaining spectroscopy of these will clearly require very low noise detectors, even on the largest telescopes, including ELTs. For instance, when observing the 88 minute period double white dwarf WD0957-666 [27],on the 3.9 m Anglo-Australian Telescope, our exposure times of 500 seconds were a compromise between smearing the spectra over too much of the orbit versus suffering too much readout noise, and this was on a relatively bright target with $V = 14.6$. This is an explicit example of the constraints discussed for Doppler tomography in section 2.2.

A similarly challenging application for high-speed spectroscopy emerges from the eclipses of white dwarfs in detached white dwarf/main-sequence binaries. These are relatively easily studied photometrically [5], but the extra light losses and dispersion of spectroscopy make them a much tougher subject for spectroscopy and I am not aware of any spectroscopic studies which target the eclipses in these objects. There is nevertheless a compelling reason to do so which is that spectroscopy of the white dwarf as it goes in and out of eclipse has the potential to determine its rotation rate by measurement of the radial velocity shifts induced by the eclipse. The rotation rate of white dwarfs in such systems has implications for the stability of double white dwarf binary stars [25]. This requires taking spectra every 5 seconds or so (preferably less) and once again, for typical systems, telescopes and instruments, moves us into the realm of readout noise.

This is an appropriate point to change topic and look at the technical difficulties of high-speed CCD spectroscopy.

3 CCD Spectroscopy

3.1 Standard CCDs

There are two key issues which make the use of standard CCDs difficult for high-speed work. First of all CCDs are usually slow to read out. It is not unusual for CCDs to be read out at ∼100 kHz pixel rates, so that an entire chip can take several tens of seconds to be read out. Many chips can be read out faster, but then one suffers worse read noise, which can more than offset any advantage of high-speed. The amount of telescope time spent reading out CCDs is potentially frightening: consider a telescope which spends 8 hours per night, 365 night/year devoted to taking spectra of 8 one hour-long spectra per night together with calibration arcs before and after each one. If each spectrum takes 1 minute to read out, then *18 solid nights* would be spent reading out the CCDs. Worse still, any programme that required exposure times shorter than 60 seconds would run at below 50% efficiency. As some of the examples of Sect. 2 showed, this is not as rare as one might imagine. Luckily there are ways around slow readouts. First one can group pixels (bin) and read out sub-sections of the chip (window). These are both often very effective. More radical is to use a frame transfer CCD allowing one half of the CCD to be read out while the other is being exposed. The EEV 47-20 CCDs used by ULTRACAM (Dhillon, this volume) allow most observations to be carried out with a deadtime of only 24 milliseconds using this technique. This would be more than adequate for the vast majority of feasible spectroscopic applications.

This brings us to the more fundamental issue of readout noise. This plays a much more important role in optical spectroscopy than it does in photometry. The key quantity to have in mind is the variance V on a given pixel which is given by

$$V = R^2 + C, \qquad (4)$$

where R is the RMS readout noise in electrons (typically ~ 3) and C is the number of electrons in the pixel, and is equal to the number of photons detected (which in general is the sum of target flux, sky background and dark counts, although the latter can usually be neglected). This is the simplest possible version of this relation and assumes perfect flat-fielding, but this is more often than not a reasonable approximation in the case of high-speed work. Once C drops below $R^2 \sim 9$, one is starting to lose out significantly to readout noise. There are techniques to alleviate this: binning again is important. If one is observing a point source, then it makes no sense to over-resolve the spatial profile, and in fact under-sampling of the spatial profile is not always much a drawback except for spotting cosmic rays, therefore binning in the spatial direction is often very useful. I have found that people often do not realise how significant an improvement binning spatially can make, but it not hard to demonstrate. Consider a case where two pixels with a total count $C = R^2$ are binned into one, and assume that C is dominated by the object (as opposed to sky background). Then the ratio of signal-to-noises, binned to unbinned, is $(3/2)^{1/2}$, equivalent to a 50% increase of exposure time. Apart from this, the only other option may be to increase the exposure time: consider again the marginal case with $C = R^2$ when one decides to double the exposure time. Then it is straightforward to show that the improvement in signal-to-noise corresponds to the improvement that would be obtained with an $8/3 \sim 2.7$-fold rather than simply two-fold increase in exposure time in the zero readout (Poisson limited) case. Of course increasing the exposure time may not be an option; after all, one is trying to resolve intrinsic variability in a target. The only other capability one then has to affect the final signal-to-noise is not to make things worse by poor reduction; it is well worth using extraction techniques designed to optimise the signal-to-noise ratio [16, 21] if at all possible.

3.1.1 When is One Readout Noise Limited?

Despite the various ways in which one can combat readout noise, there is in the end no getting around it apart from changing the detector design which I look at next. Before doing so I pause to consider the parameter space where readout noise is important. Consider a telescope of diameter D, feeding a spectrograph of resolution R_λ leading to a detector with N_d pixels per resolution element in the dispersion direction and N_s pixels across the spatial profile (loose definitions, but it is the orders of magnitude that matter here). If ϵ is the total throughput of atmosphere, instrument and detector in terms of photons detected versus those actually incident upon the atmosphere, then an exposure of t seconds of a target of AB magnitude m will produce C counts given by

$$C = \frac{11853\epsilon}{N_d N_s R_\lambda} \left(\frac{D}{1\,\mathrm{m}}\right)^2 \left(\frac{t}{1\,\mathrm{s}}\right) 10^{(16.4-m)/2.5}. \tag{5}$$

Assuming $N_d = N_s = 3$, $R_\lambda = 2000$, $D = 8\,\mathrm{m}$, $\epsilon = 0.2$, then to have $C > R^2$ for $R = 3$ electrons implies the following relation between m and t:

$$m < 16.33 - 2.5 \log\left(\frac{t}{1\,\mathrm{s}}\right). \tag{6}$$

One is therefore readout noise limited for a target of 20th magnitude target and 30 second exposures on a typical spectrograph on (currently) the world's largest telescopes. Moving to an echelle with, say, $R_\lambda = 60,000$ and $\epsilon = 0.1$, then this limit must be shifted up by 4.5 mags. A simulation of the effect of readout noise is shown in Fig. 5 for a star of considerable current interest, RX J0806+1527. This is potentially the shortest period binary known, and a strong gravitational wave source, consisting of two white dwarfs with an orbital period of just 321 seconds [33, 19]. This has yet to be proven however, and spectroscopy is by far the most promising avenue for testing it. It is however, as Fig. 5 shows, pushing the capabilities of even

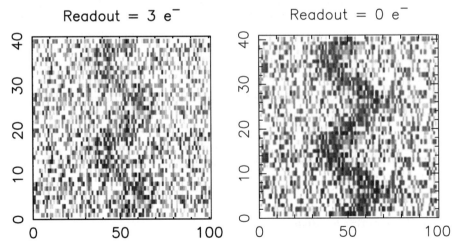

Fig. 5 A simulation of the trailed spectrum of the $V=21$ ultra-compact binary star, RX J0806+1527, as taken with 1 night using the VLT and the FORS2 spectrograph (600 grism) with (left) and without (right) readout noise. Time advances upwards; the x-range is centred on a single emission line. The exposure time was taken to be 30 seconds in order to resolve the presumed 321 second orbital period of the object, but detector readout time was assumed to be zero in each case. The emission line was taken to be about as strong as seen in the average spectrum of [19]

the VLT. The fact that as of mid-2006 this observation had yet to be performed is testament to its difficulty. Many would consider $V = 21$ to be "bright", or at least, not especially faint, but it certainly is when one is forced to take short exposures. Observations of this sort are exactly those needed for the gravitational wave sources, Sect. 2.3, except they will in general be more difficult still.

Having seen the limitations imposed by readout noise, I now turn to how it can potentially be avoided.

3.2 Electron-Multiplying CCDs

The readout noise that so seriously limits normal CCDs is added by the amplifier which converts the small charge on each pixel into a voltage. To some extent readout noise can be reduced by taking longer over the double-correlated sampling used in low-noise CCD readouts (i.e. spending longer integrating the voltage levels before and after the charge on each pixel is cleared), but $1/f$ noise limits the extent to which one can push this and one cannot in any case take too long reading each pixel if one is interested in high speed. Electron-multiplying CCDs (hereafter EMCCDs, but also known as "low light level" or L3CCDs) get round this limitation in a clever way. In these CCDs, an extra series of stages is added to the serial register *before* the charge reaches the amplifier. These stages can be clocked with higher than normal voltages which creates a significant probability that one electron will generate another as the charges are moved from stage to stage. With enough stages, a single detected electron can lead to an avalanche of several hundred or even thousands of electrons. These then dwarf the readout noise added by the amplifier. The formulae for signal-to-noise from these devices are more complex than for standard CCDs [1], but it is necessary to review them here in order to understand the advantages and limitations of these devices.

To first order, the avalanche gain register can be modelled as a series of stages at each of which there is a probability p that any electron will spawn another. If there are N such stages in total then the mean gain g, can be shown to be[1]

$$g = (1+p)^N. \qquad (7)$$

The value of p, and therefore g is a parameter that can be controlled by adjustment of the driving voltages. For example if we take $N = 536$ (appropriate to the CCD97 manufactured by the company, e2v) and $p = 0.011$, then $g = 352$. This gain is itself variable however (after it all it can in principle lie anywhere from 1 to $2^N \gg g$ although the extremes are unlikely) which increases the variance of the output over that expected from pure Poisson noise. Thus in a normal CCD readout, a mean detection rate of C electrons leads to the variance contribution of C in (4), while in

---

[1] Do not confuse this gain with the usual "gain" of CCDs which is simply a conversion factor between electrons and recorded counts or ADU. To avoid confusion I only ever talk here in terms of electrons not ADU.

an EMCCD the signal is amplified to gC on average and the corresponding variance is $(g^2 + \sigma^2)C$ where σ^2 is the variance of the gain and is given by

$$\sigma^2 = \frac{1-p}{1+p}\left(g^2 - g\right). \tag{8}$$

Had there been no dispersion in the gain then the variance would have been g^2C, exactly what one would expect if simply multiplying C by a constant; the variance in the gain thus adss extra noise. If, as is the case in practice, $p \ll 1$ and $g \gg 1$, then $\sigma^2 \approx g^2$, and the variance is a factor of 2 larger than a constant gain would have produced. Thus (4) is modified to

$$V = R^2 + 2g^2C, \tag{9}$$

and the signal-to-noise ratio for one pixel is given by

$$\frac{C}{\sqrt{(R/g)^2 + 2C}}, \tag{10}$$

compared to

$$\frac{C}{\sqrt{R^2 + C}}, \tag{11}$$

for a normal CCD. The extra variance can be thought of as being equivalent to a 50% drop in QE. Put this way it sounds bad, but comparing the above two relations for signal-to-noise, and assuming that g is so large that R/g can be neglected, one can see that the EMCCD gains once $C < R^2$. In other words once $C \sim R^2$, normal CCDs have also effectively lost a factor 2 in QE, but unlike EMCCDs normal CCDs carry on getting worse as C drops. This analysis also shows that (5) marks the dividing line between normal CCDs versus EMCCDs running in "linear" mode with counts proportional to the voltage of the amplifier.

At very low count rates, the full QE can be recovered by operating in a photon counting mode. In this case rather than take the output divided by the mean gain as an estimate of the counts, one defines a threshold T such that an output $> T$ is recorded as one photon, while an output $< T$ is recorded as zero. Provided that one is operating in a regime where the chance of more than 1 e$^-$/pixel/exposure is small, this adds no extra noise to the output, and the signal-to-noise becomes

$$\frac{C}{\sqrt{(R/g)^2 + C}}. \tag{12}$$

If g is large, this implies ideal, Poisson-limited performance. Such a device could give us data looking like the right- rather than the left-hand side of Fig. 5.

3.2.1 Problems: Thresholds, Non-Linearity and Clock-Induced Charges

EMCCDs suffer from some drawbacks, that apply to normal CCDs but which are not usually apparent above readout noise. First of all, given the variable gain, which can in principle be as low as 1, a finite threshold must imply some loss of counts and therefore QE. To know how much, one needs to know the full probability distribution of the gain. For a single electron input, low p, high g case, the output probability distribution function (PDF) for n counts is quite well approximated by $g^{-1} \exp -n/g$ [1]. Figure 6 shows a comparison between an exact calculation of the PDF and this simple approximation, showing it to be good. This immediately implies that a threshold T leads to a decrease in the QE by factor of $\exp(-T/g)$. One therefore wants $T \ll g$ to avoid too large a loss in QE, but at the same time T must not be so low that readout noise alone leads to spurious counts, i.e. $T > 4-5$ times R, so that for Gaussian noise there is a very small chance of counts induced artificially by readout noise.

The next problem is a common feature of photon counting devices, which is non-linearity at "high" count rates. In this case significant non-linearity will set in when the mean count rate rises above $\sim$e0.2/pixel/exposure. Thus one can be limited by the rate at which pixels can be clocked out. This is exactly like the IPCS except that one can elect in the case of EMCCDs to work in the linear mode if it is clear that the count rates are too high for reliable photon counting.

The third and worst problem is a new phenomenon, or at least one that is only revealed by the low noise capabilities of EMCCDs: "Clock Induced Charges" or CICs. These are electrons which are spontaneously produced during clocking. Their statistics are complex and I leave a detailed discussion of them to the appendix where I include calculations that as far as I am aware have not been published before.

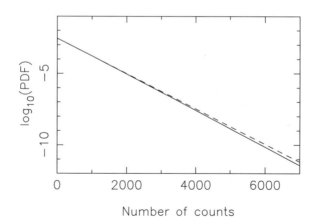

Fig. 6 The solid line shows an exact computation of the output probability distribution function of a 536 stage electron multiplying register with $p = 0.011$ given an input of 1 electron. The dashed line shows the approximation $\exp(-n/g)/g$ from [1] where n is the number of counts and the mean gain $g = 352$ in this case. The approximation only deviates significantly for values which are very rare

As far as we are concerned they are equivalent to a readout noise and lead to a variance (in the photon counting case) of the form

$$V = \left(\frac{R}{g}\right)^2 + R_C + C, \qquad (13)$$

where R_C is the number of CICs/pixel/exposure, equivalent to a readout noise before avalanche amplification of $R_C^{1/2}$. The C here should be interpreted as including the $\exp(-T/g)$ factor discussed above, and the R_C rate will similarly be threshold dependent in this case. The CICs are better thought of as a source of readout noise than background counts because they are incurred per exposure, not per unit time.

The three possible operating modes of CCDs are summarised in Fig. 7.

For EMCCDs the focus is often on photon counting mode and the need to clock the pixels out fast to retain linearity. Figure 7 shows that there remains a role for the linear mode, the fraction of parameter space it occupies depending upon the CIC rate. Note that I have extended the regime of the linear mode in Fig. 7 towards lower counts than might seem justified by the curves; this is because of the non-linearity of photon counting. In principle it can be corrected to some extent, and this is built into Fig. 7, however in practice non-linearity corrections do not work as well as one might hope and I have restricted the photon counting mode to rates less than 0.3 e^-/pixel/exposure.

Clearly the CIC rate is of central importance in how well these devices will perform in practice, as has been recognised before [11]. Rates in the range 0.004 to 0.1

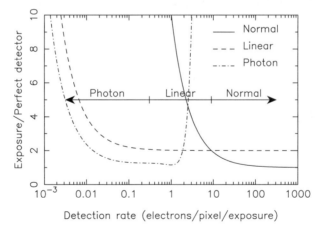

Fig. 7 The three operating modes of CCDs are shown in terms of the exposure time taken to reach a given signal-to-noise compared to a "perfect" detector of identical QE. Normal mode (solid line) is best at high count levels drops off sharply once readout noise dominates. Photon counting mode (dot-dashed line) is best at very low count levels but cannot go past 1 e^-/pixel/exposure owing to non-linearity. In between, linear or "proportional" mode (dashed line) is best despite the 50% QE drop caused by the variable avalanche gain. The rise at very low counts is caused by CICs, with a rate of 0.01 e^-/pixel/exposure assumed here

CICs/pixel/exposure have been quoted [40]. One can hope for manufacture-driven improvements with time, and controller improvements have a significant role to play too [20], but one should not lose sight of the fact that already even the high rate of 0.1 CICs/pixel/exposure is equivalent to an extremely low readout noise and can allow the count rate C to drop by a factor of 100 compared to normal CCDs before the noise floor is hit; this is an impressive potential gain and exactly what it needed to carry out high-time resolution spectroscopy on faint objects. Having said that, CICs do negate the QE advantages of CCDs compared to alternatives, in particular GaAs image intensifiers [11].

3.2.2 OPTICON-Funded L3CCD for Spectroscopy

Given the uncertainties over CIC rates, thresholds, pixel clocking rates and the like, EMCCDs are very much in the development phase for astronomy. We don't yet know indeed whether they are reliable for spectroscopy in the sense that they can return accurate atomic line profiles, equivalent widths and radial velocities. Thus as part of the OPTICON programme of the EU's Framework Programme 6 (FP6), the Universities of Sheffield and Warwick, and the Astronomy Technology Centre, Edinburgh have started a programme to characterise an EMCCD for spectroscopic work. We have purchased a CCD201 which is a frame transfer device with a 1k × 1k imaging area and with one normal and one avalanche readout. As of mid-2006, the device is mounted in a cryostat and producing test data. We will be using hardware developed for the high-speed camera ULTRACAM [12]. In terms of imaging area, this device is not competitive with existing detectors on spectrographs. On top of this the ULTRACAM controller will not push the device to its limits and so clear improvements are possible even now. However, it will allow us to see whether these devices are a promising route to explore for future development of high-speed spectroscopy.

4 Conclusion

High-speed optical spectroscopy offers a number of unique diagnostics of extreme astrophysical environments and will become only more necessary as larger surveys discover more unusual objects. Unfortunately standard CCDs, the workhorses of optical astronomy, are not well suited to the high-speed and low noise required when light is dispersed. This may be changing with recent developments of avalanche gain CCDs, but they must be tested on celestial objects before we know whether this is truly the case. This is the aim of a programme funded under the EU FP6's OPTICON consortium.

Acknowledgments I thank PPARC and OPTICON for funding during this work, and Vik Dhillon, Andy Vick, Derek Ives and Dave Atkinson for their help and collaboration in this work.

References

1. A. G. Basden, C. A. Haniff, and C. D. Mackay. Photon counting strategies with low-light-level CCDs. MNRAS, 345:985–991, November 2003.
2. A. Boksenberg. UCL Image Photon Counting System. In *Auxiliary Instrumentation for Large Telescopes, Proc. of the ESO-CERN Conference*, ed. S.Lausten, A.Reiz, page 295, 1972.
3. A. Bragaglia, L. Greggio, A. Renzini, and S. D'Odorico. Double degenerates among da white dwarfs. ApJ, 365:L13–L17, December 1990.
4. P. Brassard, G. Fontaine, M. Billères, S. Charpinet, J. Liebert, and R. A. Saffer. Discovery and Asteroseismological Analysis of the Pulsating sdB Star PG 0014+067. ApJ, 563:1013–1030, December 2001.
5. C. S. Brinkworth, T. R. Marsh, V. S. Dhillon, and C. Knigge. Detection of a period decrease in NN Ser with ULTRACAM: evidence for strong magnetic braking or an unseen companion. MNRAS, 365:287–295, January 2006.
6. J. Casares, T. Muñoz-Darias, I. G. Martínez-Pais, R. Cornelisse, P. A. Charles, T. R. Marsh, V. S. Dhillon, and D. Steeghs. Echo Tomography of Sco X-1 using Bowen Fluorescence Lines. In L. Burderi, L. A. Antonelli, F. D'Antona, T. di Salvo, G. L. Israel, L. Piersanti, A. Tornambè, and O. Straniero, editors, *AIP Conf. Proc. 797: Interacting Binaries: Accretion, Evolution, and Outcomes*, pages 365–370, October 2005.
7. S. Charpinet, G. Fontaine, P. Brassard, M. Billères, E. M. Green, and P. Chayer. Structural parameters of the hot pulsating B subdwarf Feige 48 from asteroseismology. A&A, 443:251–269, November 2005.
8. S. Charpinet, G. Fontaine, P. Brassard, E. M. Green, and P. Chayer. Structural parameters of the hot pulsating B subdwarf PG 1219+534 from asteroseismology. A&A, 437:575–597, July 2005.
9. J. C. Clemens, M. H. van Kerkwijk, and Y. Wu. Mode identification from time-resolved spectroscopy of the pulsating white dwarf G29-38. MNRAS, 314:220–228, May 2000.
10. A. Cooray, A. J. Farmer, and N. Seto. The Optical Identification of Close White Dwarf Binaries in the Laser Interferometer Space Antenna Era. ApJ, 601:L47–L50, January 2004.
11. O. Daigle, J.-L. Gach, C. Guillaume, C. Carignan, P. Balard, and O. Boisin. L3CCD results in pure photon-counting mode. In J. D. Garnett and J. W. Beletic, editors, *Optical and Infrared Detectors for Astronomy. Edited by James D. Garnett and James W. Beletic. Proceedings of the SPIE, Volume 5499, pp. 219–227 (2004).*, pages 219–227, September 2004.
12. V. Dhillon and T. Marsh. ULTRACAM - studying astrophysics on the fastest timescales. *New Astronomy Review*, 45:91–95, January 2001.
13. S. Falter, U. Heber, S. Dreizler, S. L. Schuh, O. Cordes, and H. Edelmann. Simultaneous time-series spectroscopy and multi-band photometry of the sdBV PG 1605+072. A&A, 401:289–296, April 2003.
14. P. J. Groot. Evolution of Spiral Shocks in U Geminorum during Outburst. ApJ, 551:L89–L92, April 2001.
15. D. Hils, P. L. Bender, and R. F. Webbink. Gravitational radiation from the galaxy. ApJ, 360:75–94, September 1990.
16. K. Horne. An optimal extraction algorithm for CCD spectroscopy. PASP, 98:609–617, June 1986.
17. K. Horne and T. R. Marsh. Emission line formation in accretion discs. MNRAS, 218:761–773, February 1986.
18. R. Ishioka, S. Mineshige, T. Kato, D. Nogami, and M. Uemura. Line-Profile Variations during an Eclipse of a Dwarf Nova, IP Pegasi. PASJ, 56:481–485, June 2004.
19. G. L. Israel, W. Hummel, S. Covino, S. Campana, I. Appenzeller, W. Gässler, K.-H. Mantel, G. Marconi, C. W. Mauche, U. Munari, I. Negueruela, H. Nicklas, G. Rupprecht, R. L. Smart, O. Stahl, and L. Stella. RX J0806.3+1527: A double degenerate binary with the shortest known orbital period (321s). A&A, 386:L13–L17, May 2002.

20. C. Mackay, A. Basden, and M. Bridgeland. Astronomical imaging with L3CCDs: detector performance and high-speed controller design. In J. D. Garnett and J. W. Beletic, editors, *Optical and Infrared Detectors for Astronomy. Edited by James D. Garnett and James W. Beletic. Proceedings of the SPIE, Volume 5499, pp. 203–209 (2004).*, pages 203–209, September 2004.
21. T. R. Marsh. The extraction of highly distorted spectra. PASP, 101:1032–1037, November 1989.
22. T. R. Marsh. Doppler Tomography. *LNP Vol. 573: Astrotomography, Indirect Imaging Methods in Observational Astronomy*, pages 1–26, 2001.
23. T. R. Marsh, V. S. Dhillon, and S. R. Duck. Low-Mass White Dwarfs Need Friends - Five New Double-Degenerate Close Binary Stars. MNRAS, 275:828–+, August 1995.
24. T. R. Marsh and K. Horne. Images of accretion discs. ii - doppler tomography. MNRAS, 235:269–286, November 1988.
25. T. R. Marsh, G. Nelemans, and D. Steeghs. Mass transfer between double white dwarfs. MNRAS, 350:113–128, May 2004.
26. P. F. L. Maxted, T. R. Marsh, and C. K. J. Moran. The mass ratio distribution of short-period double degenerate stars. MNRAS, 332:745–753, May 2002.
27. C. Moran, T. R. Marsh, and A. Bragaglia. A detached double degenerate with a 1.4-h orbital period. MNRAS, 288:538–544, June 1997.
28. C. Motch, S. A. Ilovaisky, and C. Chevalier. Discovery of fast optical activity in the X-ray source GX 339-4. A&A, 109:L1–L4, May 1982.
29. R. Napiwotzki, L. Yungelson, G. Nelemans, T. R. Marsh, B. Leibundgut, R. Renzini, D. Homeier, D. Koester, S. Moehler, N. Christlieb, D. Reimers, H. Drechsel, U. Heber, C. Karl, and E.-M. Pauli. Double degenerates and progenitors of supernovae type Ia. In *ASP Conf. Ser. 318: Spectroscopically and Spatially Resolving the Components of the Close Binary Stars*, pages 402–410, November 2004.
30. R. E. Nather and B. Warner. Observations of rapid blue variables. I. Techniques. MNRAS, 152:209, 1971.
31. G. Nelemans, L. R. Yungelson, and S. F. Portegies Zwart. The gravitational wave signal from the Galactic disk population of binaries containing two compact objects. A&A, 375:890–898, September 2001.
32. G. Nelemans, L. R. Yungelson, and S. F Portegies Zwart. Short-period AM CVn systems as optical, X-ray and gravitational-wave sources. MNRAS, 349:181–192, March 2004.
33. G. Ramsay, P. Hakala, and M. Cropper. RX J0806+15: the shortest period binary? MNRAS, 332:L7–L10, May 2002.
34. E. L. Robinson, T. M. Mailloux, E. Zhang, D. Koester, R. F. Stiening, R. C. Bless, J. W. Percival, M. J. Taylor, and G. W. van Citters. The pulsation index, effective temperature, and thickness of the hydrogen layer in the pulsating DA white dwarf G117-B15A. ApJ, 438:908–916, January 1995.
35. A. D. Schwope, K. H. Mantel, and K. Horne. Phase-resolved high-resolution spectrophotometry of the eclipsing polar hu aquarii. A&A, 319:894–908, March 1997.
36. D. Steeghs and J. Casares. The Mass Donor of Scorpius X-1 Revealed. ApJ, 568:273–278, March 2002.
37. D. Steeghs, E. T. Harlaftis, and K. Horne. Spiral structure in the accretion disc of the binary IP Pegasi. MNRAS, 290:L28–L32, September 1997.
38. D. Steeghs, K. O'Brien, K. Horne, R. Gomer, and J. B. Oke. Emission-line oscillations in the dwarf nova V2051 Ophiuchi. MNRAS, 323:484–496, May 2001.
39. A. Stroeer, A. Vecchio, and G. Nelemans. LISA Astronomy of Double White Dwarf Binary Systems. ApJ, 633:L33–L36, November 2005.
40. S. M. Tulloch. Photon counting and fast photometry with L3 CCDs. In A. F. M. Moorwood and M. Iye, editors, *Ground-based Instrumentation for Astronomy. Edited by Alan F. M. Moorwood and Iye Masanori. Proceedings of the SPIE, Volume 5492, pp. 604–614 (2004).*, pages 604–614, September 2004.
41. M. H. van Kerkwijk, J. C. Clemens, and Y. Wu. Surface motion in the pulsating DA white dwarf G29-38. MNRAS, 314:209–219, May 2000.

Appendix

A Clock Induced Charge Statistics

In this section we detail some of the statistical properties of CICs. CICs come in two varieties: "pre-register" and "in-register" events. Pre-register events suffer the full amplification of the avalanche stage and will have identical statistics to electrons generated from genuine signals. In-register events are amplified by a variable amount depending upon where in the avalanche register they are first produced. This leads to an extremely skewed distribution for these events with lots of low values but a tail extending to very high values as well as shown in Fig. 8. The skewed distribution makes it hard to define CIC rates in a general way. For example, Fig. 8 shows that for the particular parameters chosen about 25% of pre-register CICs would on output fall below a threshold of 100, while 90% of the in-register CICs will do so. Lowering the threshold could increase the pre-register rate by at most 25%, but could increase the in-register rate by up to 10 times. When quoting rates, it is important to define how they were measured. In this case a threshold of 100 will lead to a rate of 0.1 CIC/pixel/exposure, at the high end of total CIC rates (in- plus pre-register) [40], and so presumably the probability of a CIC being generated at any one stage of the avalanche register is usually less than I have assumed.

If the probability of a CIC being generated at any one stage of the avalanche register is p_C, then one can show that the mean output value, given zero input μ_C is given by

$$\mu_C = \frac{p_C}{p}(g-1), \tag{14}$$

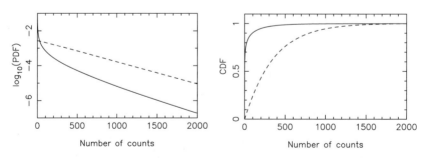

Fig. 8 In the left panel the solid line shows the distribution of in-register CICs for a 536 stage electron multiplying register with $p = 0.011$ and CIC probability/stage of $p_c = 0.0011$, i.e. this is the output PDF given a zero electron input. The dashed line shows the PDF given an input of 1 electron with p_c set to zero. The right-hand panel shows the equivalent cumulative distribution functions, and illustrates that the parameters chosen here will give a CIC above 100 counts in about 1 in 10 pixels

while the variance is

$$\sigma_C^2 = \left(\frac{2/(1+p) - p_c}{2+p}(g+1) - \frac{1-p}{1+p}\right)\mu_C. \tag{15}$$

For $p \ll 1$, $p_C \ll 1$, $g \gg 1$, this boils down to $V_C \approx g\mu_C$. For the example shown in Fig. 8, $\mu_C = 35$ while $\sigma_C = 110$. In proportional mode, this is effectively equivalent to a readout noise component of $R = \sigma_C/g = 0.3$ entering the variance equation $V = R^2 + 2C$ (ignoring the amplifier readout noise). A threshold of 100 leads to a very similar number for photon counting mode.

There may be some room for optimisation of the threshold in the presence of significant numbers of in-register CICs. In the example shown, a threshold of 100 leads to a 25% reduction in the true event rate and a 10% CIC rate, so that the signal-to-noise would be

$$\frac{0.75P}{\sqrt{0.1 + 0.75P}}, \tag{16}$$

where P is the mean number of photo-electrons per pixel per exposure. If the threshold is raised to 200, this becomes

$$\frac{0.57P}{\sqrt{0.05 + 0.57P}}, \tag{17}$$

which is better for $P < 0.057$.

The threshold cannot be optimised for pre-register events in the same way since they have identical statistics to genuine events. The relative numbers of in-register and pre-register events appears to depend upon the controller [40, 20]. My suspicion, which is possibly borne out by one of these studies [20], would be that the high voltages required for the avalanche register will make it harder to reduce the in-register compared to pre-register event rates.

Photonic Astronomy and Quantum Optics

Dainis Dravins

Abstract Quantum optics potentially offers an information channel from the Universe beyond the established ones of imaging and spectroscopy. All existing cameras and all spectrometers measure aspects of the first-order spatial and/or temporal coherence of light. However, light has additional degrees of freedom, manifest in the statistics of photon arrival times, or in the amount of photon orbital angular momentum. Such quantum-optical measures may carry information on how the light was created at the source, and whether it reached the observer directly or via some intermediate process. Astronomical quantum optics may help to clarify emission processes in natural laser sources and in the environments of compact objects, while high-speed photon-counting with digital signal handling enables multi-element and long-baseline versions of the intensity interferometer. Time resolutions of nanoseconds are required, as are large photon fluxes, making photonic astronomy very timely in an era of large telescopes.

1 What Is Observed in Astronomy?

Almost all of astronomy depends on the interpretation of properties of electromagnetic radiation ("light") from celestial sources. The sole exceptions are neutrino detections; gravitational-wave searches; analyses of cosmic rays, meteorites, and other extraterrestrial materials; and in-situ studies of solar-system bodies. All our other understanding of the Universe rests upon observing and interpreting more or less subtle properties in the light reaching us from celestial bodies.

Astronomical telescopes are equipped with myriads of auxiliary instruments which, on first sight, may give an impression of being vastly different. However, a closer examination of the underlying physical principles reveals that they all are measuring either the spatial or temporal [first-order] coherence of light (or perhaps some combination of these). All imaging devices (cameras, interferometers) are studying aspects of the spatial coherence (in various directions, and for different angular extents on the sky). All spectrally analyzing devices measure aspects of the

Dainis Dravins
Lund Observatory, Box 43, SE-22100 Lund, Sweden

temporal coherence (with different temporal/spectral resolution, and in the different polarizations). Although a gamma-ray satellite may superficially look different from a long-baseline radio interferometer, the basic physical property they are measuring is the same.

For centuries, optical astronomy has developed along with optical physics: Galileo's telescope and Fraunhofer's spectroscope were immediately applied to astronomical problems. However, the frontiers of 21st century optical physics have now moved toward photonics and quantum optics, studying individual photons, and photon streams. Those can be complex, carrying information beyond imaging, spectroscopy, or polarimetry. Different physical processes in the generation of light may cause quantum-statistical differences (e.g., different degrees of photon bunching in time) between light with otherwise identical spectrum, polarization, intensity, etc., and studies of such non-classical properties of light are actively pursued in laboratory optics.

Since almost all astronomy is based upon the interpretation of subtleties in the light from astronomical sources, quantum optics appears to have the potential of becoming another information channel from the Universe, fundamentally different from imaging and spectroscopy. What astronomy could then be possible with quantum optics?

2 What is Not Observed in Astronomy?

Figure 1 illustrates a type of problem often encountered in astrophysics: Light from various sources has been created through different (but typically unknown) physical processes: thermal radiation, stimulated emission, synchrotron radiation, etc. Now, assume one is observing these sources through "filters", adjusted so that all sources

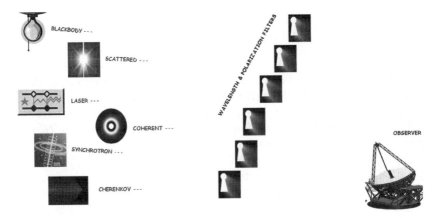

Fig. 1 An observational problem beyond current imaging and spectroscopy: Various sources emit light created through different processes. If spatial and spectral filters give to all sources the same apparent size and shape, same intensity, same spectrum, and the same polarization, how can one then tell the difference?

have the same size and shape, same intensity, same spectrum, and the same polarization. How can one then tell the difference when observing the sources from a great distance?

For sources as defined, it actually is *not possible, not even in principle*, to segregate them using *any* classical astronomical instrument. Telescopes with imaging devices (cameras or interferometers) would record the same spatial image (of the keyhole aperture), and any spectrometer would find the same spectrum. Still, the light from those sources can be physically different since photons have more degrees of freedom than those relevant for mere imaging by telescopes or for spectroscopic analysis.

Identical images are produced by photons arriving from identical directions, and identical spectra by similar distributions of photon energies. However, a photon stream has further degrees of freedom, such as the *temporal statistics of photon arrival times*, giving a measure of ordering (entropy) within the photon-stream, and its possible deviations from "randomness". Such properties are reflected in the second- (and higher-) order coherence of light, observable as correlations between pairs (or a greater number) of photons, illustrated by a few examples in Fig 2.

Clearly, the differences lie in *collective properties* of groups of photons, and cannot be ascribed to any one individual photon. The information content lies in the correlation in time (or space) between successive photons in the arriving photon stream (or the volume of a "photon gas"), and may be significant if the photon emission process has involved more than one photon.

The integer quantum spin of photons ($S=1$) make them bosons, and the density and arrival-time statistics of a photon gas in a maximum-entropy ("chaotic")[1] state of thermodynamic equilibrium then obey a Bose-Einstein distribution, analogous to the Maxwell-Boltzmann distribution for classical particles in thermodynamic equilibrium. The statistics implies a certain bunching of photons, i.e., an enhanced probability of successive photons to arrive in pairs. This is probably the best-known non-classical property of light, first measured by Hanbury Brown and Twiss in those

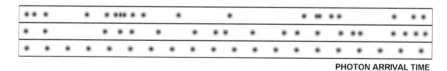

PHOTON ARRIVAL TIME

Fig. 2 Statistics of photon arrival times in light beams with different entropies (different degrees of "ordering"). The statistics can be "quantum-random", as in maximum-entropy black-body radiation (following a Bose-Einstein distribution with a characteristic "bunching" in time; top), or may be quite different if the radiation deviates from thermodynamic equilibrium, e.g. for anti-bunched photons (where photons tend to avoid one another; center), or a uniform photon density as in stimulated emission from an idealized laser (bottom). The characteristic fluctuation timescales are those of the ordinary first-order coherence time of light (on order of only picoseconds for a 1 nm passband of optical light, but traces of which are detectable also with slower time resolutions). Figure in part adapted from Loudon [98]

---

[1] In quantum optics, the term 'chaotic' denotes light from randomly fluctuating emitters, not to be confused with dynamically 'chaotic' states in mechanics.

experiments that led to the astronomical intensity interferometer [55]. However, a photon gas that is in a different entropy state does not have to obey that particular distribution, just like classical particles do not necessarily have to obey a Maxwell-Boltzmann distribution.

A well-known example is that of stimulated emission from a laser: in the ideal case its photons display no bunching whatsoever, rather arriving in a uniform stream. This is a non-chaotic state of light, far from a maximum-entropy condition. For such emission to occur, there must be at least two photons involved: one that is the stimulating one, and another that has been stimulated to become emitted. The emerging light then contains (at least) two photons that are causally connected and mutually correlated. One single photon cannot alone carry the property of stimulated emission: that requires multiple photons for its description. On a fundamental level, the phenomenon is enabled by the boson nature of photons which permits multiple particles to occupy the same quantum state. Another related example is superfluidity in liquid helium (of the isotope He^4). A volume of superfluid helium-4 obtains very special properties; however a single helium atom cannot alone carry the property of "superfluidity"; that is a *collective* property of many helium atoms, revealed by their mutual correlations. The state of superfluidity in a volume of liquid helium-4 is quite analogous to the state of stimulated emission in a volume of photon gas; both are Bose-Einstein condensations, where the respective boson fluid has a uniform particle density throughout. A practical difference is that, while the liquid helium is stationary inside its container[2], the photon gas if flowing past any detector at the speed of light, and requires detectors with very rapid response for its density fluctuations to be revealed.

These quantum fluctuations in light are fully developed over timescales equal to the inverse bandwidth of light. For example, the use of a 1 nm bandpass optical filter gives a frequency bandwidth of 10^{12} Hz, and the effects are then fully developed on timescales of 10^{-12} seconds. While instrumentation with continuous data processing facilities with such resolutions is not yet available, it is possible to detect the effects, albeit with a decreased amplitude, also at the more manageable nanosecond timescales.

By contrast, particles with half-integer quantum spin (e.g., electrons with $S=\frac{1}{2}$) are fermions, and obey different quantum statistics. In particular, groups of electrons suffer from the Pauli exclusion principle, which prohibits them to occupy the same quantum state. Thermal electrons therefore tend to avoid one another, analogous to the center illustration in Fig. 2. Photons enjoy much more freedom, and can appear either as preferentially bunched together in space and time, anti-bunched, or almost any other distribution.

Various processes studied in the laboratory, such as [multiple] scattering (in astrophysical jargon: "frequency redistribution"), passage through beamsplitters (in astrophysical jargon: "angular redistribution") modifies the relative amount of bunching, in principle carrying information of the events a photon stream has experienced since its creation (e.g., [3]). Likewise, light created under special con-

---

[2] except, perhaps, climbing its walls!

ditions (free-electron laser, resonance fluorescence, etc.; in astrophysical jargon: "non-LTE") may have their own characteristic photon statistics, as discussed below. Thus (at least in principle), quantum statistics of photons may permit to segregate circumstances such as whether the Doppler broadening of a spectral line has been caused by motions of those atoms that emitted the light, or by those intervening atoms that have scattered the already existing photons.

3 The Intensity Interferometer

Although no existing astronomical instrument is capable of directly segregating the sources in Fig. 1, there actually has been one which in principle has had that capability: the stellar *intensity interferometer*, developed years ago in Australia by Hanbury Brown, Twiss, et al., for the original purpose of measuring stellar sizes [55]. Today this would be considered a quantum-optical instrument. At the time of its design, the understanding of its functioning was the source of considerable confusion, with numerous published papers questioning its basic principles. Indeed, at that time (and even now!), the intensity interferometer has been an instrument whose functioning is challenging to intuitively comprehend.[3]

To begin with, the name itself is sort of a misnomer: there actually is nothing interfering in the instrument; rather its name was chosen for its analogy to the ordinary [phase] interferometer, whose scientific aims the original intensity interferometer was replicating (in measuring stellar diameters). Two telescopes are simultaneously measuring the random and very rapid quantum fluctuations in the light from some particular star. When the telescopes are placed next to one another, both measure the same signal, but when moving them apart, the fluctuations gradually become decorrelated; how rapidly this occurs as the telescopes are moved apart gives a measure of the size of the star. The spatial baselines are the same as would be needed in ordinary phase interferometry, but the signal observed is thus the *correlation* between the fluctuations electronically measured in each of the two telescopes, and how this correlation gradually decreases as the telescopes are moved apart from one another.

One reason for the incomprehensibility was that, at the time the optical version of the instrument was developed (following earlier radio experiments), the quantum optical properties of light were still incompletely known, and novel light sources such as the laser were not yet fully developed. Especially confusing was to comprehend how *two* telescopes could simultaneously observe the "same" light from a star. Given that light consisted of photons, how could then the same photon be simultaneously detected by two different telescopes?

It was to be some years until a good explanation and theory could be developed. Not only did this describe the functioning of that instrument, but it also led to a more complete quantum mechanical description of the nature of light. In particular, the

---

[3] The author actually has heard the late Robert Hanbury Brown himself make the remark that *"this must be the most incomprehensible instrument in all of astronomy"*.

resulting quantum theory of optical coherence shows that an arbitrary state of light can be specified with a series of coherence functions essentially describing one-, two-, three-, etc. -photon-correlations. While the ordinary (first-order) coherence can be manifest as the interference in amplitude and phase between light waves, the second-order coherence is manifest as either the correlation between photon arrival rates at different locations at any one given time (second-order *spatial* coherence, the effect exploited in the intensity interferometer), or as correlations in photon arrival rates between different delay times along the light beam, at any one given spatial location. The latter gives the second-order *temporal* coherence, an effect exploited in photon-correlation spectroscopy for measuring the wavelength width of scattered light. A key person for the development of the quantum theory of optical coherence was Roy Glauber [47–51], whose pioneering efforts were rewarded with the 2005 Nobel Prize in physics.

The functioning of the intensity interferometer is now well documented, and will not be repeated here. For details of the original instrument, see the original papers by Hanbury Brown et al., as well as retrospective overviews [55–62]. The principles are explained in various textbooks and reference publications, e.g., [52, 98, 102], and very lucidly in Labeyrie et al. [83].

The significant observational advantage is that telescopic optical quality and control of atmospheric path-lengths need only to correspond to some [reasonable] fraction of the electronic resolution. For a realistic value of 10 ns, say, the corresponding light-travel distance is 3 m, and optical errors of maybe one tenth of this

Fig. 3 The only previous astronomical instrument which could have segregated among the light sources of Fig. 1. The original *intensity interferometer* at Narrabri, Australia, observed one star simultaneously with two telescopes, whose mutual distance was gradually changed. (University of Sydney; reprinted with permission from its School of Physics)

can be accepted, enabling coarse flux collectors to be used (rather than precise telescopes), and avoiding any sensitivity to atmospheric seeing (thus enabling very long baselines). The price to be paid is that large photon fluxes are required (leading to bright limiting magnitudes) and that (for a two-telescope system) only the modulus of the visibility function is obtained, giving a measure of the extension of the source, but not its full phase-correct image.

The quantity obtained from an intensity interferometer is the normalized *second-order correlation function* of light with its time-variable intensity $I(t)$:

$$g^{(2)}(\tau) = \frac{\langle I(t)I(t+\tau)\rangle}{\langle I(t)\rangle^2},$$

where τ is the correlation time-delay, t is time, and $\langle \rangle$ denotes long-term averaging.

Although its function was perhaps not fully appreciated at the time it was developed, an intensity interferometer works correctly only for sources whose light is in the maximum entropy, thermodynamic equilibrium state. For such light, simple relations exist between the modulus of the first-order coherence function (the visibility measured in ordinary phase interferometers) and higher-order correlation functions. The relation between the normalized second-order correlation $g^{(2)}$ and the ordinary first-order quantity is $g^{(2)} = |g^{(1)}|^2$, so that the modulus of the visibility $g^{(1)}$ can be deduced, yielding stellar diameters from intensity correlation measurements.[4]

For a stable wave (such as an idealized laser), $g^{(2)}(\tau) = g^{(2)}(0) = 1$ for all time delays; while for chaotic, maximum-entropy light $g^{(2)}(0) = 2$, a value reflecting the degree of photon bunching in the Bose-Einstein statistics of thermodynamic equilibrium. In the laboratory, one can follow how the physical nature of the photon gas gradually changes from chaotic ($g^{(2)} = 2$) to ordered ($g^{(2)} = 1$) when a laser is "turned on" and the emission gradually changes from spontaneous to stimulated (although the spectrum does not change). Measuring $g^{(2)}$ and knowing the laser parameters involved, it is possible to deduce the atomic energy-level populations, which is an example of a parameter of significance to theoretical astrophysics ("non-LTE departure coefficient") which cannot be directly observed with any classical measurements of one-photon properties. Chaotic light that has been scattered against a Gaussian frequency-redistributing medium obtains a higher degree of photon bunching: $g^{(2)}(0) = 4$; while fully antibunched light has $g^{(2)}(0) = 0$. The latter state implies that whenever there is one photon at some time $= t$ [then $I(t) = 1$], there is none immediately afterwards, i.e., $I(t+\tau) = 0$ for sufficiently small τ. Experimentally, this can be produced in, e.g., resonance fluorescence, and is seen through sub-Poissonian statistics of recorded photon counts, i.e. narrower distributions than would be expected in a "random" situation. Corresponding relations exist for higher-order correlations, measuring the properties for groups of three, four, or a

---

[4] The expressions given here are simplified ones; a full quantum-optical description involves photon annihilation operators, etc.

greater number of photons. For introductions to the theory of such quantum-optical phenomena, see, e.g., [42, 97, 98, 102, 105, 121]. Experimental procedures are described by Bachor & Ralph [3], Becker [6], and Saleh [132].

The functioning of the intensity interferometer implicitly assumes the photon statistics of starlight to correspond to the Bose-Einstein distribution characteristic for thermodynamic equilibrium, as from a black-body radiator (top plot in Fig. 2). A hypothetical star shining with idealized laser light (as bottom plot in Fig. 2) would produce no intensity fluctuations whatsoever: no matter over what baseline one would place the telescopes, the instrument would not yield any meaningful signal from which to deduce the angular size of the star. The intensity interferometer was not designed for such purposes, but it could in principle have been used to search for possible deviations from randomness in the photon streams. For example, if the stellar diameter is determined independently (using ordinary phase interferometry, say), the difference between that value, and that deduced from the second-order coherence could be interpreted as due to a difference in photon statistics and in the entropy of light.

4 Quantum Phenomena in Astronomy

While the existence *in principle* of various quantum-optical phenomena cannot be avoided, it is less obvious which astronomical objects will in practice be sources of measurable amounts of non-trivial photon-stream properties. In this Section, we review various studies of radiation processes where such phenomena could be expected.

What is the quantum nature of the light emitted from a volume with departures from thermodynamic equilibrium of the atomic energy level populations? Will a spontaneously emitted photon stimulate others, so that the path where the photon train has passed becomes temporarily deexcited and remains so for perhaps a microsecond until collisions and other effects have restored the balance? Does then light in a spectral line perhaps consist of short photon showers with one spontaneously emitted photon leading a trail of others emitted by stimulated emission? Such amplified spontaneous emission ("partial laser action") might occur in atomic emission lines from extended stellar envelopes or stellar active regions. Predicted locations are mass-losing high-temperature stars, where the rapidly recombining plasma in the stellar envelope can act as an amplifying medium [84, 158, 159]. Analogous effects could exist in accretion disks [38]. In the infrared, several cases of laser action are predicted for specific atomic lines [39, 54, 117]. Somewhat analogous situations (corresponding to a laser below threshold) have been studied in the laboratory. The radiation from "free" clouds (i.e. without any laser resonance cavity) of excited gas with population inversion is then analyzed. One natural mode of radiative deexcitation indeed appears to be the emission of "photon showers" triggered by one spontaneously emitted photon which is stimulating others along its flight vector out from the volume (Fig. 4).

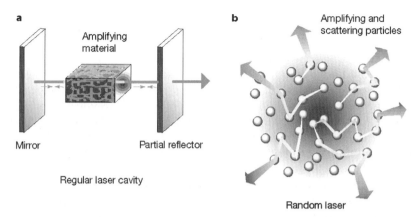

Fig. 4 Different geometries producing laser emission in the laboratory, and in space ("random laser"): Wiersma [163]. Reprinted with permission from Macmillan Publishers Ltd: Nature **406**, 132, © 2000

Geometric differences between a laboratory laser, and an astrophysical one include that the laboratory one is normally enclosed inside an elongated cavity, whose end mirrors have the purpose of both increasing the path-length traversed by light inside the amplifying medium, and serving as a wavelength filter. Cosmic lasers involve the same basic processes, but of course do not have mirrors in resonance cavities, rather relying on sheer path-length to achieve amplification. An astrophysical laser must be of a more "random" nature, presumably producing monochromatic flashes which are initiated by "lucky" photons inside the amplifying volume. The beams are emitted in narrow angles, and we only observe those beams that happen to be directed at the Earth; the vast majority which do not point towards us, remain invisible.

Following an early theoretical prediction [93], and an experimental realization in the laboratory [86] random lasers are currently receiving significant attention, not least in trying to understand their statistical properties and intrinsic intensity fluctuations [16, 40, 90, 106, 156].

At first sight, it might appear that light from a star should be nearly chaotic because of the very large number of independent radiation sources in the stellar atmosphere, which would randomize the photon statistics. However, since the time constants involved in the maintenance of atomic energy level overpopulations (e.g. by collisions) may be longer than those of their depopulation by stimulated emission (speed of light), there may exist, in a given solid angle, only a limited number of radiation modes reaching the observer in a given time interval (each microsecond, say) and the resulting photon statistics might then well be non-chaotic. Proposed mechanisms for pulsar emission include stimulated synchrotron and curvature radiation ("free-electron laser") with suggested timescales of nanoseconds, over which the quantum statistics of light again would be non-chaotic. Obviously, the list of potential astrophysical targets can be made longer [27]. In general, photon statistics for the radiation from any kind of energetic source could convey something about the processes where the radiation was liberated.

4.1 Mechanisms Producing Astrophysical Lasers

The possibility of *laser action*, i.e., an enhanced fraction of stimulated emission in certain spectral lines due to *amplified spontaneous emission*, has been suggested for a number of spectral lines in different sources. Already long ago it was realized that deviations from thermodynamic equilibrium in atomic energy-level populations could lead to such emission, although the possibility was not taken seriously before the construction of the laboratory microwave maser in 1954, and the visible-light laser in 1960. However, after the discovery in 1965 of the first celestial OH maser at λ 18 cm, the astrophysical possibilities become apparent [126].

Laser action normally requires a population inversion in atomic or molecular energy levels. Such deviations from thermodynamic equilibrium can be achieved through selective radiative excitation, or electronic recombination in a cooling plasma. Selective excitation, where another (often ultraviolet) emission line of a closely coinciding wavelength excites a particular transition, overpopulating its upper energy level, is often referred to as *Bowen fluorescence*. Following Bowen [12], plausible combinations of atomic and molecular lines have been studied by several people, e.g., [13, 43].

Already long ago, a few authors did touch upon the possibility of laser action, but dismissed it as unrealistic or only a mathematical curiosity. Thus, Menzel [103], when discussing situations outside of thermodynamic equilibrium, concluded "... *the condition may conceivably arise when the value of the integral* [of absorbed radiation] *turns out to be negative. The physical significance of such a result is that energy is emitted rather than absorbed... as if the atmosphere had a negative opacity. This extreme will probably never occur in practice.*" (ApJ **85**, p 335, 1937). However, 33 years later the same Donald Menzel [104] was a pioneer in proposing laser action in non-LTE atmospheres.

If a strongly ionized plasma is rapidly cooled, population inversion occurs during the subsequent electronic recombination cascade, permitting laser action. This scheme, known as plasma laser or recombination laser, has been studied in the laboratory (e.g., supersonic expansion of helium plasma from nozzles into low-density media), and is applicable to stars with mass loss. If the electron temperature is sufficiently low, the recombination occurs preferentially into the upper excited states of the ion, which then decay by a radiative and/or collisional cascade to the ground state. Within the cascade, population inversions may form among the excited states, depending upon the relative transition probabilities. Also, stepwise ionization implies the moving of electrons to higher energy levels, likewise tending to invert the energy-level populations.

Another mechanism for creating non-equilibrium populations of energy levels is an external X-ray illumination of stellar atmospheres, originating from a hot component in a close binary system [129].

Possibly, the earliest modeling of an astrophysical source in terms of laser action, was for the λ 190.9 nm C III intercombination line in the Wolf-Rayet star γ Vel by West in 1968 [162], although the interpretation was later doubted [19]. Several infrared atomic transitions with astrophysically favorable parameters for population inversion were identified by Smith [140], while Menzel [104] suggested laser action in non-LTE atmospheres by analyzing the microscopic form of the radiative transfer equation. Jefferies [70] showed that external pumping may not be required for close-lying atomic energy levels with energy separations ΔE much smaller than kT, with T the kinetic temperature of the gas. Lavrinovich & Letokhov [84] suggested population inversions for lines such as O I 844.6 nm in the atmospheres of hot Be stars, while Peng & Pradhan [117] modeled infrared lasers in active galactic nuclei and novae. Selective excitation appears to cause the bright emission of ultraviolet Fe II lines in gas ejecta close to the central star of η Car [77] as well as in the symbiotic star RR Tel [64]. Masing in the forbidden [Fe XI] 6.08 μm line may be common, caused by peculiarities in the atomic energy structure [39]. Other authors have discussed infrared lasers in planetary atmospheres, active galactic nuclei, accretion systems and stellar envelopes; a review is by Johansson and Letokhov [76].

For longer wavelengths, there is a broad literature on millimeter and radio masers, especially molecular ones. Millimeter recombination lines of high levels in hydrogen (H30 α, H50 β, and others) are seen towards η Car, apparently affected by maser emission [22], as well as in other sources.

Theoretical treatments of astrophysical radiative transfer including laser action have been made by several authors: [24, 38, 150, 151, 158, 159]. A common conclusion seems to be that while laser action seems possible under many diverse conditions, the amplification of any specific line requires a sufficiently extended region within a rather narrow range of parameter space (pressure, temperature, electron density).

Generally, radio and microwave masers can be produced more easily (gyrosynchrotron emission, curvature radiation) than those at shorter wavelengths [148],

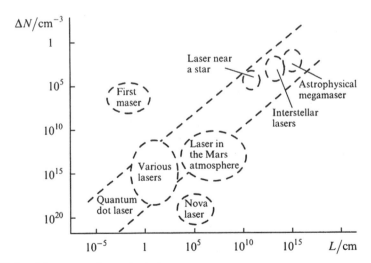

Fig. 5 Various laboratory and astrophysical masers and lasers span very many orders of magnitude in the active-medium particle density—active-medium size diagram. ΔN denotes the density of atomic population inversions: Letokhov [94]. Reprinted from Quant Electr **32**, 1065 (2002), with permission from Turpion Ltd

although there are suggestions for astrophysical lasers even in the X-ray regime, perhaps a recombination laser in hydrogen-like ions around accreting neutron stars [14, 44].

In the absence of feedback, the amplifying medium of an astrophysical laser represents an amplifier of spontaneous emission. Because of the large size of astrophysical lasers, the radiation pattern is determined by the geometry of the amplifying region. In the case of a nearly spherical geometry, the radiation should be isotropic.

4.2 Radiation from Luminous Stars

Eta Carinae, the most luminous star known in our Galaxy, is some 50–100 times more massive than our Sun and 5 million times as luminous (M_{bol} approx $= -12$). This star is highly unstable, undergoing giant outbursts from time to time; one in 1841 created the bipolar Homunculus Nebula. At that time, and despite the comparatively large distance (3 kpc), η Car briefly became the second brightest star in the night sky. It is now surrounded by nebulosity expanding at around 650 km s^{-1}.

Spectra of the bright condensations ("Weigelt blobs") in the η Car nebulosity display distinct Fe II emission lines, whose formation has been identified as due to stimulated emission: [73, 74]; Fig. 6. It is argued that laser amplification and stimulated emission must be fairly common for gaseous condensations in the vicinity of bright stars, caused by an interplay between fast radiative and slow collisional relaxation in these rarefied regions. These processes occur on highly different time

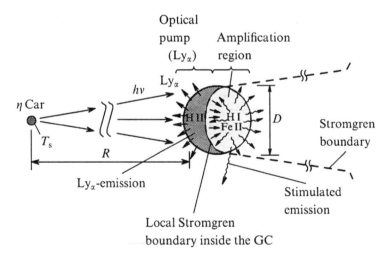

Fig. 6 Outline of a compact gas condensation (GC) near the hot luminous star η Car with the Strömgren boundary between the photoionized (H II) and neutral (H I) regions. Intense Ly α radiation from H II regions is creating laser action through a pumping of Fe II atomic energy levels. The region of laser amplification is marked: Johansson & Letokhov [72]. Reprinted from JETP Lett **75**, 495, © 2002

scales, radiative relaxation operating over $10^{-9} - 10^{-3}$ s, and collisional over seconds. In case of excitation of some high-lying electronic levels of an atom or ion with a complex energy-level structure, radiative relaxation follows as a consequence of spontaneous emission, in the course of which an inverse population develops of some pair(s) of levels, producing laser action. Such emission is a diagnostic of non-equilibrium and spatially non-homogeneous physical conditions as well as a high brightness temperature of Ly α in ejecta from eruptive stars (Fig. 7).

A somewhat similar geometry is found in symbiotic stars, close binary systems where a hot star ionizes part of an extended envelope of a cooler companion, leading to complex radiative mechanisms. The combined spectrum shows the superposition of absorption and emission features together with irregular variability. In conditions where strong ultraviolet emission lines of highly ionized atoms (e.g., O VI λ 103.2, 103.8 nm) irradiate high-density regions of neutral hydrogen (with the Ly β line at λ 102.6 nm), Raman-scattered lines may be observed. Such have been identified at λ 682.5 and 708.2 nm in the symbiotic star V1016 Cyg [135].

Stimulated emission does not necessarily require population inversions of atomic energy levels: Sorokin & Glownia [144] suggest "lasers without inversion" to explain the emission of a few narrow emission lines that dominate the visible or ultraviolet spectra of certain objects such as some symbiotic stars. Although the electronic level structures of the atoms/ions producing these emissions preclude the maintenance of population inversions, there are other ways to produce stimulated

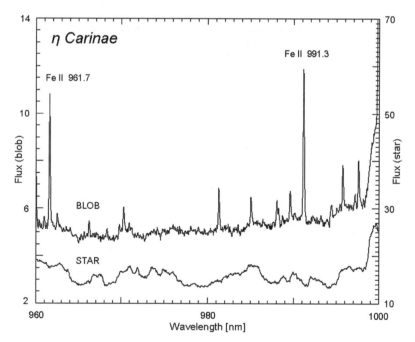

Fig. 7 Spectra of the central star of η Car (flux scale at right) and the bright condensation in the nebulosity outside (Weigelt blob B; left scale), with the two lasing Fe II lines λ 961.7 and 991.3 nm: Johansson & Letokhov [74]. Reprinted from A&A **428**, 497 with permission from its publisher

4.3 Hydrogen Lasers in Emission-Line Objects

MWC 349 is a peculiar hot emission-line star, often classified as B[e], with a B0 III companion. It is a very bright infrared source and an extremely strong radio star. The estimated mass is ≈ 30 M_{Sun} and its distance ≈ 1.2 kpc. It possesses a dense neutral Keplerian circumstellar disk (extent ≈ 300 AU), and an ionized wind with very low terminal velocity (≈ 50 km/s). Proposed models include a quasi-spherical outflow of ionized gas in the outer part of the circumstellar envelope, and a differentially rotating disk viewed edge-on in its inner part [53].

Perhaps the most distinct features are strong hydrogen recombination line masers at mm and infrared wavelengths (H10α to H40α; [128]). The emission appears to be created by photoionization in a region with large electron density in a Keplerian disk at a distance ≈ 20–30 AU from the star [23]; Figs. 8–9.

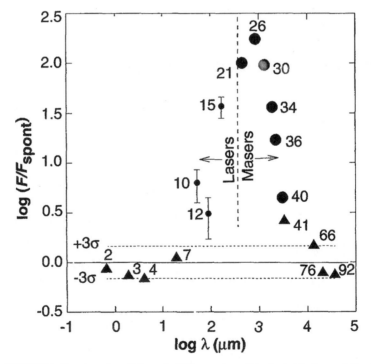

Fig. 8 MWC349A: Log-log plot of the ratio F/F_{spont} where F is the total observed flux in successive hydrogen recombination lines, and F_{spont} the estimated contribution from spontaneous emission. Large dots indicate masing mm and sub-mm lines; small dots are infrared detections. The numbers are the principal quantum numbers for each line's lower level: Strelnitski et al. [149]. Reprinted with permission from Science **272**, 1459, © 1996 AAAS

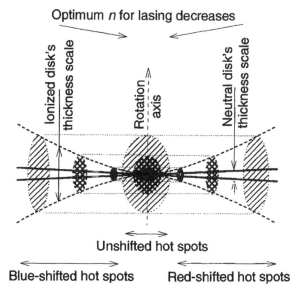

Fig. 9 Proposed structure of the lasing and masing circumstellar disk in MWC349A. This edge-on view is presumably as we observe it: Strelnitski et al. [149]. Reprinted with permission from Science **272**, 1459, © 1996 AAAS

4.4 Quantum Optics in the Earliest Universe

The physical processes exemplified by the lasers and masers in MWC349A may have been quite fundamental already in the *very* early Universe, soon after the Big Bang, at the epochs of recombination and reionization. The physical properties of masing hydrogen recombination lines at redshift $Z \approx 1000$ were modeled by Spaans & Norman [145]. Given small-scale overdense regions, maser action is possible owing to the expansion of the Universe and the low ambient temperature at radiative decoupling. Due to the redshift, these spectra would today be observable in the long-wavelength radio region (e.g., with the planned Square Kilometer Array, SKA), but the basic physical processes would be analogous to those observable at shorter wavelengths in today's local Universe.

4.5 CO_2 Lasers in Planetary Atmospheres

Laser/maser emission is not a property that is exclusive to extreme deep-space environments, but can be found already in our planetary-system neighborhood (and even in the Earth's upper atmosphere!), in particular due to an interplay between solar ultraviolet radiation, and the energy levels in molecules such as CO_2.. Their observation ideally requires spectral resolutions in the range 10^6 to 10^7 [143].

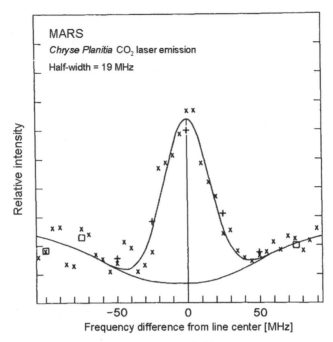

Fig. 10 Spectra of a Martian CO_2 laser emission line ($^{12}C^{16}O_2$ R8 line, $\lambda 10.33\mu m$) as a function of frequency difference from line center (in MHz). Lower curve is the total emergent intensity in the absence of laser emission; the emission peak is modeled with a Gaussian fit: Mumma et al [107]. Reprinted with permission from Science **212**, 45, © 1981 AAAS

4.6 How Short (and Bright) Pulses Exist in Nature?

The nanosecond structure observed within giant pulses from radio pulsars represents the currently most rapid fluctuations found in astronomical sources. Soglasnov et al. [142] reported on observations of giant radio pulses from a millisecond pulsar at 1.65 GHz. Pulses were observed with widths ≤15 ns, corresponding to a brightness temperature of $T_b \geq 5 \times 10^{39}$ K, the highest so far observed in the Universe. Some 25 giant pulses are estimated to be generated during each neutron-star revolution. Their radiation energy density can exceed 300 times the plasma energy density at the surface of the neutron star and can even exceed the magnetic field energy density at that surface, constraining possible mechanism for their production.

Radiation mechanisms generating pulsar emission structure on very short timescales have been suggested already earlier [45]. If—as in classical textbooks—one would assume that the source size corresponds to the light-travel-time during the source variability, the structures responsible for pulses a few ns wide must be less than one meter in extent [63] (the speed of light is 30 cm ns^{-1}).

Giant pulses have so far been observed only in some radio pulsars [36, 80, 82, 124], while theoretical ideas include energy release in nonlinear plasma turbulence

Photonic Astronomy and Quantum Optics

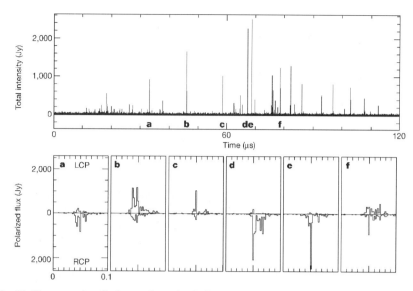

Fig. 11 Nanosecond radio bursts from the Crab pulsar. Top: Details of a pulse seen with 2 ns resolution. Bottom: 100-ns sections, showing the left-, and right-circularly polarized flux from six of the nanopulses: Hankins et al. [63]. Reprinted with permission from Macmillan Publishers Ltd: Nature **422**, 141, © 2003

[161], stimulated Compton scattering [118, 119], as well as effects from angular beaming arising due to relativistic motion of the radiating sources [46].

Whether pulsars emit also *optical* flashes on nanosecond timescales is not clear. Searches for such pulses from the Crab pulsar were made already soon after its

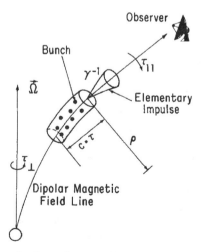

Fig. 12 Possible mechanism for fluctuations on sub-microsecond timescales of pulsar emission, originating from relativistically moving particle bunches: Gil [45]. Reprinted from Ap&SS **110**, 293, with kind permission of Springer Science and Business Media, © 1985

discovery, but without success. However, Shearer et al. [139] detected a correlation between optical and giant radio pulse emission from the Crab. Optical pulses coincident with the giant radio pulses were on average 3 % brighter than those coincident with normal radio pulses. This correlation suggests that there might also exist optical nanopulses, however, possibly requiring extremely large telescopes for their study.

4.7 Photon Bubbles

When the intensity of light becomes very high, a volume filled with light takes on properties of a three-dimensional photon gas. Having a low mass density, it becomes buoyant, while exerting a pressure to its sides which may balance the surrounding material-gas pressure. In situations of very intense photon flux (e.g., in very hot stars) the development of *photohydrodynamic turbulence* involving the formation of "bubbles" of photon gas is expected. The bubbles would be filled with light, and

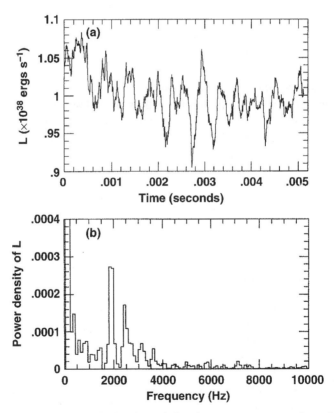

Fig. 13 Theoretically simulated luminosity variations in a neutron-star accretion column due to photon-bubble oscillations: Klein et al. [79]. Reproduced from ApJ **457**, L85 by permission of the AAS

would rise through the stellar surface, giving off photon bursts [66, 125, 138, 146]. Such phenomena are predicted to generate sub-millisecond fluctuations in accretion onto compact objects and can enable radiation fluxes orders of magnitude above the Eddington "limit" [7, 8, 79].

4.8 Emission in Magnetic Fields of Magnetars

Certain supernova explosions—those where the original star was spinning fast enough and had a strong enough magnetic field—can produce magnetars, highly magnetic neutron stars. Such objects may be most interesting targets for studying exotic radiation processes, including coherent laser effects. Normal pulsars can emit coherent radiation in the radio range, with a frequency corresponding to the plasma frequency in the pulsar magnetosphere; analogous processes in magnetar magnetospheres would have the plasma frequency appropriately scaled up, generating radiation in the optical or infrared.

A maser mechanism has been proposed to explain the optical pulsations from anomalous X-ray pulsars. Those would be generated by a curvature-drift-induced maser in the magnetar polar caps [99]. Another mechanism for coherent optical emission could be magnetospheric currents and plasma instabilities above magnetar surfaces [34]. Such coherent emission in the infrared and optical would probably be beamed, pulsed and polarized. Possibly, magnetospheric plasma could be produced in episodic sparking, on scales down to nanoseconds.

Magnetars are also among objects that may be suspected of hosting cosmic freeelectron lasers (not involving inverted atomic energy-level populations but rather relativistic electrons in strong magnetic fields). Such lasers, having their own characteristic photon statistics, are a topic of wide current interest for various laboratory experiments [130, 136, 137].

5 Searching for Laser/Maser Effects

Circumstantial indicators for optical laser action include high line-to-continuum ratios; "anomalous" intensity ratios among lines in the same multiplet; rapid time variability (perhaps uncorrelated among different lines); small spatial extent, and narrow spectral lines (caused by the amplification of radiation at the line center being higher than in the wings), coupled with theoretical expectations.

Since any region radiating into a specific direction at any moment in time must be very much smaller than the full stellar atmosphere, there will exist a large number of independent emission regions, and Doppler shifts from stellar rotation or expanding atmospheres will smear out the integrated line profile, masking the spectrally narrow contributions from individual emission regions. Also, if only a small fraction of the source is lasing, spontaneous emission from the rest may overwhelm the small laser "hot-spots" when observed with insufficient spatial resolution. Still, suitable interferometry could reveal the existence of such hot-spots, even if they are continually changing [84, 85].

In radio, masers have been studied in great detail, and have become an important part of radio astrophysics for both local sources [35], and for the distant Universe [96]. Some examination has been made also of the statistics of radiation amplitudes. In radio, where photon-counting is not (yet) feasible, this corresponds to photon statistics in the optical. Ordinarily, however, the radiation is assumed to arise through processes in which individual particles radiate independently, and the resulting radiation fields then become chaotic (Gaussian). This assumption greatly simplifies the radiative transfer equations, although a few attempts to detach from this assumption have been made [165, 166]: in stimulated emission, the nonlinear gain characteristics may well disturb the Gaussian nature of the original field.

Early observational efforts to search for deviations from Gaussian statistics in coherent radio sources were made by Evans et al. [37] who studied OH masers, concluding that the statistics were Gaussian within at least some percent. Likewise, Paschenko et al. [115], and Lekht et al. [89] found the statistics of OH sources to have noiselike signal properties. DeNoyer & Dodd [26] noted that a way to detect the non-Gaussian electromagnetic fields in saturated masers would be to perform statistical tests in individual frequency channels; some studies of OH masers, however, did not reveal any such signals.

A claimed detection of non-Gaussian statistics was reported in the radio emission from some pulsars by Jenet et al. [71]. The large Arecibo telescope was used to measure the statistics from four pulsars using high time resolution (100 ns), detecting temporally coherent emission in three of these. However, Smits et al. [141], in repeating similar measurements, concluded that the coherent signatures more likely originate from interstellar scintillation, rather than being intrinsic to the pulsars.

6 Modeling Astrophysical Photon Statistics

Analogous to past observational ventures into previously unexplored parameter domains, the search for phenomena of photon statistics in astronomical light is an explorative scheme where one does not know beforehand exactly what to expect, nor in which sources. Also—judging from past experience—any more detailed development of relevant astrophysical theory is unlikely to occur until after actual observations are available. Nevertheless, some authors have made efforts towards this direction, and somewhat analogous physical situations may illustrate what challenges to expect.

The theoretical problem of light scattering in a macroscopic turbulent medium such as air is reasonably well studied. In particular, the equations of transfer for I^2 and higher-order moments of intensity I have been formulated and solved: e.g., [155]. A familiar result implies that stars twinkle more with increasing atmospheric turbulence. The value of $\langle I \rangle$, i.e. the total number of photons transmitted per unit time may well be constant, but $\langle I^2 \rangle$ increases with greater fluctuations in the medium. The quantum-optical problem of scattering light against atoms is somewhat related, except that the timescales involved are now those of the coherence time of light, not those of terrestrial windspeeds.

Theoretical treatments of astrophysical radiative transfer have so far almost exclusively treated the first-order quantities of intensity, spectrum and polarization, and not the transfer of I^2 and higher-order terms. There are a few notable exceptions, however, like the analytical solution of the higher-order moment equation relevant for radio scintillations in the interstellar medium [88, 91, 92] and attempts to formulate the quantum mechanical description of the transfer of radiation, including non-Markovian effects (i.e. such referring to more than one photon at a time) in a photon gas [100, 101]; the transfer equation for the density matrix of phase space cell occupation number states [112, 134], the need to introduce concepts from non-linear optics [166], and other relevant formulations [21, 147, 160]; further there must be many pertinent papers in laboratory quantum optics. However, as opposed to the laboratory case, the often complex nature of astronomical sources may not lend itself to such simple treatments, and there does not yet appear to exist any theoretical predictions for specific astronomical sources of any spectral line profiles of higher-order than one (i.e. the ordinary intensity versus wavelength).

It appears that the equations for radiative transfer must be written in a *microscopic* form, i.e. considering each radiation, excitation, and scattering process on a photon[s]-to-atom[s]-to-photon[s] level, rather than treating only statistically averaged light intensities, opacities, etc. While it could be argued that the many radiation sources in any astronomical object would tend to randomize the emerging radiation (erasing the signatures of non-equilibrium processes), the number of independent sources that contribute to the observed photon signal in any given short time interval is actually very limited. If 1000 photons per millisecond are collected, say, those can originate from at most 1000 spatially distinct locations, which during that particular millisecond happened to beam their radiation towards the Earth. Perhaps, such 1000 photons could originate from only 990 sources, say, in which case some insight in the source physics could be gained through measurements of photon statistics.

There exist other types of quantum phenomena known in the laboratory, which quite conceivably might be detected in photon statistics and correlations from also some astronomical sources. These include emission cascades which could show up in cross-correlations between emission successively appearing in different spectral lines in the same deexcitation cascade, or various collective effects of light-emitting matter ("superradiance", i.e. several *atoms* emitting together).

7 Photon Orbital Angular Momentum

Photons have surprisingly many properties: One single photon arriving from a given direction, of any given wavelength, still can have hundreds of different states since it may carry different amounts of orbital angular momentum! The linear momentum of electromagnetic radiation is associated with radiation pressure, while the angular momentum is associated with the polarization of the optical beam. Only quite recently has it been more widely realized that radiation may in addition carry also *photon orbital angular momentum*, POAM. There were some pioneering laboratory experiments already in the 1930's [10], but only recently has it become possible to measure POAM for also individual photons [87]. The amount of POAM is characterized by an integer ℓ in units of the Planck constant $\hbar$, so that an absorber placed in the path of such a beam will, for each photon absorbed, acquire an angular momentum $\ell\hbar$. The integer ℓ gives the POAM states of the photon and also determines—in a quantum information context—how many bits of information that can be encoded in a single photon. Since, in the laboratory, photons can now be prepared with ℓ up to at least the order of 300, it implies that single photons may carry [at least] 8 bits of information, of considerable interest for quantum computing, and a main reason for the current interest in these phenomena [15].

The POAM phenomenon can be illustrated on a microscopic level: a small particle can be made to orbit around the light beam's axis while, at the same time, the beam's angular momentum (circular polarization) causes the particle to rotate on its own axis [114]. Some astronomical applications were already suggested by Harwit [65], although it is not verified whether there exist actual astronomical sources where a significant fraction of the photons carry such properties.

However, POAM manipulation at the telescope may permit high-contrast imaging inside an "optical vortex", with possible applications in exoplanet imaging [41, 152, 153]. This utilizes that the cross-sectional intensity pattern of all POAM beams has an annular character (with a dark spot in the center) which persists no matter how tightly the beam is focused. In a sense, such light thus acts as its own coronagraph, extinguishing the light from the central star, while letting through that from a nearby exoplanet.

The spatial characteristics of POAM make it sensitive to spatial disturbances such as atmospheric seeing [116], although it is not obvious whether that sensitivity could also be somehow exploited to extract information about spatial structure in the light source itself.

Analogous phenomena can be recorded also in the radio range, where suitably designed three-dimensional antenna configurations can detect the wavefront curvature of helical radio waves, offering observables beyond those conventionally measured [9, 17, 18].

8 Role of Large Optical Telescopes

High time resolution astrophysics, quantum optics and large telescopes are closely connected since the sensitivity to rapid variability and higher-order coherence increases *very rapidly* with telescope size.

Light-curves basically become useless for resolutions below microseconds where typical time intervals between successive photons may even be longer than the time resolution, and where timescales of variability are both irregular and unknown. Instead studies have to be of the statistics of the arriving photon stream, such as its correlations or power spectra. All such statistical functions depend on (at least) the second power of the measured source intensity. Figure 14 compares the observed signal (I), its square and fourth powers, for telescopes of different size. The signal

Telescope diameter	Intensity ⟨I⟩	Second-order correlation ⟨I²⟩	Fourth-order photon statistics ⟨I⁴⟩
 3.6 m	1	1	1
 8.2 m	5	27	720
 4 × 8.2 m	21	430	185,000
 50 m	193	37,000	1,385,000,000
100 m	770	595,000	355,000,000,000

Fig. 14 Comparisons between the observed signal of source intensity I, its square and fourth powers, for telescopes of different size. The signal for classical quantities increases with the intensity I; the signal in power spectra and similar functions suitable for variability searches, as I^2, and that of four-photon correlations as I^4, as relevant for some quantum-statistical studies. The advent of very large telescopes enormously increases the potential for quantum optics, and high time resolution astrophysics in general

for classical quantities increases with the intensity I; the signal in power spectra and second-order (two-photon) intensity correlation as I^2; while the signal for a four-photon correlation equals the probability that four photons are recorded in rapid succession, and thus increases with the fourth power of the intensity. This very steep dependence makes the largest telescopes *enormously more sensitive* for high-speed astrophysics and quantum optics.

These large numbers may appear unusual when compared to the more modest gains expected for classical types of instruments, and initially perhaps even difficult to believe. Such numbers are, however, well understood among workers in non-linear optics. The measured $<I^4>$ is proportional to the conditional probability that four photons are recorded within a certain time interval. $<I^4>$ itself is, strictly speaking, not a physical observable: either one detects a photon in a time interval, or one does not. $<I^4>$ therefore has the meaning of successive intensity measurements in rapid succession: $<I(t) \cdot I(t+dt) \cdot I(t+2dt) \cdot I(t+3dt)>$. In an experiment where one is studying the multi-step ionization of some atomic species, where four successive photons have to be absorbed in rapid succession, one indeed sees how a doubling of the light intensity causes a 16-fold increase in the ionization efficiency. Or, for light of identical intensity, how the efficiency increases if the illuminating light source is changed to another of the same intensity but with different statistical properties, i.e. a different value of $<I^4>$.

But it does not stop there. The prospect of improved detectors will further increase the efficiency in a multiplicative manner. An increased quantum efficiency in the visual of a factor 3, say, or in the near infrared of a factor 10, will mean factors of 10 and 100 in second-order quantities, while the signal in fourth-order functions will improve by factors 100 and 10,000, respectively. These factors should thus be *multiplied* with those already large numbers to give the likely gains for very large telescopes equipped with future photon-counting detectors, as compared to present ones.

Due to analogous steep dependences on intensity, the research field of non-linear optics was opened up by the advent of high-power laboratory lasers. In a similar vein, the advent of very large telescopes could well open up the field of very-high-speed astrophysical variability, and bring astronomical quantum optics above a detection threshold.

It is worth noting that—contrary to the situation in imaging—effects of atmospheric or telescopic *seeing* should not compromise measurements of intensity correlations. (This insensitivity to effects of atmospheric turbulence was indeed one of the main advantages of the original intensity interferometer.) To first order, atmospheric seeing induces only phase distortions and angular dislocations for the incoming light but does not affect the amount of illumination falling onto the telescope aperture. That, however, is modulated by the second-order effect of atmospheric *scintillation*. Characteristic timescales for scintillation fall in the range of milliseconds with only very little power remaining on scales of microseconds or shorter [28–30]. Nevertheless, any more precise search for very rapid astrophysical fluctuations will have to quantitatively check and calibrate the possible effects of the atmosphere, of any high-speed adaptive-optics systems used to compensate seeing, and of various transient atmospheric phenomena (such as meteors or Cherenkov radiation from energetic cosmic-ray particles).

9 Synergy with Large Radio Telescopes

SKA, the *Square Kilometer Array* is the large international radio telescope now being planned for longer radio wavelengths. One of its science drivers is strong-field tests of gravity using pulsars and black holes. The radio emission normally comes from lower-density regions in the magnetospheres around such objects, and expected studies of strong gravitational fields will focus on studying the orbit evolution of pulsars in orbit around other compact objects. The *SKA*'s sensitivity will likely enable the discovery of more than 10,000 pulsars in our Galaxy, including more than 1,000 millisecond ones. Many of these will probably be optically detectable with extremely large telescopes, and the *SKA* identifications will offer an excellent list of new targets for optical study.

Hydrogen recombination lasers/masers at the epoch of atomic recombination in the *very* early Universe ($Z \approx 1000$) were mentioned in Sect. 4.4 above. SKA might be able to observe these emission lines, strongly redshifted into the long-wavelength radio, even if those detections are likely to be difficult. On the other hand, optical large telescopes should permit studies of the equivalent physical processes of recombination lasers in nearby Galactic sources such as MWC 349A. This should place the interpretation of the cosmological emissions on a firmer physical basis, and could become an excellent example of the synergy between great telescopes operating in widely different wavelength regions.

Quantum effects are more readily reached at the longer radio wavelengths, where the number of photons per coherence volume can be very large. Indeed, the manifestation of the bunching of photons in the optical corresponds to "wave noise" in the radio. Such aspects also more readily produce conditions for maser emission in astronomical sources. Mechanisms such as the electron-cyclotron maser may operate in numerous classes of astronomical sources, from blazar jets near black holes, to exoplanets [154]. The very shortest nanosecond radio bursts seen from a few pulsars imply extremely high brightness temperatures and essentially reach the quantum limit (Fig. 15).

Thus, radio observations offer powerful tools that are complimentary to those in the optical. The optical probes higher-density regions close to stars and compact objects (rather than low-density clouds); optical spectral lines are those of atomic transitions (rather than those of low-temperature molecules), and the optical permits full studies of quantum phenomena that require the counting of individual photons, not yet practical at the longer radio wavelengths.

Radio observations permit various studies related to quantum statistics of radiation (some of which have already been pursued), e.g., the statistics of microwave

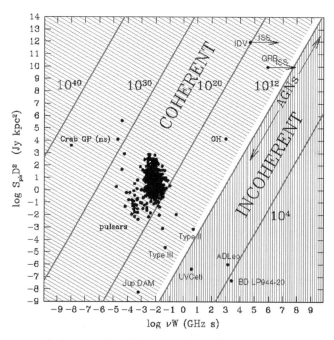

Fig. 15 Phase space for known and anticipated transient radio signals. The horizontal axis is the product of transient duration W and radio frequency ν while the vertical axis is the product of the observed peak flux density S_{pk} and the square of the distance D, thus proportional to luminosity. Lines of constant brightness temperature [K] are shown. A quantum limit is set by the uncertainty principle which requires signals to have a duration of at least one cycle of electromagnetic radiation. With frequency ν given in units of 10^9 Hz, this requires physical signals to be to the right of $\log \nu W = -9$: Wilkinson et al., [164]. Reprinted from New Astron Rev **48**, 1551, © 2004, with permission from Elsevier

maser radiation. Among future challenges remains the study of photon statistics of the cosmic microwave background; possibly this could be one future task in observational cosmology, once its currently sought polarization has been mapped?

10 Intensity Interferometry of Non-Photons

In an astronomical context, one naturally thinks of intensity interferometry in terms of correlations between streams of stellar photons. However, it may be illustrative to realize that in science and industry there exist widespread applications of this interferometric method pioneered by Hanbury Brown & Twiss, sometimes just referred to as "HBT-interferometry". The property being exploited, i.e., the "bunching" in time of photons applies not only to experiments in scattering of optical light [20] or of X-rays [167], but also to any other class of *bosons* (particles with integer spin), a phenomenon being applied in high-energy physics. The observed source is then not a star but perhaps the particles (e.g., pions, kaons) produced in an interaction region between high-energy particle beams. The angular size of the source can then be inferred from the cross-correlation of intensity fluctuations observed by suitably placed detectors. The effort that has been invested in particle applications of intensity interferometry (hundreds of papers) is very much greater than that in the past Hanbury Brown–Twiss experiment for astronomy. Of course, in the laboratory, there are additional degrees of freedom, such as three-dimensional intensity interferometry of time-variable sources. For reviews, see Boal et al. [11] and Alexander [2].

A somewhat different situation applies to fermions, i.e. particles with half-integral-valued spin, such as electrons. These obey Fermi-Dirac statistics and follow the Pauli exclusion principle which prohibits two or more particles to occupy the same quantum state. Thus, a gas of electrons or other fermions shows "antibunching", a property opposite to the bunching characteristic for thermal photons (Fig. 2 center). This is seen experimentally in the laboratory as a certain anticorrelation between nearby electrons [78].

It may be essential to appreciate these quantum properties of electrons when planning quantum-optical experiments. The many quantum-statistical properties, and different degrees of bunching in photon streams are permitted precisely by their boson nature and the more "permissive" Bose-Einstein statistics.

Although, using common language, one often speaks of photon "detection", *photons* as such are actually *never* directly detected! Rather, what is being detected and studied is some electrical signal from *photo-electrons* that results from some photon interaction inside the detector. It is a sobering thought that quantum-statistical properties to be measured, e.g., the bunching of several photons in the same quantum state, is a property that can not even in principle be possessed by these photo-electrons. One may then wonder how it is possible to at all study boson properties through a medium (electrons) that cannot, not even in principle, carry such properties? For photocathode detectors, the explanation apparently is the very short time (femtoseconds?) required for a photoelectron to exit a photocathode and then be detected as an individual particle.

However, semiconductor detectors are more complex, and have longer timescales for the relaxation of their inner energy levels [95]. A further discussion of possible ensuing complications is outside the scope of this paper (but see, e.g., [127]), though it is worth remembering that measurements of the quantum statistics of the incoming light may require an adequate understanding also of the quantum statistical properties of the detector through which these measurements are to be made: in a quantum world it may not always be possible to separate the observable from the observer.

11 Modern Intensity Interferometry

Astronomical observations with single or multiple large telescopes, involving the precise electronic timing of arriving photons within intense light fluxes enable various classes of intensity interferometry, and intensity correlation experiments. Given that contemporary electronic techniques are much more powerful than those originally used in the pioneering Narrabri instrument, one can envision various modern variants, and using only one single telescope aperture, spectroscopic measurements become feasible.

11.1 Photon-Correlation Spectroscopy

Photon-correlation (also called intensity-fluctuation, intensity-correlation or self-beating) spectroscopy is the temporal equivalent of spatial intensity interferometry. The cross-correlation of the optical fluctuations is then measured at one and the same spatial location, but with a variable temporal baseline (as opposed to the same instant of time but with a variable spatial baseline in the spatial intensity interferometer).

In any spectroscopic apparatus, the spectral resolution is ultimately limited by the Heisenberg uncertainty principle: $\Delta E \Delta t \geq \hbar/2$. Thus, to obtain a small value of the uncertainty in energy, ΔE, the time to measure it, Δt, must be relatively great. Methods to increase Δt include the use of larger diffraction gratings, and tilting them parallel to the direction of light propagation in order to increase the time light spends inside the instrument (this is the fundamental reason why echelle gratings give higher resolution than those in lower diffraction orders). Interference filters inside which light travels back and forth many times give high resolution (e.g., Fabry-Perot interferometers with a large finesse). Or—as in photon-correlation spectroscopy—instead of mechanical devices, one can use electronic timing of the light along its direction of propagation. Since this can be made for temporal delays up to perhaps one second, this enables a spectral resolution corresponding to that of a hardware instrument with physical size equal to one light-second! This makes possible spectral resolutions of 1 Hz, equivalent to $R \approx 10^{14}$, many orders of magnitude beyond those feasible with classical spectrometers (Table 1).

Table 1 Spectrometer length and equivalent light-travel-time requirements for different resolving powers at λ 600 nm

Spectral resolution R	Length	Time
100,000	6 cm	200 ps
1,000,000	60 cm	2 ns
10,000,000	6 m	20 ns
100,000,000	60 m	200 ns
1,000,000,000	600 m	2 μs

Somewhat paradoxically, the lower the spectral resolution, the more stringent the temporal requirements. The reason is that the timescale of intensity fluctuations is set by the self-beating time of the light within the detected passband. If this is 1 MHz, say, the characteristic time is its inverse (i.e., 1 μs), but for an astronomically more realistic optical passband of 1 nm, one is down to the order of ps, facing the same challenges as in spatial intensity interferometry. The method has become widespread for various laboratory applications to measure the very small Doppler broadenings caused by, e.g., exhaust fumes, molecular suspensions, or blood cells in living organisms undergoing random motions on a level of perhaps only mm s^{-1}.

Analogous to the spatial information extracted from intensity interferometry, photon-correlation spectroscopy does not reconstruct the full shape of the source spectrum, but "only" gives the power spectrum of the source with respect to the baseline over which it has been observed. In the case of the intensity interferometer, information is obtained on the relative power of spatial frequencies covered by the spatial baselines. In the case of the photon-correlation spectrometer, the result is the power of temporal [electromagnetic] frequencies covered by the temporal baselines, thus yielding the spectral linewidth but not automatically reconstructing the precise shape of a possibly asymmetric line profile. (Although the latter might actually be possible using more elaborate correlation schemes over multiple temporal baselines.)

The signal-to-noise ratios follow similar relations as in intensity interferometry, i.e., while generally being expensive in terms of photon flux, sources of high brightness temperature (such as narrow emission-line components) are easier to measure. Again similar to intensity interferometry, the method formally assumes Gaussian (chaotic; thermal; maximum-entropy) photon statistics, and will not work with purely coherent light. However, the analysis of superposed coherent and chaotic radiation could yield even more information [1, 157].

Technical discussions are in various papers of which several appeared around the time laser light scattering became established as a laboratory technique: [67–69, 81, 131–133]. For more general introductions to intensity-correlation spectroscopy principles and applications see Becker [6], Chu [20], Degiorgio & Lastovka [25], Oliver [113], Pike [122, 123], and Saleh [132].

Photon-correlation spectroscopy does not appear to have been used in astronomy, although there are analogs with [auto]correlation spectrometers used for radio astronomy. While the basic physical principles are related, the latter build upon the detection of the phase and amplitude of the radio wave, an option not feasible for optical light due to its very much higher electromagnetic frequencies.

11.2 Spectroscopy with Resolution R = 100,000,000

Photon-correlation spectroscopy is especially suitable for the study of very narrow emission features. To resolve the infrared emission profiles from CO_2 lasers on Mars requires spectral resolving powers $R = \lambda/\Delta\lambda$ in the range of $10^6 - 10^7$ (Fig. 10), realized though various heterodyne techniques. The optical laser emission components around η Carinae are theoretically expected to have spectral widths $\Delta \nu$ on order 30–100 MHz [73, 75], thus requiring $R \approx 10^8$ ($\approx$10 MHz) to be resolved, far beyond the means of classical spectroscopy. However, such emission lines will be self-beating on timescales on the order of $\Delta t \approx 100$ ns, well within reach of photon-correlation spectroscopy.

There might be a particular advantage for photon-correlation spectroscopy in that it is insensitive to (the probably rapidly variable) wavelength shifts of the emission line components due to local velocities in the source; one does not need to know exactly at which wavelength within the observed passband there might, at any one instant of time, exist some narrow emission-line components: they will all contribute to the correlation signal.

Possibly, the earliest laboratory measurement of a narrow emission line through photon-correlation spectroscopy was by Phillips et al. [120] who, not long after the original experiments by Hanbury Brown & Twiss, deduced linewidths on the order of 100 MHz, corresponding to correlation times of some nanoseconds.

11.3 Optical e-interferometry

Using two or multiple telescopes (or telescope apertures), a precise electronic timing of arriving photons within intense light fluxes, combined with digital signal storage and handling, enables various new modes of electronic digital intensity interferometry, with degrees of freedom not available in the past. Discussions of such e-interferometry are in Dravins et al. [31], who especially point at the potential of electronically combining multiple subapertures of extremely large telescopes, mainly for observations at short optical wavelengths. Pairs or groups of telescopes no longer need to be mechanically movable to track a star across the sky: the mechanical displacements can be replaced by continuously changing electronic delays. Ofir & Ribak [109–111] evaluate concepts for multidetector intensity interferometers. By digitally combining the signals across all possible pairs and triples of baselines (reminiscent of that in long baseline radio interferometry), not only is the noise level reduced and fainter sources accessible but, utilizing various proposed algorithms, one might achieve also the reconstruction of actual two-dimensional images and not merely their power spectra, as the case in classical two-element operations.

Since the required mechanical accuracy of the optics and stability of the atmosphere is on the order of centimeters (rather than a fraction of an optical wavelength, as in phase interferometry), long baselines may become feasible. Very long baseline optical interferometry (VLBOI; perhaps over 1–100 km) might thus be achieved by such electronic means; possible targets could include either nearby phenomena such as the granulation structure on the surfaces of nearby stars, or distant sources such as quasar cores. Initial experiments in VLBOI need not have dedicated telescopes but could be carried out with existing telescopes distributed across the grounds of adjacent observatories, or with flux collectors built primarily for measuring optical Cherenkov radiation in air from high-energy gamma rays, or perhaps even the coarser ones built for solar energy research.

Faster detectors are being developed, enabling higher sensitivity for such e-interferometry. However, faster electronics can only be exploited up to some point since there is a matching requirement on the optomechanical systems. A time resolution of 1 ns corresponds to 30 cm light travel distance, and the mechanical and optical paths then need to be controlled to only some reasonable fraction of this (perhaps 3 cm). However, a timing improvement to 100 ps, say, would require mechanical accuracies on mm levels, going beyond what is achieved in flux collectors, and beginning to approach the levels of seeing-induced fluctuations in atmospheric path-length differences.

Sources that are most advantageous to observe with intensity interferometry are those of small spatial extent with high brightness temperatures. These include suspected sites of stellar laser emission, where possible correlation studies were suggested already earlier [84, 85]. A recent paper [75] suggests to combine intensity interferometry with optical heterodyne spectroscopy (mixing the astronomical signal with a frequency-sweeping laboratory laser) to simultaneously identify spatial and spectral structures in the lasers around η Car.

12 Instrumentation for Extremely Large Telescopes

The practical detectability of higher-order effects in photon streams requires large photon fluxes as does indeed any study of very rapid variability. To identify nanosecond-scale variability will likely require sustained photon-count rates of many megahertz. This is one reason why the optical and infrared regions are most promising in searches for very rapid astrophysical variability. For many interesting sources, required count rates will be reached with foreseen optical telescopes, while the limited sizes of space instruments preclude such count rates for X-rays. While radio observations can be very sensitive, the absence of photon-counting could limit some classes of experiments; also the statistics of radiation can be strongly affected by the propagation through the ionized interstellar medium. Such considerations, as well as the dramatic increase in sensitivity to short-term variability (Fig. 14) indicate that high time resolution astrophysics and quantum optics are most suitable tasks for the coming generation of extremely large optical telescopes.

As part of the conceptual design studies of instruments for ESO's planned extremely large telescope, an instrument, *QuantEYE*, was conceived as the highest time-resolution instrument in optical astronomy [4, 31–33, 108], also described elsewhere in this volume [5]. *QuantEYE* is designed to explore astrophysical variability in the yet unexplored micro- and nanosecond domains, reaching the realm of quantum-optical phenomena. Foreseen targets include millisecond pulsars, variability close to black holes, surface convection on white dwarfs, non-radial oscillation spectra in neutron stars, fine structure on neutron-star surfaces, photon-gas bubbles in accretion flows, and possible free-electron lasers in the magnetic fields around magnetars. Measuring statistics of photon arrival times, *QuantEYE* is able to perform photon-correlation spectroscopy at resolutions $\lambda/\Delta\lambda > 100,000,000$, opening up for the first "extreme-resolution" optical spectroscopy in astrophysics, as well as to search for deviations from thermodynamic equilibrium in the quantum-optical photon bunching in time.

Irrespective of whether an instrument is eventually built following the precise concepts of *QuantEYE* (or whether it is overtaken by ongoing developments in detector technologies), its evaluation demonstrates how one can push the envelope of the parameter domains available to observational astronomy. During recent decades, these domains were mainly expanded through the opening up of successive new wavelength regions, each giving a new and different view of the Universe, thus supplementing the previous ones. However, now that almost all wavelength regions are accessible, the time domain becomes a bold new frontier for astrophysics, enabled by the forthcoming generation of extremely large optical telescopes, equipped with sensitive and fast photon-counting detectors, and supported by the now-established quantum theory of optical coherence!

13 Conclusions

During recent decades is has been realized, theoretically and in the laboratory that both individual photons, and groups of them, can be much more complex, and carry much more information than was commonly believed in the past. That even a single photon, of any given wavelength and polarization, and coming from any given direction, still can have hundreds of different states has come as a surprise to many who believed that photon properties were already well understood.

For astronomy, this poses both an opportunity and a challenge: since our understanding of the Universe is based upon the delicate decoding of information carried by light from celestial sources, we should exploit every opportunity to extract additional information, even if moving into uncharted and unknown territory, and not knowing beforehand what type of information will be conveyed. Quantum optics offers a new tool for studying extreme physical conditions in space, and the forthcoming extremely large optical telescopes promise to offer adequate sensitivity for exploiting quantum optics as another information channel from the Universe.

Acknowledgments This work was supported by the Swedish Research Council, The Royal Physiographic Society in Lund, and the European Southern Observatory. During the design study of *QuantEYE*, numerous constructive discussions were held with colleagues at Lund Observatory, at the University of Padova, and at ESO Garching. Valuable comments on the manuscript were received from Henrik Hartman and Sveneric Johansson in Lund, Vladilen Letokhov (Lund & Moscow), and from an anonymous referee.

References

1. Aleksandrov EB, Golubev YuM, Lomakin AV, Noskin VA (1983) *Intensity-fluctuation spectroscopy of optical fields with non-Gaussian statistics.* Sov Phys Usp **26**:643–663 = Usp Fiz Nauk **140**:547–582
2. Alexander G (2003) *Bose-Einstein and Fermi-Dirac interferometry in particle physics.* Rep Progr Phys **66**:481–522
3. Bachor HA, Ralph TC (2004) *A Guide to Experiments in Quantum Optics,* Wiley-VCH, Weinheim
4. Barbieri C, Dravins D, Occhipinti T, Tamburini F, Naletto G, Da Deppo V, Fornasier S, D'Onofrio M, Fosbury RAE, Nilsson R, Uthas H (2006a) *Astronomical applications of quantum optics for extremely large telescopes.* J Mod Opt, in press
5. Barbieri C, Naletto G, Tamburini F, Occhipinti T, Giro E, D'Onofrio M (2006b) *From QuantEYE to AquEYE – Instrumentation for astrophysics on its shortest timescales.* This volume
6. Becker W (2005) *Advanced Time-Correlated Single Photon Counting Techniques.* Springer, Berlin
7. Begelman MC (2001) *Super-Eddington atmospheres that do not blow away.* ApJ **551**:897–906
8. Begelman MC (2006) *Photon bubbles and the vertical structure of accretion disks.* ApJ **643**:1065–1080
9. Bergman J, Carozzi T, Karlsson R (2003) *Multipoint antenna device.* International Patent Publication WO 03/007422
10. Beth RA (1936) *Mechanical Detection and Measurement of the Angular Momentum of Light.* Phys Rev **50**:115–125
11. Boal DH, Gelbke CK, Jennings BK (1990) *Intensity interferometry in subatomic physics.* Rev Mod Phys **62**:553–602
12. Bowen IS (1927) *The origin of the chief nebular lines.* PASP **39**:295–297
13. Bowen IS (1947) *Excitation by line coincidence.* PASP **59**:196–198
14. Burdyuzha VV, Shelepin LA (1990) *The possibility of an X-ray recombination laser in hydrogen-like ions for the conditions of accreting neutron stars.* AdSpR **10**:163–165
15. Calvo GF, Picón A, Bagan E (2006) *Quantum field theory of photons with orbital angular momentum.* Phys Rev A **73**, id 013805 (10 pp)
16. Cao H (2003) *Lasing in random media.* Waves Random Media **13**:R1–R39
17. Carozzi TD, Bergman JES (2006) *Real irreducible sesquilinear-quadratic tensor concomitants of complex bivectors.* J Math Phys **47**, id 032903 (7 pp)
18. Carozzi T, Karlsson R, Bergman J (2000) *Parameters characterizing electromagnetic wave polarization.* Phys Rev E **61**:2024–2028
19. Castor JI, Nussbaumer H (1972) *On the excitation of C III in Wolf-Rayet envelopes.* MNRAS **155**:293–304
20. Chu, B (1991) *Laser Light Scattering. Basic Principles and Practice, 2nd ed.* (1991) Academic Press, Boston
21. Cooper J, Ballagh RJ, Burnett K, Hummer DG (1982) *On redistribution and the equations for radiative transfer.* ApJ **260**:299–316
22. Cox P, Martin-Pintado J, Bachiller R, Bronfman L, Cernicharo J, Nyman LÅ, Roelfsema PR (1995) *Millimeter recombination lines towards η Carinae.* A&A **295**:L39–L42
23. Danchi WC, Tuthill PG, Monnier JD (2001) *Near-infrared interferometric images of the hot inner disk surrounding the massive young star MWC 349A.* ApJ **562**:440–445
24. Das Gupta S, Das Gupta SR (1991) *Laser radiation in active amplifying media treated as a transport problem - Transfer equation derived and exactly solved.* Ap&SS **184**:77–142
25. Degiorgio V, Lastovka JB (1971) *Intensity-correlation spectroscopy.* Phys Rev A **4**:2033–2050
26. DeNoyer LK, Dodd JG (1982) *Detecting saturation in astrophysical masers.* Bull AAS **14**:638
27. Dravins D (1994) *Astrophysics on its shortest timescales.* ESO Messenger no **78**:9–19

28. Dravins D, Lindegren L, Mezey E, Young AT (1997a) *Atmospheric Intensity Scintillation of Stars. I. Statistical Distributions and Temporal Properties.* PASP **109**:173–207
29. Dravins D, Lindegren L, Mezey E, Young AT (1997b) *Atmospheric intensity scintillation of stars. II. Dependence on optical wavelength.* PASP **109**: 725–737
30. Dravins D, Lindegren L, Mezey E, Young AT (1998) *Atmospheric intensity scintillation of stars. III. Effects for different telescope apertures.* PASP **110**:610–633, erratum **110**:1118
31. Dravins D, Barbieri C, Da Deppo V, Faria D, Fornasier S, Fosbury RAE, Lindegren L, Naletto G, Nilsson R, Occhipinti T, Tamburini F, Uthas H, Zampieri L (2005) *QuantEYE. Quantum Optics Instrumentation for Astronomy. OWL Instrument Concept Study.* ESO doc OWL-CSR-ESO-00000–0162, 280 pp
32. Dravins D, Barbieri C, Fosbury RAE, Naletto G, Nilsson R, Occhipinti T, Tamburini F, Uthas H, Zampieri L (2006a) *QuantEYE: The quantum optics instrument for OWL.* In: Herbst T (ed) *Instrumentation for Extremely Large Telescopes*, MPIA spec publ **106**, pp 85–91
33. Dravins D, Barbieri C, Fosbury RAE, Naletto G, Nilsson R, Occhipinti T, Tamburini F, Uthas H, Zampieri L (2006b) *Astronomical quantum optics with extremely large telescopes.* In: Whitelock P, Dennefeld M, Leibundgut B (eds) *The Scientific Requirements for Extremely Large Telescopes*, IAU Symp **232**:502–505
34. Eichler D, Gedalin M, Lyubarsky Y. (2002) *Coherent emission from magnetars.* ApJ **578**:L121–L124
35. Elitzur M (1992) *Astronomical Masers.* ARA&A **30**:75–112
36. Ershov AA, Kuzmin AD (2005) *Detection of giant pulses in pulsar PSR J1752+235.* A&A **443**:593–597
37. Evans NJ, Hills RE, Rydbeck OEH, Kollberg E (1972) *Statistics of the radiation from astronomical masers.* Phys Rev A **6**:1643–1647
38. Fang LZ (1981) *Stimulated recombination emission from rapidly cooling regions in an accretion disc and its application to SS 433.* MNRAS **194**:177–185
39. Ferland GJ (1993) *A masing [Fe XI] line.* ApJS **88**:49–52
40. Florescu L, John S (2004) *Theory of photon statistics and optical coherence in a multiple-scattering random-laser medium.* Phys Rev E **69**, id. 046603 (16 pp)
41. Foo G, Palacios DM, Swartzlander GA (2005) *Optical vortex coronagraph.* Opt Lett **30**:3308–3310
42. Fox M (2006) *Quantum Optics. An Introduction.* Oxford University Press, Oxford
43. Gahm GF, Lindgren B, Lindroos KP (1977) *A compilation of fluorescent molecular lines originating in or around stellar objects with strong atomic emission lines.* A&AS **27**: 277–283
44. Ghosh TK, Das AK, Mukherjee TK, Mukherjee PK (1995) *The $2p^5 3l$ configurations of highly stripped Ne-like ions: Possibility of X-ray laser emission.* ApJ **452**:949–953
45. Gil J (1985) *Fluctuations of pulsar emission with sub-microsecond time-scales.* Ap&SS **110**:293–296
46. Gil J, Melikidze, G I (2005) *Angular beaming and giant subpulses in the Crab pulsar.* A&A **432**:L61–L65
47. Glauber RJ (1963a) *Photon correlations.* Phys Rev Lett **10**:84–86
48. Glauber RJ (1963b) *The quantum theory of optical coherence.* Phys Rev **130**:2529–2539
49. Glauber RJ (1963c) *Coherent and incoherent states of the radiation field.* Phys Rev **131**:2766–2788
50. Glauber RJ (1965) *Optical coherence and photon statistics.* In: DeWitt C, Blandin A, Cohen-Tannoudji C (eds) *Quantum Optics and Electronics*, Gordon and Breach, New York, pp 65–185
51. Glauber RJ (1970) *Quantum theory of coherence.* In Kay SM, Maitland A (eds) *Quantum Optics*, Academic Press, London, pp 53–125
52. Goodman JW (1985) *Statistical Optics.* Wiley, New York
53. Gordon MA, Holder BP, Jisonna LJ, Jorgenson RA, Strelnitski VS (2001) *3 year monitoring of millimeter-wave radio recombination lines from MWC 349.* ApJ **559**:402–418

54. Greenhouse MA, FeldmanU, Smith HA, Klapisch M, Bhatia AK, Bar-Shalom A (1993) *Infrared coronal emission lines and the possibility of their laser emission in Seyfert nuclei.* ApJS **88**:23–48; *erratum* **99**:743 (1995)
55. Hanbury Brown R (1974) *The Intensity Interferometer.* Taylor & Francis, London
56. Hanbury Brown R (1985) *Photons, Galaxies and Stars (selected papers).* Indian Academy of Sciences, Bangalore
57. Hanbury Brown R (1991) *Boffin. A Personal Story of the Early Days of Radar, Radio Astronomy and Quantum Optics.* Adam Hilger, Bristol
58. Hanbury Brown R, Twiss RQ (1956a) *Correlation between photons in two coherent beams of light.* Nature **177**:27–29
59. Hanbury Brown R, Twiss RQ (1956b) *A test of a new type of stellar interferometer on Sirius.* Nature **178**:1046–1048
60. Hanbury Brown R, Twiss RQ (1958) *Interferometry of the intensity fluctuations in light III. Applications to astronomy.* Proc Roy Soc London Ser A **248**:199–221
61. Hanbury Brown R, Davis J, Allen RL (1967a) *The stellar interferometer at Narrabri Observatory – I. A description of the instrument and the observational procedure.* MNRAS **137**:375–392
62. Hanbury Brown R, Davis J, Allen LR, Rome JM (1967b) *The stellar interferometer at NarrabriObservatory – II. The angular diameters of 15 stars.* MNRAS **137**: 393–417
63. Hankins TH, Kern JS, Weatherall JC, Eilek JA (2003) *Nanosecond radio bursts from strong plasma turbulence in the Crab pulsar.* Nature **422**:141–143
64. Hartman H, Johansson S (2000) *Ultraviolet fluorescence lines of Fe II observed in satellite spectra of the symbiotic star RR Telescopii.* A&A **359**:627–634
65. Harwit M (2003) *Photon orbital angular momentum in astrophysics.* ApJ **597**:1266–1270
66. Hsieh SH, Spiegel EA (1976) *The equations of photohydrodynamics.* ApJ **207**: 244–252
67. Hughes AJ, Jakeman E, Oliver CJ, Pike ER (1973) *Photon-correlation spectroscopy: Dependence of linewidth error on normalization, clip level, detector area, sample time and count rate.* J Phys A: Math Nucl Gen **6**:1327–1336
68. Jakeman E (1970) *Statistical accuracy in the digital autocorrelation of photon counting fluctuations.* J Phys A: Gen Phys **3**:L55–L59
69. Jakeman E (1972) *The effect of heterodyne detection on the statistical accuracy of optical linewidth measurements.* J Phys A: Gen Phys **5**:L49–L52; *corrigendum* **5**:1738
70. Jefferies JT (1971) *Population inversion in the outer layers of a radiating gas.* A&A **12**:351–362
71. Jenet FA, Anderson SB, Prince TA (2001) *The first detection of coherent emission from radio pulsars.* ApJ **558**:302–308
72. Johansson S, Letokhov VS (2002) *Laser action in a gas condensation in the vicinity of a hot star.* Pis'ma Zh Éksp Teor Fiz **75**:591–594 = JETP Lett **75**:495–498
73. Johansson S, Letokhov VS (2004a) *Anomalous Fe II spectral effects and high H I Ly-alpha temperature in gas blobs near Eta Carinae.* Pis'ma Astron Zh **30**:67–73 = Astron Lett **30**:58–63; *erratum* **30**:433
74. Johansson S, Letokhov VS (2004b) *Astrophysical lasers operating in optical Fe II lines in stellar ejecta of Eta Carinae.* A&A **428**:497–509
75. Johansson S, Letokhov VS (2005) *Possibility of measuring the width of narrow Fe II astrophysical laser lines in the vicinity of Eta Carinae by means of Brown-Twiss-Townes heterodyne correlation interferometry.* New Astron **10**:361–369
76. Johansson S, Letokhov VS (2006) *Astrophysical lasers and nonlinear optical effects in space.* New Astron Rev, submitted
77. Johansson S, Davidson K, Ebbets D, Weigelt G, Balick B, Frank A, Hamann F, Humphreys RM, Morse J, White R L (1996) *Is there a dichromatic UV laser in Eta Carinae?* In: Benvenuti P, Macchetto FD, Schreier EJ (eds) *Science with the Hubble Space Telescope-II*, Space Telescope Science Institute, Baltimore, pp 361–365
78. Kiesel H, Renz A, Hasselbach F (2002) *Observation of Hanbury Brown–Twiss anticorrelations for free electrons.* Nature **418**:392–394

79. Klein RI, Arons J, Jernigan G, Hsu JJL (1996) *Photon bubble oscillations in accretion-powered pulsars.* ApJ **457**:L85–L89
80. Knight HS, Bailes M, Manchester RN, Ord SM, Jacoby BA (2006) *Green Bank telescope studies of giant pulses from millisecond pulsars.* ApJ **640**:941–949
81. Koppel DE (1974) *Statistical accuracy in fluorescence correlation spectroscopy.* Phys Rev A **10**:1938–1945
82. Kuzmin AD, Ershov AA (2004) *Giant pulses in pulsar PSR B0031–07.* A&A **427**:575–579
83. Labeyrie A, Lipson SG, Nisenson P (2006) *An Introduction to Optical Stellar Interferometry.* Cambridge Univ Press, Cambridge
84. Lavrinovich NN, Letokhov VS (1974) *The possibility of the laser effect in stellar atmospheres.* Zh Éksp Teor Fiz **67**:1609–1620 = Sov Phys–JETP **40**:800–805 (1975)
85. Lavrinovich NN, Letokhov VS (1976) *Detection of narrow "laser" lines masked by spatially inhomogeneous broadening in radiation emitted from stellar atmospheres.* Kvant Elektron **3**:1948–1954 = Sov J Quantum Electron **6**:1061–1064
86. Lawandy NM, Balachandran RM, Gomes ASL, Sauvain E (1994) *Laser action in strongly scattering media.* Nature **368**:436–438
87. Leach J, Padgett MJ, Barnett SM, Franke-Arnold S, Courtial V (2002) *Measuring the orbital angular momentum of a single photon.* Phys Rev Lett **88**, id 257901 [4 pp]
88. Lee LC, Jokipii JR (1975) *Strong scintillations in astrophysics. III - The fluctuations in intensity.* ApJ **202**:439–453
89. Lekht EE, Rudnitskii GM, Franquelin O, Drouhin JP (1975) *Statistical properties of the emission of OH maser sources.* Pis'ma Astron Zh **1**:29–32 = Sov Astron Lett **1**:37–38
90. Lepri S, Cavalieri S, Oppo GL, Wiersma DS (2006) *Statistical regimes of random laser fluctuations.* Phys Rev E, submitted = arXiv:physics/0611059
91. Lerche I (1979a) *Scintillations in astrophysics. I – An analytic solution of the second-order moment equation.* ApJ **234**:262–274
92. Lerche I (1979b) *Scintillations in astrophysics. II – An analytic solution of the fourth-order moment equation.* ApJ **234**:653–668
93. Letokhov VS (1967) *Generation of light by a scattering medium with negative resonance absorption.* Zh Éksp Teor Fiz **53**:1442–1452 = Sov Phys JETP **26**:835–840 (1968)
94. Letokhov VS (2002) *Astrophysical lasers.* Kvant Elektron **32**:1065–1079 = Quant Electr **32**:1065–1079
95. Lim TS, Chern JL, Otsuka K (2002) *Higher-order photon statistics of single-mode laser diodes and microchip solid-state lasers.* Opt Lett **27**:2197–2199
96. Lo, KY (2005) *Mega-masers and galaxies.* ARA&A **43**:625–676
97. Loudon R (1980) *Non-classical effects in the statistical properties of light.* Rep Prog Phys **43**:913–949
98. Loudon R (2000) *The Quantum Theory of Light,* 3rd ed. Oxford Univ Press, Oxford
99. Lu Y, Zhang SN (2004) *Maser mechanism of optical pulsations from anomalous X-ray pulsar 4U 0142+61.* MNRAS **354**:1201–1207
100. Macháček M (1978) *A quantum mechanical description of the transfer of radiation. I – The radiation processes.* Bull Astron Inst Czechosl **29**:268–277
101. Macháček M (1979) *The quantum mechanical description of the transfer of radiation. II – The equation of transfer.* Bull Astron Inst Czechosl **30**:23–28
102. Mandel L, Wolf E (1995) *Optical Coherence and Quantum Optics.* Cambridge Univ Press, Cambridge
103. Menzel DH (1937) *Physical processes in gaseous nebulae. I.* ApJ **85**:330–339
104. Menzel DH (1970) *Laser action in non-LTE atmospheres.* In: Groth HG, Wellmann P (eds) *Spectrum Formation in Stars with Steady-State Extended Atmospheres.* Nat Bur Stds Spec Publ **332**:134–137
105. Meystre P, Sargent M (1990) *Elements of Quantum Optics.* Springer, Berlin
106. Mishchenko EG (2004) *Fluctuations of radiation from a chaotic laser below threshold.* Phys Rev A **69**, id 033802 [6 pp]

107. Mumma MJ, Buhl D, Chin G, Deming D, Espenak F, Kostiuk T (1981) *Discovery of natural gain amplification in the 10-micrometer carbon dioxide laser bands on Mars - A natural laser.* Science **212**:45–49
108. Naletto G, Barbieri C, Dravins D, Occhipinti T, Tamburini F, Da Deppo V, Fornasier S, D'Onofrio M, Fosbury RAE, Nilsson R, Uthas H, Zampieri L (2006) *QuantEYE: A quantum optics instrument for extremely large telescopes.* In: McLean IS, Iye M (eds) *Ground-Based and Airborne Instrumentation for Astronomy.* SPIE Proc **6269**:635–643
109. Ofir A, Ribak EN (2006a) *Offline, multidetector intensity interferometers – I. Theory.* MNRAS **368**:1646–1651
110. Ofir A, Ribak EN (2006b) *Offline, multidetector intensity interferometers – II. Implications and applications.* MNRAS **368**:1652–1656
111. Ofir A, Ribak EN (2006c) *Micro-arcsec imaging from the ground with intensity interferometers.* In: Monnier JD, Danchi WC (eds) *Advances in Stellar Interferometry.* SPIE proc **6268**:1181–1191
112. Ojaste J, Sapar A (1979) *Statistics of photons and its recording.* Publ Tartu Astrof Obs **47**:93–102
113. Oliver CJ (1978) *The extraction of spectral parameters in photon-correlation spectroscopy.* Adv Phys **27**:387–435
114. Padgett M, Courtial J, Allen L (2004) *Light's orbital angular momentum.* Phys Today **57(5)**:35–40
115. Paschenko M, Rudnitskij GM, Slysh VI, Fillit R (1971) *A measurement of the one-dimensional distribution function of the signal from some Galactic OH radio sources.* Astr Tsirk No.626 [3 pp.]
116. Paterson C (2005) *Atmospheric turbulence and orbital angular momentum of single photons for optical communication.* Phys Rev Lett **94**, id 153901 [4 pp]
117. Peng J, Pradhan AK (1994) *Laser action in far-infrared astrophysical sources* ApJ **432**:L123–L126
118. Petrova SA (2004a) *Toward explanation of microstructure in pulsar radio emission.* A&A **417**:L29–L32
119. Petrova SA (2004b) *On the origin of giant pulses in radio pulsars.* A&A **424**:227–236
120. Phillips DT, Kleiman H, Davis SP (1967) *Intensity-correlation linewidth measurement.* Phys Rev **153**:113–115
121. Picinbono B, Bendjaballah C (2005) *Characterization of nonclassical optical fields by photodetection statistics.* Phys Rev A **71**, id 013812 [12 pp]
122. Pike ER (1970) *Optical spectroscopy in the frequency range $1–10^8$ Hz.* Rev Phys Techn **1**:180–194
123. Pike ER (1976) *Photon correlation spectroscopy.* In Smith RA (ed) *Very High Resolution Spectroscopy.* Academic Press, London, pp 51–73
124. Popov MV, Kuzmin AD, Ul'yanov OM, Deshpande AA, Ershov AA, Zakharenko VV, Kondratiev VI, Kostyuk SV, Losovskii BYa, Soglasnov VA (2006) *Instantaneous radio spectra of giant pulses from the Crab pulsar from decimeter to decameter wavelengths.* Astron Zh **83**:630–637 = Astron Rep **50**:562–568
125. Prendergast KH, Spiegel EA (1973) *Photon bubbles.* Comm Astrophys Space Phys **5**:43–49
126. Robinson BJ, McGee RX (1967) *OH Molecules in the interstellar medium.* ARA&A **5**:183–212
127. Rohde PP, Ralph TC (2006) *Modelling photo-detectors in quantum optics.* J Mod Opt **53**:1589–1603
128. Quirrenbach A, Frink S, Thum C (2001) *Spectroscopy of the peculiar emission line star MWC349.* In Gull TR, Johansson S, Davidson K (eds) *Eta Carinae and Other Mysterious Stars: The Hidden Opportunities of Emission Spectroscopy.* ASP Conf Ser **242**: 183–186
129. Sakhibullin NA, ShimanskyVV (2000) *Non-LTE effects for Na I lines in X-ray illuminated stellar atmospheres.* Pis'ma Astr Zh **26**:369–379 = Astron Lett **26**:309–318
130. Saldin EL, Schneidmiller EA, Yurkov MV (1998) *Statistical properties of radiation from VUV and X-ray free electron laser.* Opt Comm **148**:383–403.

131. Saleh BA (1973) *Statistical accuracy in estimating parameters of the spatial coherence function by photon counting techniques.* J Phys A: Math Nucl Gen **6**:980–986
132. Saleh BA (1978) *Photoelectron Statistics.* Springer, Berlin
133. Saleh BEA, Cardoso MF (1973) *The effect of channel correlation on the accuracy of photon counting digital autocorrelators.* J Phys A: Math Nucl Gen **6**:1897–1909
134. Sapar A (1978) *Transfer equation for the density matrix of phase space cell occupation number states.* Publ Tartu Astrof Obs **46**:17–32
135. Schmid HM (1989) *Identification of the emission bands at* $\lambda\lambda$ *6830, 7088.* A&A **211**:L31–L34
136. Schroeder CB (2002) *Photon statistics of the SASE FEL.* Nucl Instr Meth Phys Res A **483**:499–503
137. Schroeder CB, Pellegrini C, Chen P (2001) *Quantum effects in high-gain free-electron lasers.* Phys Rev E **64**, id 056502 [10 pp]
138. Shaviv N (2005) *Exceeding the Eddington limit.* In: Humphreys R, Stanek K (eds) *The Fate of the Most Massive Stars.* ASP Conf Ser **332**:183–189
139. Shearer A, Stappers B, O'Connor P, Golden A, Strom R, Redfern M, Ryan O (2003) *Enhanced optical emission during Crab giant radio pulses.* Science **301**:493–495
140. Smith HA (1969) *Population inversions in ions of astrophysical interest.* ApJ **158**:371–383
141. Smits JM, Stappers BW, Macquart JP, Ramachandran R, Kuijpers J (2003) *On the search for coherent radiation from radio pulsars.* A&A **405**:795–801
142. Soglasnov VA, Popov MV, Bartel N, Cannon W, Novikov AYu, Kondratiev VI, Altunin VI (2004) *Giant pulses from PSR B1937+21 with widths $\leq$ 15 nanoseconds and $T_b \geq 5 \times 10^{39}$ K, the highest brightness temperature observed in the Universe.* ApJ **616**:439–451
143. Sonnabend G, Wirtz D, Vetterle V, Schieder R (2005) *High-resolution observations of Martian non-thermal CO_2 emission near 10 μm with a new tuneable heterodyne receiver.* A&A **435**:1181–1184
144. Sorokin PP, Glownia JH (2002) *Lasers without inversion (LWI) in Space: A possible explanation for intense, narrow-band, emissions that dominate the visible and/or far-UV (FUV) spectra of certain astronomical objects.* A&A **384**:350–363
145. Spaans M, Norman CA (1997) *Hydrogen recombination line masers at the epochs of recombination and reionization.* ApJ **488**:27–34
146. Spiegel EA (1976) *Photohydrodynamic instabilities of hot stellar atmospheres.* In: Cayrel R, Steinberg M (eds) *Physique des Mouvements dans les Atmosphères Stellaires.* Editions du CNRS, Paris, pp 19–50
147. Streater A, Cooper J, Rees DE (1988) *Transfer and redistribution of polarized light in resonance lines. I – Quantum formulation with collisions.* ApJ **335**:503–515
148. Strelnitski V (2002) *The puzzle of natural lasers.* In: Migenes V, Reid MJ (eds) *Cosmic Masers: From Protostars to Blackholes,* IAU Symp **206**:479–481
149. Strelnitski V, Haas MR, Smith HA, Erickson EF, Colgan SWJ, Hollenbach DJ (1996a) *Far-infrared hydrogen lasers in the peculiar star MWC 349A.* Science **272**:1459–1461
150. Strelnitski VS, Ponomarev VO, Smith HA (1996b) *Hydrogen masers. I. Theory and prospects.* ApJ **470**:1118–1133
151. Strelnitski VS, Smith HA, Ponomarev VO (1996c) *Hydrogen masers. II. MWC349A.* ApJ **470**:1134–1143
152. Swartzlander GA (2001) *Peering into darkness with a vortex spatial filter.* Opt Lett **26**:497–499
153. Tamburini F, Anzolin G, Umbriaco G, Bianchini A, Barbieri C (2006) *Overcoming the Rayleigh criterion limit with optical vortices.* Phys Rev Lett **97**, id 163903 [4 pp]
154. Treumann, RA (2006) *The electron-cyclotron maser for astrophysical applications.* Astron Astrophys Rev **13**:229–315
155. Uscinski BJ (1977) *The Elements of Wave Propagation in Random Media.* McGraw-Hill, New York
156. van der Molen KL, Mosk AP, Lagendijk A (2006) *Intrinsic intensity fluctuations in random lasers.* Phys Rev A **74**, id 053808 [6 pp]

157. Vannucci G, Teich MC (1980) *Computer simulation of superposed coherent and chaotic radiation.* Appl Opt **19**:548–553
158. Varshni YP, Lam CS (1976) *Laser action in stellar envelopes.* Ap&SS **45**:87–97
159. Varshni YP, Nasser RM (1986) *Laser action in stellar envelopes. II – He I.* Ap&SS **125**:341–360
160. Verga AD (1982) *Irreversible thermodynamics in a radiating fluid.* ApJ **260**: 286–298
161. Weatherall JC (1998) *Pulsar radio emission by conversion of plasma wave turbulence: Nanosecond time structure.* ApJ **506**:341–346
162. West DK (1968) [*Conference discussion on possible laser action in C III λ 1909 Å line of γ Vel*]. In: Gebbie KB, Thomas RN (eds) *Wolf-Rayet Stars.* Nat Bur Stds Spec Publ **307**:221–227
163. Wiersma D (2000) *Laser physics: The smallest random laser.* Nature **406**:132–133
164. Wilkinson PN, Kellermann KI, Ekers RD, Cordes JM, Lazio TJW (2004) *The exploration of the unknown.* New Astron Rev **48**:1551–1563
165. Wu YC (1992) *Statistical nature of astrophysical maser radiation for linear masers.* Can J Phys **70**:432–440; *erratum* **71**:403 (1993)
166. Wu YC (1993) *Nonlinear optics in celestial natural maser-laser environment.* Ap&SS **209**:113–121
167. Yabashi M, Tamasaku K, Ishikawa T (2004) *Intensity interferometry for the study of X-ray coherence.* Phys Rev A **69**, id 023813 [9 pp]

ULTRACAM: An Ultra-Fast, Triple-Beam CCD Camera for High-Speed Astrophysics

V. S. Dhillon

In this paper, I describe the scientific motivation, design and performance of ULTRACAM—a three-colour CCD camera for imaging photometry at high temporal resolutions.

1 Introduction

Charge-Coupled Devices (CCDs) revolutionised astronomy when they became available in the 1970's. CCDs are linear, stable, robust, low-power devices, with large formats, small pixels and excellent sensitivity over a wide range of wavelengths and light levels; in fact, they are almost perfect detectors, suffering only from poor time resolution and readout noise compared to the photon-counting detectors that they replaced. These limitations of CCDs are inherent to their architecture, in which the photo-generated electrons must first be extracted (or *clocked*) from the detection site and then digitised.

There are ways in which the readout noise of CCDs can be eliminated—see the article by Tom Marsh in this volume. There are also ways in which CCDs can be made to read out faster. First, the clocking rate can be increased and the digitisation time decreased, but only if the CCD is of sufficient quality to allow these adjustments to be made without an unacceptable increase in charge-transfer inefficiency and readout noise, respectively. Second, the duty cycle of the exposures can be increased; for example, *frame-transfer* CCDs provide a storage area into which photo-generated charge can be clocked. This charge is then digitised whilst the next exposure is taking place and, because digitisation generally takes much longer than clocking, the dead time between exposures is significantly reduced. Third, the data acquisition hardware and software can be designed in such a way that the rate at which it is able to archive the data (e.g. to a hard disk) is always greater than the rate at which data is digitised by the CCD—a situation I shall refer to as *detector-limited* operation.

V. S. Dhillon
Department of Physics & Astronomy, University of Sheffield, Sheffield S3 7RH, UK
e-mail: vik.dhillon@sheffield.ac.uk

We have employed all three of the techniques described above to harness the greater sensitivity and versatility of CCDs (compared to photon-counting detectors) for high-speed optical photometry. The resulting instrument, known as ULTRACAM (for ULTRA-fast CAMera), was commissioned on the 4.2 m William Herschel Telescope (WHT) on 16 May 2002. Other contributions to this volume detail the scientific results obtained with ULTRACAM. In this paper, I describe the scientific motivation behind ULTRACAM, present an outline of its design and report on its measured performance.

2 Scientific Motivation

It is widely recognised that temporal resolution is a relatively unexplored region of observational parameter space (see, for example, [9]), particularly in the optical part of the spectrum. There is a simple reason for this: CCDs are the dominant detector on all of the largest ground-based telescopes and the resources of the observatories have understandably been targeted towards increasing the area and sensitivity of their detectors rather than prioritising high-speed readout. As a result, it has been difficult for astronomers to achieve frame rates of higher than of order one frame per minute using CCDs, although there are a few notable exceptions (see Table 1 for details).

Aside from the serendipitous value of exploring a new region of observational parameter space, it is the study of compact objects such as white dwarfs, neutron stars and stellar-mass black holes which benefits the most from high-speed observations. The dynamical timescale, t_{dyn}, of a star of mass, M, and radius, R, is given by

Table 1 Instruments/modes for high-speed astrophysics with CCDs—this is not intended to be an exhaustive list, but an indication of the variety of ways in which CCDs can be used for high-speed optical observations. Those listed with a dagger ($\dagger$) symbol are not continuous modes and data-taking has to be periodically stopped for archiving. The frame rates listed are estimates of the maximum values possible with each mode. Abbreviations: F-T—frame-transfer; MCP—microchannel plate; EMCCD—electron-multiplying CCD

Instrument/mode	Telescope	Detector	Frame rate	Reference
Low-smear drift (LSD) mode†	WHT	CCD	4 Hz	[35]
Time-series mode†	AAT	CCD	100 Hz	[42]
Phase-binning mode	Hale	CCD	1000 Hz	[18], [19]
Stroboscopic mode	GHO	CCD	7.5 Hz	[44], [20]
Freerun mode	Mayall	CCD+MCP	24000 Hz	[12], [13]
Continuous-clocking mode	Keck-II	CCD	14 Hz	[31], [40]
High time-resolution (HIT) mode†	VLT	CCD	833 Hz	[6]
ULTRACAM	WHT/VLT	CCD	500 Hz	This paper
UCT photometer	SAAO	F-T CCD	1 Hz	[32], [47]
Acquisition camera	Gemini South	F-T CCD	7 Hz	[15]
JOSE camera†	NOT	F-T CCD	150 Hz	[41], [2]
SALTICAM	SALT	F-T CCD	10 Hz	[33]
LuckyCam	NOT	EMCCD	30 Hz	[21]

$$t_{dyn} \sim \sqrt{\frac{2R^3}{GM}}, \qquad (1)$$

and ranges from seconds in white dwarfs to milliseconds in neutron stars and black holes. This means that the rotation and pulsation of these objects or material in close proximity to them (e.g. in an accretion disc) tends to occur on timescales of milliseconds to seconds. Other areas of astrophysics which benefit from observations obtained on these timescales are studies of eclipses, transits and occultations, where increased time-resolution can be used to (indirectly) give an increased spatial resolution.

Studying the above science requires a photometer with the following capabilities:

1. *Short exposure times (from milliseconds to seconds).* There is little point in going much faster than milliseconds, as the gravity of compact objects does not allow for bulk motions on timescales below this. There are exceptions to this rule, however; for example, microsecond time-resolution would allow the study of magnetic instabilities in accreting systems, and nanosecond time-resolution would open up the field of quantum optics (see [9]).
2. *Negligible dead time between exposures.* It takes a finite amount of time to archive a data frame and start the next exposure. To preserve time resolution, it is essential that this dead time is a small fraction of the exposure time.
3. *Multi-channels (3 or more) covering a wide wavelength range.* Ideally, one would obtain spectra at high temporal resolution in order to fully characterise the source of variability (e.g. to determine its temperature). The faintness of most compact objects precludes spectroscopy, unfortunately. At a minimum, therefore, at least three different pass-bands covering as wide a portion of the optical spectrum as possible need to be observed, as this would allow a blackbody spectrum to be distinguished from a stellar spectrum. It is particularly important that one of the three channels is sensitive to the far blue (i.e. between approximately 3000–4000 Å), as the flickering and oscillations observed in many accreting binaries is much more prominent at these wavelengths (see, for example, [27]).
4. *Simultaneous measurement of the different wavelength bands.* The requirement for multi-channel photometry must be tied to a requirement that each channel is observed simultaneously. A single-channel instrument with a filter wheel, for example, could obtain data in more than one colour by changing filters between each exposure or obtaining a full cycle in one filter and then a second cycle in another filter. The first technique results in poor time resolution and both the first and second techniques are observationally inefficient (relative to an instrument which can record multi-channel data simultaneously). In addition, the non-simultaneity of both techniques makes them unsuitable for the study of colour variations which occur on timescales shorter than the duration of observations in a single filter.
5. *Imaging capability.* A photometer with an imaging capability is essential to observe variable sources. This is because, unless the atmospheric conditions

are perfectly photometric, it is necessary to simultaneously measure the target, comparison stars and sky background so that any variability observed can be unambiguously assigned to the correct source. Although non-imaging photometers based on multiple photomultiplier tubes can work around this, they still suffer from their dependency on a fixed aperture which is larger than the seeing disc and which therefore increases the sky noise substantially. This is not a problem with imaging photometers, in which the signal-to-noise of the extracted object counts can be maximised using optimal photometry techniques (e.g. [30]).

6. *High efficiency and portability.* General studies of variability, i.e. the study of any object which eclipses, transits, occults, flickers, flares, pulsates, oscillates, erupts, outbursts or explodes, involves observing objects which span a huge range in brightness. The faintest (e.g. the pulsars) are extremely dim and observing them at high temporal resolution is a photon-starved application on even the largest aperture telescopes currently available. Others (e.g. cataclysmic variables) are relatively bright and can be observed at sub-second time-resolution at excellent signal-to-noise on only a 2 m class telescope. Any instrument designed specifically for high-speed observations must therefore be portable, so that an appropriate aperture telescope can be selected for the project at hand, and highly efficient at recording incident photons, to combat the short exposure times required.

3 Design

None of the instruments listed in Table 1 meet all six of the scientific requirements given at the end of Sect. 2. Here I present the design of ULTRACAM, an instrument which does meet these requirements.

3.1 Optics

The starting point for the ULTRACAM optical design was the required field of view. The field of view has to be large enough to give a significant probability of finding a comparison star of comparable brightness to our brightest targets. The probability of finding a comparison star of a given magnitude depends on the search radius and the galactic latitude of the star. Using the star counts listed by [39], the probability of finding a star of magnitude $R = 12$ at a galactic latitude of $30°$ (the all-sky average) is 80% if the search radius is 5 arcminutes.[1] Most of the targets discussed in Sect. 2 are significantly fainter than $R = 12$, so a field of view of 5 arcminutes virtually guarantees the presence of a suitable comparison star.

---

[1] An on-line comparison star probability calculator can be found at http://www.shef.ac.uk/physics/people/vdhillon/ultracam.

The next consideration in the ULTRACAM optical design was the required pixel scale. Arguably the best compromise in terms of maximising spatial resolution whilst minimising the contributions of readout noise and intra-pixel quantum efficiency variations (e.g. [17]) is to use a pixel scale which optimally samples the median seeing at the telescope. On the WHT, the median seeing ranges from 0.55 to 0.73 arcseconds depending on the method of measurement [46], so a pixel scale of 0.3 arcseconds provides a good compromise. The E2V 47-20 CCDs used in ULTRACAM (see Sect. 3.2) were selected because their 1024 pixels on a side delivers a field of view of 5 arcminutes at a scale of 0.3 arcseconds/pixel, matching the requirements listed above perfectly.

There are a total of 30 optical elements in ULTRACAM, as shown in Fig. 1, all of which have a broad-band anti-reflection coating with transmission shown in Fig. 2. The maximum number of optical elements encountered by a photon passing through ULTRACAM is 14. Light from the telescope is first collimated and then split into three beams by two dichroic beamsplitters. Each beam is then re-imaged by a camera onto a detector, passing through a filter and CCD window along the way. Each of these components is discussed in turn below.

3.1.1 Collimator

The collimator is based on a Petzval lens with a very remote re-imaged pupil. As well as feeding collimated light to the re-imaging cameras, the collimator in ULTRACAM corrects for the aberrations in the telescope it is mounted on. As the first optical element, the collimator also has to have a high transmission in the scientifically important $\sim$3000–4000 Å region (see Sect. 2) without introducing

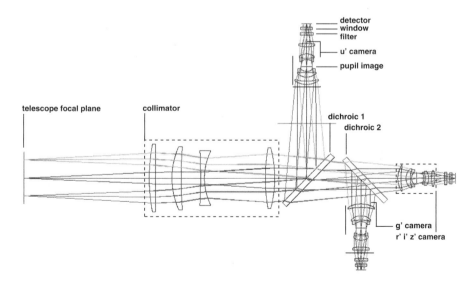

Fig. 1 Ray-trace through the ULTRACAM optics, showing the major optical components: the collimator, dichroics, cameras, filters and detector windows

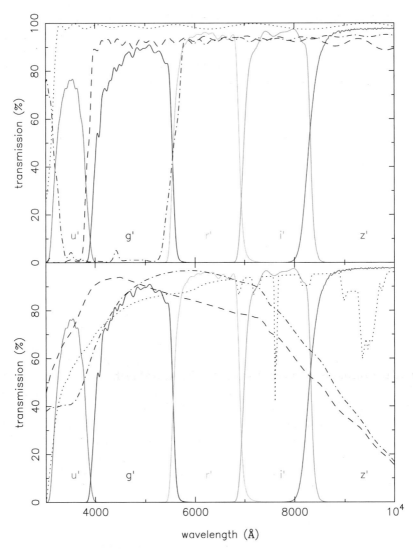

Fig. 2 Top: Transmission profiles of the ULTRACAM SDSS filter-set (solid lines), the anti-reflection coating used on the ULTRACAM lenses (dotted line), and the two dichroics (dashed line and dashed-dotted line). Bottom: Transmission profiles of the ULTRACAM SDSS filter-set (solid lines) and the atmosphere for unit airmass (dotted line). Also shown are the quantum efficiency curves of the u' and g' CCDs (dashed line) and the $r'i'z'$ CCD (dashed-dotted line)

significant chromatic aberration, restricting the glasses it is made from to N-PSK3, CaF_2 and LLF1.

The collimator shown in Fig. 1 is used when mounting ULTRACAM on the WHT, but it can be replaced with another collimator designed to accommodate the different optical characteristics of another telescope. For example, we already have a second collimator for use with the 2.3 m Aristarchos Telescope in Greece, and a third for use at the Visitor Focus of the 8.2 m Very Large Telescope (VLT) in Chile (on which ULTRACAM was commissioned in May 2005). Each of these collimators have identical mounting plates on their lens barrels, so it is a simple matter to switch between them on the opto-mechanical chassis (Sect. 3.3). Note that the rest of the ULTRACAM optics remain unchanged when moving between different telescopes.

3.1.2 Dichroics

Two dichroic beamsplitters divide the light from the collimator into three different beams, which shall hereafter be referred to as the 'red', 'green' and 'blue' channels. Each dichroic consists of a UV-grade fused silica substrate with a long-wave pass (LWP) coating that reflects incident light with wavelengths shorter than the cut-point whilst transmitting longer wavelengths. Such LWP dichroics generally have a higher throughput than SWP dichroics and their throughput has been further enhanced by coating the back surface of each dichroic with a broad-band anti-reflection coating. The ULTRACAM dichroics were manufactured by CVI Technical Optics, Isle of Man.

To keep the cost of the custom optics as low as possible, whilst still retaining maximum throughput and image quality, it was decided to dedicate the blue channel of ULTRACAM to the scientifically-essential SDSS u'-band. As a result, the 50% cut point of the first dichroic was set to 3870 Å (see Fig. 2). A similar argument forced us to dedicate the green channel of ULTRACAM to the SDSS g'-band, giving a 50% cut point for the second dichroic of 5580 Å. In both cases, the cut-points are measured using incident light which is randomly polarised.

It should be noted that the ULTRACAM dichroics operate in a collimated beam. This has the great advantage that ghosts produced by reflections off the back surface of the dichroics are in focus (but slightly aberrated) and fall more-or-less on top of the primary image. This means that the light in the ghosts, which is typically less than 0.001% of the primary image, is included in the light of the target when performing photometry, thereby maximising the signal-to-noise of the observations and removing a potential source of systematic error.

3.1.3 Re-Imaging Cameras

The three re-imaging cameras are based upon a four-element double-Gauss type lens, with two glass types per camera to minimise costs. The glass thickness has also been kept to a minimum, particularly in the blue arm, to keep the throughput as high as possible. The cameras have been optimised for an infinite object distance, with a field angle dictated by the required field size and the need for the telescope pupil to be re-imaged onto the camera stop (shown by the horizontal/vertical bars near the centre of each camera in Fig. 1) at sufficient distance from the last surface of the collimator to allow space for the dichroics. The re-imaging cameras and collimators used in ULTRACAM were manufactured by Specac Ltd., UK.

3.1.4 Filters and Windows

The Sloan Digital Sky Survey (SDSS) photometric system [14] was adopted as the primary filter set for ULTRACAM. This system is becoming increasingly prevalent in astrophysics and consists of five colour bands (u', g', r', i' and z') that divide the entire range from the atmospheric cut-off at ~3000 Å to the sensitivity limit of CCDs at ~11 000 Å into five essentially non-overlapping pass-bands. The fact that SDSS filters show only negligible overlap between their pass-bands is one of the main reasons they have been adopted for use in ULTRACAM, as the dichoic cut-points then have only a negligible effect on the shape of the filter response. The alternative Johnson-Morgan-Cousins system (U, B, V, R_C, I_C) suffers from overlapping pass-bands which would be substantially altered when used in conjunction with the ULTRACAM dichroics. The thinned CCDs used in ULTRACAM would also suffer from fringing when used with R_C due to the extended red tail of this filter's bandpass, but this is eliminated by the sharp red cut-off in SDSS r'. The SDSS filters used in ULTRACAM were procured from Asahi Spectra Ltd., Tokyo.

As well as SDSS filters and clear filters (for maximum throughput in each channel), ULTRACAM also has a growing set of narrow-band filters, including CIII/NIII+HeII (central wavelength 4662 Å, FWHM 108 Å), NaI (central wavelength 5912 Å, FWHM 312 Å), Hα (central wavelength 6560 Å, FWHM 100 Å), red continuum (central wavelength 6010 Å, FWHM 118 Å) and blue continuum (central wavelength 5150 Å, FWHM 200 Å). All ULTRACAM filters are 50×50 mm^2 and approximately 5 mm thick, but have been designed so that their optical thicknesses are identical, so that their differing refractive indices are compensated by slightly different thicknesses, making the filters interchangeable without having to significantly refocus the instrument. Clearly, it is only possible to use these filters in combinations compatible with the red, green and blue channels of ULTRACAM, such as $u'g'r'$, $u'g'i'$, $u'g'z'$, $u'g'$ clear, u' CIII/NIII+HeII red-continuum, $u'g'$ NaI, u' blue-continuum Hα.

The final optical element encountered by a photon in ULTRACAM is the CCD window. This allows light to fall on the chip whilst retaining the vacuum seal of the CCD head (see Sect. 3.2). The windows are plane, parallel discs made of UV grade fused silica and coated with the same anti-reflection coating as the lenses (see Fig. 2).

3.2 Detectors

ULTRACAM uses three E2V 47-20 CCDs as its detectors. These are frame transfer chips with imaging areas of 1024×1024 pixels^2 and storage areas of 1024×1033 pixels^2, each pixel of which is 13 μm on a side. To improve quantum efficiency, the chips are thinned, back-illuminated and anti-reflection coated with E2V's enhanced broad-band astronomy coating (in the case of the blue and green chips) and standard mid-band coating (in the case of the red chip)—see Fig. 2. The chips have a single serial register which is split into two halves, thereby doubling the frame rate, and each of these two channels has a very low noise amplifier at its end, delivering readout noise of only ~$3.5\,e^-$ at pixel rates of 11.2 μs/pixel (and ~$5\,e^-$ at 5.6 μs/pixel). The devices used in ULTRACAM are grade 0 devices, i.e. they are of the highest cosmetic quality available. The full well capacity of these devices is ~$100\,000\,e^-$, and the devices are hence operated at approximately unity gain with the 16-bit analogue-to-digital converters in the CCD controller (see Sect. 3.4).

In order for dark current to always be a negligible noise source, it must be significantly lower than the number of photons received from the sky. In the worst case, in which an ULTRACAM observation is being performed in the u'-band, when the Moon is new, on a dark site (such as La Palma), and on a small aperture telescope (such as a 1 m), the number of photons from the sky incident on ULTRACAM would be only ~$0.3\,e^-$/pixel/s. In this most pessimistic scenario, therefore, the dark current would need to be below ~$0.1\,e^-$/pixel/s to be a negligible noise source. Fortunately, running the E2V 47-20 in inverted mode at 233 K delivers a dark current of only ~$0.05\,e^-$/pixel/s, beating our requirement by a factor of two. This is a relatively high temperature, so the chips are cooled by a three-stage Peltier cooler. To maintain cooling stability, and hence dark current stability, the hot side of the Peltier is itself maintained at a constant temperature of 283 K using a recirculating water chiller. Note that this chiller is also used to cool the CCD controller (see Sect. 3.4), which is located at the base of the instrument and can become extremely hot during operation.

The great advantage of using Peltier cooling as opposed to liquid nitrogen, for example, is that the resulting CCD heads are very small and lightweight. With three such heads in ULTRACAM, this has dramatically decreased both the mass and volume of the instrument. To prevent condensation on the CCDs whilst observing, the heads have been designed to hold a vacuum of below 10^{-3} Torr for several weeks, and we also blow dry air or nitrogen gas across the front of the head to prevent condensation from forming on the CCD window.

3.3 Mechanics

Small-aperture telescopes generally have much smaller size and mass constraints on instruments than large-aperture telescopes. Given the requirement for portability detailed in Sect. 2, we therefore designed the mechanical structure of ULTRACAM assuming the space envelope and mass limits of a typical 2 m telescope.

The mechanical structure of ULTRACAM is described in detail by [43]. Briefly, it is required to: i) provide a stable platform on which to mount the optics, CCD heads and CCD controller; ii) allow easy access to the optics and CCD heads for alignment, filter changes and vacuum pumping; iii) provide alignment mechanisms for the optics and CCD heads; iv) exhibit low thermal expansion, as all three cameras must retain parfocality; v) exhibit low flexure (less than 10 μm) at any orientation, so stars do not drift out of the small windows defined on the three chips; vi) be electrically and thermally isolated from the telescope in order to reduce pickup noise and allow efficient operation of our water chiller.

The mechanical structure chosen for ULTRACAM is a double octopod, as shown in Fig. 3, which provides a rigid and open framework meeting all of the requirements described above. The Serrurier trusses of the double octopod are made of carbon fibre, which offers similar structural strength to steel but at a five-fold reduction in mass and a low thermal expansion coefficient. All of the remaining parts of the mechanical chassis are of aluminium, giving a total mass for the instrument of only 82 kg and an overall length of just 792 mm (including the CCD controller and WHT collimator).

ULTRACAM requires a mounting collar to interface it to a telescope focus. The collar, which is constructed of steel, places ULTRACAM at the correct distance from the telescope focal plane. A layer of G10/40 isolation material is placed between the collar and the top plate of ULTRACAM to provide thermal and electrical isolation from the telescope. The mounting collar also contains a motorised focal-plane mask. This is a steel blade which can be moved in the focal plane to prevent light from falling on regions of the CCD chip outside the user-defined windows typically used for observing. Without this mask, the light from bright stars falling on the active area of the chip above the CCD windows would cause vertical streaks in the windows (see Fig. 1 of [7] for an example). The mask also prevents photons from the sky from contaminating the background in drift-mode windows (see Sect. 3.7 and [43] for details).

3.4 Data Acquisition system

In this section we provide a brief overview of the ULTRACAM data acquisition system. A much more detailed description can be found in the paper by [4].

3.4.1 Hardware

Figure 4 shows the data acquisition hardware used in ULTRACAM. Data from the three CCD chips are read out by a San Diego State University (SDSU) Generation III CCD controller ([22], [23]). The SDSU controller, which is in wide use at many of the world's major ground-based telescopes, was adopted by the ULTRACAM project due to its user programmability, fast readout, low noise and ability to operate several multiple-channel CCDs simultaneously. The SDSU controller is hosted by a rack-mounted dual-processor PC running Linux patched with RealTime Application Interface (RTAI) extensions. The use of RTAI allows one processor to be strictly controlled so as to obtain accurate timestamps from the Global Positioning System (GPS) antenna located outside the dome and connected to the PC via a serial port (see Sect. 3.5 for details).

The instrument control PC communicates with the SDSU controller via a Peripheral Component Interconnect (PCI) card and two 250 MHz optical fibres. As well as communicating through the fibres, the SDSU controller also has the ability to interrupt the PC using its parallel port interrupt line, which is required to perform

ULTRACAM: An Ultra-Fast, Triple-Beam CCD Camera

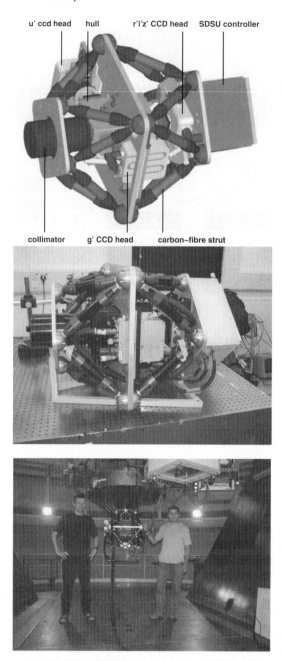

Fig. 3 Top: CAD image of the ULTRACAM opto-mechanical chassis, highlighting some of the components described in the text. Middle: Photograph of ULTRACAM in the test focal station of the WHT (courtesy Sue Worswick). Bottom: Photograph of ULTRACAM mounted on the Cassegrain focus of the WHT

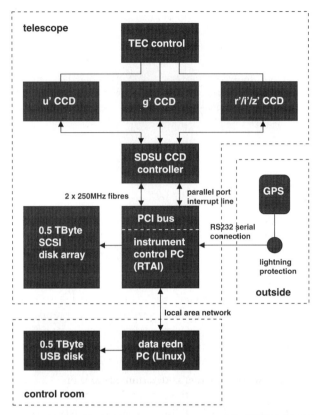

Fig. 4 Schematic showing the principal hardware components of the ULTRACAM data acquisition system. The connections between the hardware components and their locations at the telescope are also indicated

accurate timestamping (see Sect. 3.5). Data from the CCDs are passed from the SDSU PCI card to the PC memory by Direct Memory Access (DMA), from where the data is written to a high-capacity SCSI disk array. All of the real work in reading out the CCDs is performed by the SDSU controller; the PCI card merely forwards commands and data between the instrument control PC and the SDSU controller.

3.4.2 Software

The SDSU controller and PCI card both have on-board digital signal processors (DSPs) which can be programmed by downloading assembler code from the instrument control PC. A user wishing to take a sequence of windowed images with ULTRACAM, for example, would load the relevant DSP application onto the SDSU controller (to control the CCDs) and PCI card (to handle the data). The user can also modify certain parameters, such as the exposure time or binning factors (see Sect. 3.7 for a full list), by writing the new values directly to the DSP's memory.

All communication within the ULTRACAM system, including the loading of DSP applications on the SDSU controller, is via Extensible Markup Language (XML) documents transmitted using the Hyper Text Transfer Protocol (HTTP). This is an international communications standard, making the ULTRACAM data acquisition system highly portable and enabling users to operate the instrument using any interface able to send XML documents via HTTP (e.g. a web browser, perl scripts, java).

3.5 Timestamping

Stamping CCD frames with start times accurate to a small fraction of the typical exposure time is a key requirement for any astronomical instrument. In ULTRA-CAM, with frame rates of up to 500 Hz, this requirement is particularly severe, as it demands timestamping accurate to a fraction of a millisecond. Without this level of accuracy, for example, it would be impossible to determine the rate of decrease in the orbital periods of interacting binary stars as measured from eclipse timings [5], or compare the pulse arrival times in the optical and X-ray light from Anomalous X-ray Pulsars [7].

Whenever an exposure is started, the SDSU controller sends an interrupt to the instrument control PC which, thanks to the use of RTAI, *immediately* writes the current time (see below for how this is determined) to a First-In First-Out (FIFO) buffer. The data handling software then reads the timestamp from the FIFO and writes it to the header of the next buffer of raw data written to the PC memory. In this way, the timestamps and raw data always remain synchronised. Moreover, as the SDSU controller reads out all three chips simultaneously, the red, green and blue channels of ULTRACAM also remain perfectly synchronised.

The clock provided on most PC motherboards is not able to provide the current time with sufficient accuracy for our purposes, as it typically drifts by many milliseconds per second. The solution adopted by ULTRACAM is to use a PCI-CTR05 9513-based counter/timer board, manufactured by Measurement Computing Corporation, in conjunction with a Trimble Acutime 2000 smart GPS antenna. Every 10 seconds, the GPS antenna reports UTC to an accuracy of 50 nanoseconds. At the same time, the number of ticks reported by the counter board is recorded. At a later instant, when a timestamp is requested, the system records the new value of the counter board, calculates the number of ticks that have passed since the last GPS update, multiplies this by the duration of a tick (which we have accurately measured in the laboratory) and adds the resulting interval to the previous UTC value reported by the GPS. Since the counter board ticks at 1 MHz, to an accuracy of 100 ppm, this is a much more reliable method of timestamping than using the system clock of the PC.

Laboratory measurements indicate that the timestamping in ULTRACAM has a relative (i.e. frame-to-frame) accuracy of $\sim 50\,\mu s$. The absolute timing accuracy of ULTRACAM has been verified to an accuracy of ~ 1 millisecond by comparing observations of the Crab pulsar with the ephemeris of [26].

3.6 Pipeline Data Reduction System

ULTRACAM can generate up to 1 MByte of data per second. In the course of a typical night, therefore, it is possible to accumulate up to 50 GBytes of data, and up to 0.5 TBytes of data in the course of a typical observing run. To handle these high data rates, ULTRACAM has a dedicated pipeline data reduction system[2], written in C++, which runs on a linux PC or Mac located in the telescope control room and connected to the instrument control PC via a dedicated 100BaseT LAN (see Fig. 4).

Data from a run on an object with ULTRACAM is stored in two files, one an XML file containing a description of the data format, and the other a single, large unformatted binary file containing all the raw data and timestamps. This latter file may contain millions of individual data frames, each with its own timestamp, from each of the three ULTRACAM CCDs. The data reduction pipeline grabs these frames from the SCSI disk array by sending HTTP requests to a file server running on the instrument control PC.

The ULTRACAM data reduction pipeline has been designed to serve two apparently conflicting purposes. Whilst observing, it acts as a quick-look data reduction facility, with the ability to display images and generate light curves in real time, even when running at the highest data rates of up to 1 MByte per second and at the highest frame rates of up to 0.5 kHz. After observing, the pipeline acts as a fully-featured photometry reduction package, including optimal extraction [30]. To enable quick-look reduction whilst observing, the pipeline keeps many of its parameters hidden to the user and allows the few remaining parameters to be quickly skipped over to generate images and light curves in as short a time as possible. Conversely, when carefully reducing the data after a run, every single parameter can be tweaked in order to maximise the signal-to-noise of the final data.

3.7 Readout Modes

Figure 5 illustrates how the frame-transfer chips in ULTRACAM are able to deliver high frame rates. A typical observation with ULTRACAM consists of the observation of the target star in one user-defined CCD window and a nearby non-variable comparison star in another. Once the exposure is complete, the image area is shifted into the storage area, a process known as *vertical clocking*. This is very rapid, taking only 23.3 μs per row and hence ∼24 milliseconds to shift the entire 1024 rows into the storage area. As soon as the image area is shifted in this way, the next exposure begins. Whilst exposing, the previous image in the storage area is shifted onto the serial register and then undergoes *horizontal clocking* to one of two readout ports, where it is digitised. In other words, the previous frame is being read out whilst the

---

[2] Available for download at http://deneb.astro.warwick.ac.uk/phsaap/software/ultracam/html/index.html.

ULTRACAM: An Ultra-Fast, Triple-Beam CCD Camera

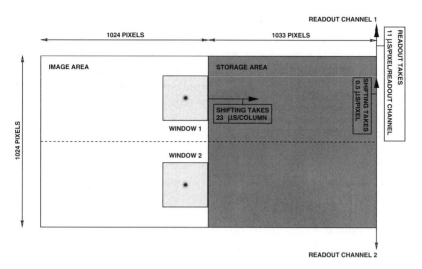

Fig. 5 Illustration of frame-transfer operation in ULTRACAM. Note that the horizontal shifting shown in the diagram is conventionally referred to as vertical clocking, and the vertical shifting is usually referred to as horizontal clocking

next frame is exposing, thereby reducing the dead time to the time it takes to shift the image into the storage area, i.e. 24 milliseconds.

ULTRACAM has no shutter—the fast shifting of data from the image area to the storage area acts like an electronic shutter, and is far faster than conventional mechanical shutters. This does cause some problems, however, such as vertical trails of star-light from bright stars, but these can be overcome in some situations by the use of a focal-plane mask (see Sect. 3.3).

Setting an exposure time with ULTRACAM is a more difficult concept than in a conventional non-frame-transfer camera. This is because ULTRACAM attempts to frame as fast as it possibly can, i.e. it will shift the image area into the storage area as soon as there is room in the storage area to do so. Hence, the fastest exposure time is given by the fastest time it takes to clear sufficient room in the storage area, which in turn depends on the number, location, size and binning factors of the windows in the image area, as well as the vertical clocking and digitisation times, all of which are variables in the ULTRACAM data acquisition system. To obtain an arbitrarily long exposure time with ULTRACAM, therefore, an *exposure delay* is added prior to the vertical clocking to allow photons to accumulate in the image area for the required amount of time. Conversely, to obtain an arbitrarily short exposure time with ULTRACAM, it is necessary to set the exposure delay to zero and adjust the window, binning and digitisation parameters so that the system can frame at the required rate. As it takes 24 ms to vertically clock the entire image area into the storage area, this provides a hard limit to the maximum frame rate of 40 Hz, with a duty cycle (given by the exposure time divided by the sum of the exposure and dead times) of less than 3%, and a useable limit of $\sim$10 Hz (with a more acceptable duty cycle of 75%).

Clearly, an alternative readout strategy is required in order to push beyond the $\sim$10 Hz frame-rate barrier and approach the desired kilohertz frame rates required to study the most rapid variability. For this purpose, we have developed *drift mode*. In this mode, the windows are positioned on the border between the image and storage areas and, instead of vertically clocking the entire image area into the storage area, only the window is clocked into the (top of) the storage area. A number of such windows are hence present in the storage area at any one time. This dramatically reduces the dead time, as now the frame rate is limited to the time it takes to clock only a small window into the storage area. For example, in the case of two windows of size 24×24 pixels^2 and binned 4×4, it is possible to achieve a frame rate of $\sim$500 Hz with a duty cycle of $\sim$75%. This is currently the highest frame rate that ULTRACAM has achieved on-sky whilst observing a science target.

Drift mode only offers the possibility of 2 windows and should only be used when frame rates in excess of 10 Hz are required. This is because the windows in drift mode spend longer on the chip and hence accumulate more dark current and, without the use of the focal plane mask, more sky photons. For frame rates of less than 10 Hz, ULTRACAM offers normal 2-window, 4-window, 6-window and full-frame readout modes (the latter reading out in approximately 3 seconds with only 24 milliseconds dead time).

For some applications, e.g. when taking flat fields or observing bright standard stars, it is desirable to use a full frame or two large windows and yet have short exposure times. This is not possible with the modes described above. To enable exposures times of arbitrarily short length, therefore, ULTRACAM offers the so-called *clear* modes. These modes, which are available in full-frame and 2-window formats, clear the chip prior to exposing for the requested amount of time. This means that any charge which has built up in the image area whilst the previous exposure is reading out is discarded. The disadvantage of this mode is that the duty cycle becomes very poor (3% in the case of a 0.1 second exposure time in full-frame-clear mode).

4 Performance

During the design phase of the project, a set of functional and performance requirements for ULTRACAM were established against which the instrument was tested during the commissioning phase. A detailed description of how ULTRACAM performed when measured against these requirements is given by [43]. Other measures of the performance of an astronomical instrument include how well the project was managed, how reliable the instrument is in operation, how much telescope time it wins, and how much science it enables. In what follows, I attempt to summarise the performance of ULTRACAM measured against these (very different) metrics.

1. ULTRACAM was delivered on budget (£ 300 000) and 3 months ahead of the three-year schedule set to design, build and commission it.

2. ULTRACAM is an extremely reliable instrument, as it has no moving parts. To date, ULTRACAM has suffered from <2% technical downtime, and this has nearly all been due to unavoidable hardware failures (such as hard disks and power supplies).
3. ULTRACAM has met nearly all of its performance requirements, including throughput (50% in g'), image quality (seeing limited at 0.6"/0.3" on WHT/VLT), flexure (10 µm), timing accuracy (50 µm) and detector-limited data throughput. Using high-quality optics, detectors and coatings, ULTRACAM has achieved zero points on the VLT of approximately $u' = 26.0$, $g' = 28.1$, $r' = 27.6$, $i' = 27.7$ and $z' = 26.7$.
4. The frame rates achieved on sky have ranged from 0.05 Hz to 500 Hz on targets ranging in brightness from 8–26 magnitude. It can be seen that the maximum frame rate is a factor of two below the required 1 kHz value set at the start of the project. This is not a serious drawback, however, as the science performed by ULTRACAM to date (see Table 2) has almost all concentrated on the 0.1–1 second regime.
5. The ULTRACAM project team has adopted a policy of open access, offering the instrument to any astronomer on a shared-risks, collaborative basis. As a result, ULTRACAM has been awarded a great deal of competitively allocated observing time: 87 nights on the WHT and 23 nights on the VLT in the 4 years since commissioning. A breakdown of how these 110 nights have been spent on different types of object is given in Table 2.
6. To date, a total of ~25 refereed papers have been published which are wholly or partly dependent on ULTRACAM data. Some examples of the research done with ULTRACAM are given in Table 2, as well as elsewhere in this volume.
7. Any instrument which is not continually enhanced and upgraded will eventually cease to be competitive. For this reason, we have always pursued a vigorous enhancements programme with ULTRACAM, focussed on the three most

Table 2 Breakdown of the percentage of time spent observing different classes of astronomical object with ULTRACAM on the WHT and VLT. The right-hand column provides references to some of the ULTRACAM papers published in each area

Target	Time	References
Cataclysmic variables/accreting white dwarfs	22%	[24], [10]
Black-hole X-ray binaries	19%	[36], [37]
sdB stars/asteroseismology	15%	[1], [16]
Kuiper belt object occultations	11%	[34]
Eclipsing white-dwarf/red-dwarf binaries	10%	[5], [29]
Pulsars	5%	[8], [7]
Ultra-compact binaries	4%	[3]
Flare stars	4%	[28]
Extrasolar planet transits	3%	
Isolated white dwarfs	2%	[38]
Isolated brown dwarfs	2%	[25]
Gamma-ray bursts	1%	[45]
Active galactic nuclei/Blazars	1%	
Titan/Pluto occultations	1%	[11], [48]

important areas for high time-resolution astrophysics: increasing the maximum frame rate, reducing the readout noise and minimizing downtime/inefficiency. The performance of the instrument has improved by at least a factor of two in these three areas in the 4 years since commissioning.

5 Conclusions

I have described the scientific requirements, design and performance of ULTRA-CAM. I have listed examples of the varied science that has been performed with the instrument, details of which can be found in some of the other contributions to this volume. As for the future, we intend to continue to operate ULTRACAM for approximately one run per year on both the VLT and the WHT. We are also about to commission a spectroscopic version of the instrument—see the contribution by Tom Marsh in this volume for details. In the longer term, we are exploring the possibility of building ULTRACAM-II, which would be dedicated for use in Chile and which would have four channels, a larger field of view and, if available, use multi-channel, split frame-transfer electron-multiplying CCDs.

Acknowledgments ULTRACAM was designed and built, and is operated and enhanced, as a joint project by me and Tom Marsh of the University of Warwick.

I would also like to thank the staff of the UK Astronomy Technology Centre in Edinburgh, the Astronomy Group and Central Mechanical Workshops at the University of Sheffield, the Isaac Newton Group of Telescopes on La Palma and the European Southern Observatory at Paranal for the essential contributions they have made to the project.

Finally, I would like to thank PPARC for providing the funding to build, operate and exploit ULTRACAM.

References

1. C. Aerts, C. S. Jeffery, G. Fontaine, V. S. Dhillon, T. R. Marsh, and P. Groot. *MNRAS*, 367:1317, 2006.
2. J. E. Baldwin, R. N. Tubbs, G. C. Cox, C. D. Mackay, R. W. Wilson, and M. I. Andersen. *A & A*, 368:L1, 2001.
3. S. C. C. Barros, T. R. Marsh, V. S. Dhillon, P. J. Groot, G. Nelemans, G. Roelofs, and D. Steeghs. *MNRAS*, submitted, 2006.
4. S. M. Beard, A. J. A. Vick, D. Atkinson, V. S. Dhillon, T. R. Marsh, S. McLay, M. J. Stevenson, and C. Tierney. In H. Lewis, editor, *Advanced Telescope and Instrumentation Control Software II.*, page 218. SPIE, 4848, 2002.
5. C. S. Brinkworth, T. R. Marsh, V. S. Dhillon, and C. Knigge. *MNRAS*, 365:287, 2006.
6. C. Cumani and K.-H. Mantel. *Experimental Astronomy*, 11:145, 2001.
7. V. S. Dhillon, T. R. Marsh, F. Hulleman, M. H. van Kerkwijk, A. Shearer, S. P. Littlefair, F. P. Gavriil, and V. M. Kaspi. *MNRAS*, 363:609, 2005.
8. V. S. Dhillon, T. R. Marsh, and S. P. Littlefair. *MNRAS*, in press (astro-ph/0607478), 2006.
9. D. Dravins. *ESO Messenger*, 78:9, 1994.
10. W. J. Feline, V. S. Dhillon, T. R. Marsh, and C. S. Brinkworth. *MNRAS*, 355:1, 2004.

11. A. Fitzsimmons, A. Zalucha, J. L. Elliot, H. B. Hammel, J. Thomas-Osip, V. S. Dhillon, T. R. Marsh, F. W. Taylor, and P. G. J. Irwin. *MNRAS*, submitted, 2006.
12. J. L. A. Fordham, H. Kawakami, R. M. Michel, R. Much, and J. R. Robinson. *MNRAS*, 319:414, 2000.
13. J. L. A. Fordham, N. Vranesevic, A. Carramiñana, R. Michel, R. Much, P. Wehinger, and S. Wyckoff. *ApJ*, 581:485, 2002.
14. M. Fukugita, T. Ichikawa, J. E. Gunn, M. Doi, K. Shimasaku, and D. P. Schneider. *AJ*, 111:1748, 1996.
15. R. I. Hynes, P. A. Charles, J. Casares, C. A. Haswell, C. Zurita, and T. Shahbaz. *MNRAS*, 340:447, 2003.
16. C. S. Jeffery, V. S. Dhillon, T. R. Marsh, and B. Ramachandran. *MNRAS*, 352:699, 2004.
17. P. Jorden, J.-M. Deltorn, and P. Oates. *Gemini*, 41:1, 1993.
18. B. Kern. PhD thesis, California Institute of Technology, 2002.
19. B. Kern and C. Martin. *Nat*, 417:527, 2002.
20. J. Kotar, S. Vidrih, and A. Čadež. *Rev Sci Instrum, 74, 3111 (astro-ph/0303368)*, 2003.
21. N. M. Law, S. T. Hodgkin, and C. D. Mackay. *MNRAS*, 368:1917, 2006.
22. R. W. Leach, F. L. Beale, and J. E. Eriksen. New-generation CCD controller requirements and an example: the San Diego State University generation II controller. In S. D'Odorico, editor, *Proc. SPIE Vol. 3355, Optical Astronomical Instrumentation*, page 512, 1998.
23. R. W. Leach and F. J. Low. CCD and IR array controllers. In M. Iye and A. F. Moorwood, editors, *Proc. SPIE Vol. 4008, Optical and IR Telescope Instrumentation and Detectors*, page 337, 2000.
24. S. P. Littlefair, V. S. Dhillon, T. R. Marsh, and B. T. Gänsicke. *MNRAS*, 371:1435, 2006.
25. S. P. Littlefair, V. S. Dhillon, T. R. Marsh, T. Shahbaz, and E. L. Martín. *MNRAS*, 370:1208, 2006.
26. A. G. Lyne, C. A. Jordan, and M. E. Roberts. Jodrell bank crab pulsar timing results. Monthly ephemeris, Jodrell Bank Observatory, University of Manchester, 2005.
27. T. R. Marsh and K. Horne. *MNRAS*, 299:921, 1998.
28. M. Mathioudakis, D. S. Bloomfield, D. B. Jess, V. S. Dhillon, and T. R. Marsh. *A & A*, 456:323, 2006.
29. P. F. L. Maxted, T. R. Marsh, L. Morales-Rueda, M. A. Barstow, P. D. Dobbie, M. R. Schreiber, V. S. Dhillon, and C. S. Brinkworth. *MNRAS*, 355:1143, 2004.
30. T. Naylor. *MNRAS*, 296:339, 1998.
31. K. O'Brien, K. Horne, B. Boroson, M. Still, R. Gomer, J. B. Oke, P. Boyd, and S. D. Vrtilek. *MNRAS*, 326:1067, 2001.
32. D. O'Donoghue. *Baltic Astr*, 4:517, 1995.
33. D. O'Donoghue et al. *MNRAS*, in press (astro-ph/0607266), 2006.
34. F. Roques, A. Doressoundiram, V. Dhillon, T. Marsh, S. Bickerton, J. J. Kavelaars, M. Moncuquet, M. Auvergne, I. Belskaya, M. Chevreton, F. Colas, A. Fernandez, A. Fitzsimmons, J. Lecacheux, O. Mousis, S. Pau, N. Peixinho, and G. P. Tozzi. *AJ*, 132:819, 2006.
35. R. G. M. Rutten, F. J. Gribbin, D. J. Ives, A. Bennett, and V. S. Dhillon. Drift-mode ccd readout. User manual, Isaac Newton Group, La Palma, 1997.
36. T. Shahbaz, V. S. Dhillon, T. R. Marsh, J. Casares, C. Zurita, P. A. Charles, C. A. Haswell, and R. I. Hynes. *MNRAS*, 362:975, 2005.
37. T. Shahbaz, V. S. Dhillon, T. R. Marsh, C. Zurita, C. A. Haswell, P. A. Charles, R. I. Hynes, and J. Casares. *MNRAS*, 346:1116, 2003.
38. R. Silvotti, M. Pavlov, G. Fontaine, T. R. Marsh, and V. S Dhillon. *Mem. Soc. Astron. Ital.*, 77:486, 2006.
39. D. A. Simons. Technical note no. 30, Gemini Observatory, 1995.
40. W. Skidmore, K. O'Brien, K. Horne, R. Gomer, J. B. Oke, and K. J. Pearson. *MNRAS*, 338:1057, 2003.
41. D. St-Jacques, G. C. Cox, J. E. Baldwin, C. D. Mackay, E. M. Waldram, and R. W. Wilson. *MNRAS*, 290:66, 1997.
42. R. A. Stathakis and H. M. Johnston. Rgo spectrograph. User manual, Anglo-Australian Observatory, 2002.

43. M. J. Stevenson. PhD thesis, University of Sheffield, 2004.
44. A. Čadež, S. Vidrih, M. Galičič, and A. Carramiñana. *A&A*, 366:930, 2001.
45. P. Vreeswijk, T. Marsh, S. Littlefair, V. Dhillon, and K. O'Brien. *GRB Coordinates Network*, 3445:1, 2005.
46. R. W. Wilson, N. O'Mahony, C. Packham, and M. Azzaro. *MNRAS*, 309:379, 1999.
47. P. A. Woudt and B. Warner. *MNRAS*, 328:159, 2001.
48. A. Zalucha, A. Fitzsimmons, J. L. Elliot, J. Thomas-Osip, H. B. Hammel, V. S. Dhillon, T. R. Marsh, F. W. Taylor, and P. G. J. Irwin. *Icarus*, submitted, 2006.

OPTIMA: A High Time Resolution Optical Photo-Polarimeter

G. Kanbach, A. Stefanescu, S. Duscha, M. Mühlegger, F. Schrey, H. Steinle, A. Slowikowska and H. Spruit

Abstract A high-speed photo-polarimeter, "OPTIMA" short for Optical Pulsar Timing Analyzer, has been designed and developed in the group for gamma-ray astronomy of the Max-Planck-Institut für Extraterrestrische Physik. This sensitive, portable detector is used to observe optical emissions of sources that radiate mainly at X- and gamma-ray energies, like pulsars and other highly variable compact sources. The single photon counting instrument is based on fiber fed avalanche photodiodes (APDs), a GPS timing receiver, a CCD camera for target acquisition and a stand-alone data acquisition and control system. Several configurations are available: for photometry a hexagonal bundle with seven channels and one fiber offset for sky background monitoring; for polarimetry a rotating polarization filter in front of the photometer or a newly developed 4-channel double Wollaston system; and for coarse spectroscopy a 4-colour prism spectrograph.

1 General Layout

The concept of aperture timing photometry is very familiar to astronomers working in high-energy X- and gamma-ray astronomy. Single photon events are located on a sky map and their arrival times are registered. The desired target photons are then selected for analysis depending on their angular distance from the target source. In the optical band we can operate in the same way if a suitable fast 2-dimensional detector with single photon sensitivity is available. Such detectors have been realized, e.g.

G. Kanbach · A. Stefanescu · S. Duscha · M. Mühlegger · F. Schrey · H. Steinle · A. Slowikowska
Max Planck Institut für Extraterrestrische Physik, Postfach 1312, 85741 Garching, Germany
e-mails: gok@mpe.mpg.de, astefan@mpe.mpg.de, sduscha@mpe.mpg.de, mmuehleg@mpe.mpg.de, fzs@mpe.mpg.de, hcs@mpe.mpg.de

H. Spruit
Max Planck Institut für Astrophysik, Karl-Schwarzschild-Str. 1, 85741 Garching, Germany
e-mail: henk@mpa-garching.mpg.de

A. Slowikowska
Nicolaus Copernicus Astronomical Center, Dept. of Astrophysics, Torun, Poland
e-mail: aga@mpe.mpg.de or aga@ncac.torun.pl

in the form of multi-anode micro-channel arrays (MAMA) on the TRIFFID camera [14] or as small, compact arrays of solid-state detectors (e.g. the Josephson junction detector S-Cam3 described by [9]). These cameras are still characterized by either low quantum efficiency or technically challenging operations. However they offer the advantages of imaging a larger field with fine resolution: event selection can be adapted to the seeing and sky background conditions and reference stars are often present in the same exposures.

Although the future will belong to advanced fast 2-D photon counting/timing cameras with high sensitivity (see e.g. the contribution by Strueder et al., these proceedings), we have chosen to first develop a simple photon counting system ("OPTIMA" the Optical Pulsar Timing Analyzer) that relies on fixed apertures and is equipped with detectors of high quantum efficiency. The initial science goals were to measure optical pulsar lightcurves and highly variable binary systems and to establish timing relations to emissions in other wavelength ranges (radio and X-rays). For this purpose accurate quantitative photometry is not of prime importance and fixed apertures of an appropriate size are sufficient. The apertures are given by optical fibers which are placed in the focal plane of a telescope and feed the light of target stars and sky background to avalanche photodiodes (APDs). To ameliorate the negative aspects of fixed aperture photometry we installed an small 'integral field unit' of apertures in the form of a hexagonal close-packed bundle of fibers and a fast read-out for the field viewing acquisition CCD camera. The schematics of the OPTIMA detector (configuration up to 2005) and a photograph of the open system are shown in Figs. 1 and 2. Since 2006 a new OPTIMA configuration (called 'OPTIMA-Burst') is available. This configuration uses basically the components of the previous experiment but has a separate box for the APD detectors. The focal plane fiber pick-ups of the earlier version are also present in OPTIMA-Burst, but there is an additional aperture that inputs light into a new double-Wollaston polarimeter. All fibers (of about 2 m length) are fed through a semi-rigid tube to the APD box. Schematics and a photograph of OPTIMA-Burst are shown in Figs. 3 and 4.

We started to develop the OPTIMA photometer in 1998 ([21],[22]) and used progressively more complete systems on a 1.3 m telescope (Mt. Skinakas, Crete), on the 3.5 m telescope (Calar Alto, Spain), on the 2.1 m Guillermo Haro telescope, Cananea, Mexico, the 2.5 m NOT on La Palma, and, in the southern hemisphere, on the 74 inch Mt. Stromlo and 2.2 m ESO/La Silla observatories. The new version OPTIMA-Burst has been used at the Skinakas observatory and early operations were described by [20] and [10].

1.1 Fiber Pick-up and Detectors

OPTIMA intercepts the image formed in the focal plane of a large telescope with a slanted mirror. The reflected light is re-imaged on a commercial CCD camera. We currently use a fast-readout Apogee AP6 camera featuring a Kodak chip (type KAF1000E, 1024×1024 pixels of 24.4 μm size, backside illuminated). A full frame

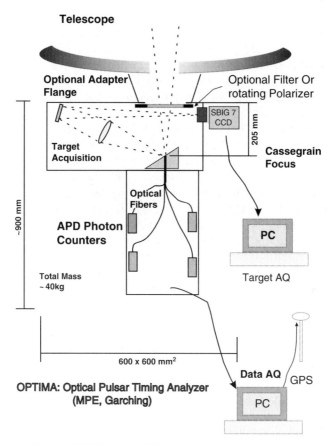

Fig. 1 Schematic layout of OPTIMA (pre-2006)

is downloaded in about 1 sec. Embedded in the slanted mirror and coincident with the focal plane are the 'photon-counting' apertures (numbers 1–4 in Fig. 5) and two small LED lightsources (labelled A and B). These LEDs can be switched on via computer command and serve to control the overall alignment of the field-viewing optics and camera. Aperture no. 1 (size $\sim$345 μm) in Fig. 5 is the diaphragm for the double-Wollaston polarimeter. Opening no. 2, with 1.7 mm diameter, contains the hexagonal fiber bundle mounted in a fine steel tube, and the apertures no. 3 (fiber input to a 4-channel prism spectrograph) and no. 4 with a fiber to record the night sky brightness near to the target. The field-viewing optics shows a region of approximately 12' $\times$ 12' (at the telescope of Mt. Skinakas, Greece, f-ratio 7.64, D = 129 cm) with some vignetting near the edges. The main task of this system is to acquire the target star and to derive the telescope-control commands to move the target into any chosen aperture. During the photon-counting measurements, when the telescope guiding is controlled by an external auto-guider, the secondary task of the OPTIMA CCD is to take serial images of the field with short integration times

Fig. 2 Photograph of pre-2006 OPTIMA with open APD box of photon counters (3). The target acquisition optics (filters, target imaging and fiber pick-up) is located in box (1) and the CCD camera is mounted externally (2)

(typically 10 sec). This series of exposures is evaluated for the atmospheric seeing and transparency conditions during the measurement.

For the photon-counting observation of faint sources it is very important to convert the highest possible fraction of incoming photons into countable signals. This efficiency should include the light losses on all optical surfaces (input to the

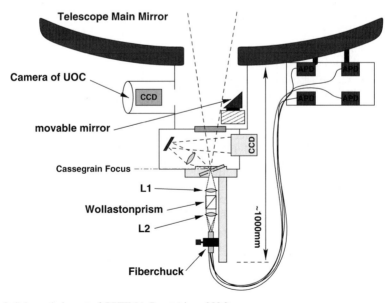

Fig. 3 Schematic layout of OPTIMA-Burst (since 2006)

Fig. 4 Photograph of OPTIMA-Burst mounted to the Cassegrain focus of the Mt. Skinakas 1.3 m telescope (UoC, Heraklion, Greece). The components are numbered like in Fig. 2

fibers through a polished surface and coupling of the fiber into the optics included in the photon counting modules, each ∼4%), absorption in the fiber (very small for the chosen quartz fibers), and losses incurred by the taper that reduces fiber diameter from about 300 μm (input) to 100 μm (output). The 'optical' efficiency of the tapered fibers is not easily calibrated. Measurements by [21] estimated the fiber efficiency to be around 90%. Finally the quantum efficiency of the detectors should be large over a wide spectral band. Most previous systems for recording single optical photons with good time resolution used vacuum electronics with photocathodes of peak quantum efficiencies of typically 20% and a narrow wavelength range. Much better quantum efficiencies can be reached with solid state detectors. OPTIMA uses Avalanche Photodiodes (APDs). These silicon devices have been produced with peak quantum efficiencies of up to 80% and a wide band of sensitivity ranging from 250 to 1100 nm. We use commercially available APD based single photon counting modules of type SPCM-AQR-15-FC produced by Perkin-Elmer. These devices operate in a Geiger counter mode where a photon initiated avalanche pulse is quenched by the instantaneous reduction of the bias voltage. The diodes have a diameter of 200 μm and are electrically cooled with Peltier elements. The selected units offer low dark count rates of typically less than 50 cps, are insensitive to electromagnetic interference and are very reliable. They can record photons up to rates of $\sim 2 \times 10^6$ cps before noticeable dead-time losses occur. The present data acquisition however can not keep up with such rates. Typical DAQ event losses around 1% are encountered for rates of 4×10^4 cps. The achieved quantum efficiency of the APD detectors is shown in Fig. 6. Although it falls short of the values mentioned above, it is still above 20% for a spectral range from 450 to 950 nm.

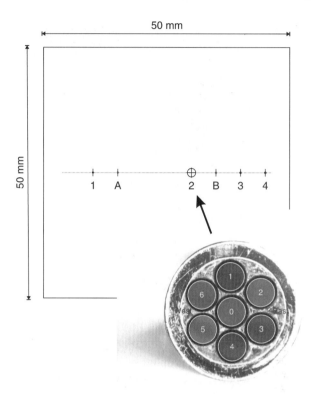

Fig. 5 Layout of the fiber input apertures in the field-viewing mirror (see text) and photograph of the hexagonal fiber bundle (input: single fiber diameter ~ 320 μm; output diameter: ~ 100 μm; length: 2 m; illuminated from output side): central fiber for the target, ring fibers for the close-by sky or nebular environment

To estimate the efficiency of OPTIMA one needs to specify the spectrum of the incident radiation. As an example we use the spectrum of the Crab pulsar as given by [18] and calculate an 'unfiltered' efficiency to convert the incoming photons at the telescope into counts. With realistic assumptions on the optical losses we arrive at an overall efficiency of about 15–20%. This efficiency was confirmed at various telescopes.

1.2 Timing, Data Acquisition, and Software

The signals provided by the global positioning system (GPS) supply a global absolute time base. We use a receiver (from Datum Inc.) which can process the clock pulses of up to six satellites simultaneously and reaches an absolute time accuracy of better than 2 μs on the "pulse per second (PPS)" GPS signal. This signal disciplines a local high frequency oscillator (250 kHz, i.e. leading to a time resolution window of 4 μs) with the same precision which provides a continuous UTC time signal to

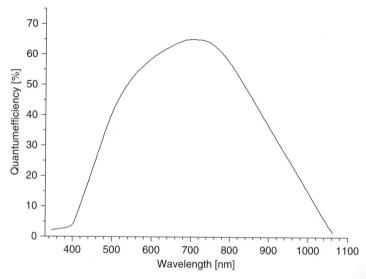

Fig. 6 Typical quantum efficiency of the Perkin-Elmer single photon counting modules SPCM-AQR-15-FC; dark noise ranging from 50 to 250 counts/sec

the system bus of the PC used for data acquisition (DAQ). The task of the DAQ unit is thus to correlate the electronic signals of the APD detector modules with the high resolution time base and assign UTC arrival times to each detected photon. The timing of the conversion cycles of the DAQ card is controlled by the GPS based oscillator, so that the transfer of the APD detector signals is running at a fixed rate. The absolute starting time of each software triggered acquisition sequence is precisely known. The controlling software counts the number of conversion cycles since the start of the sequence and stores this sequential number together with an identifier of the respective detector channel for each detected photon. Conversion cycles without detected photons are skipped. Based on the cycle number, the acquisition frequency and the absolute time of the start of the sequence, the UTC arrival time of every recorded photon can be restored during data analysis. During the long-term measurements the consistency and continuity of the time base are continously controlled. The presently used DAQ system is limited to rates below about $\sim 10^5$ cps because of pile-up in the time-resolution window. Future versions of the DAQ should be able to control higher rates, which are achievable with the photon counters (up to several MHz).

Typical count rates from the night sky in dark conditions are ~ 1–2 kHz per fiber resulting in several GBytes of data for a night of observing. Data are first staged to RAM and periodically ($\sim$ every 10 mins.) stored on HDD. Off-line data analysis includes the options to transfer the topocentric photon arrival times to the solar system barycenter. Pulsar phases and light curves can be calculated if the pulsar ephemeris is known. If unknown periodicities or irregular variations are investigated, FFT analysis and rate plots are available.

1.3 Polarimetry and Spectroscopy

Two versions of polarimetry were developed in order to extend the observational modes of the originally photometric instrument. The first polarimeter was realized with a rotating polaroid filter that modulates the complete field-of-view with all photon-counting apertures. This mode of operation is especially well suited to the measurement of a polarized source embedded in an extended polarized nebula, like the Crab pulsar, where simultaneous target and background polarization data can be taken with the fiber bundle array. The disadvantages of a rotating polaroid are the rather low transmission ($\sim$32% for unpolarized white light), the wavelength dependent modulation efficiency, and the fact that the polarization measurements are done in sequence as the filter turns. The latter property restricts the use of this system to targets with slowly variable (with respect to the filter rotation rate of about 3 Hz) or regular periodic polarization. Irregular transient sources need to be measured with a system that offers parallel simultaneous polarimeters. This is the second variant of polarimeter, a double Wollaston system, that is available for OPTIMA.

1.3.1 Rotating Polaroid Filter (RPF)

Figure 7 shows a photograph of the polarizer seen from top (telescope side) and bottom. The device can be introduced into the incoming beam above the fiber pick-up so that all fiber channels and the CCD image are fully covered. The polarizing filter (Type 10K by Spindler & Hoyer) is mounted on a precision roller bearing and is rotated with the motor visible on the bottom view. Typical rotation frequencies of up to 10 Hz can be adjusted through the supply voltage of the motor. Incoming linearly polarized light is modulated at twice the rotation frequency. In the top view a magnetic switch (magnet on the rotating filter, Hall sensor on the base plate) can be seen. The reference position of the filter is given by a signal from the Hall sensor.

Fig. 7 Photograph of the OPTIMA rotating polarization filter. Left (seen from telescope side): 2: permanent magnet on rotating filter; the polarization direction of the filter (4) is perpendicular to the radius vector to the magnet. 3: Hall sensor on base plate. The filter rotates from north to east. Right: motor and belt. Typical rotation rates are 3–4 Hz

It is registered and timed in the same way as a photon event and stored in a separate channel. This allows interpolation of the position of the polarizing filter for any event using the time difference between the Hall sensor signals. Slight irregularities in the rotation frequency of the RPF that occur on time-scales longer than fractions of a second, e.g. due to supply voltage drifts or mechanical resistance changes in the bearing and motor, can be corrected with sufficient accuracy. We have tested the RPF in the lab with unpolarized and linearly polarized light to ensure and prove that the OPTIMA fibers and detectors have no intrinsic, systematic response to polarized light.

The efficiency of modulation with the installed polaroid filter is restricted to λ <800 nm, whereas the APDs are sensitive to $\sim$1000 nm. We therefore employed in our later measurements ([15]) an additional IR blocking filter (transmission only for λ <750 nm). This filter ensures nearly complete modulation of the polarized radiation. Measurements without this IR-blocking filter have a reduced modulation depth of $\sim$60% for a typical Crab pulsar spectrum and have to be corrected.

1.3.2 Double Wollaston Polarimeter

Figure 8 shows the basic layout of the new polarimeter optics. The target star is positioned in a diaphragm (aperture of $\sim$345 µm in the field viewing mirror) and the emerging beam is collimated. Two quartz Wollaston prisms are positioned side by side in the collimated beam (separated by a thin opaque plate) so that about half of the beam falls on each prism. The polarized and symmetrically diverging output beams (divergence about 1°) are re-focussed onto a fiber pick-up where four regular

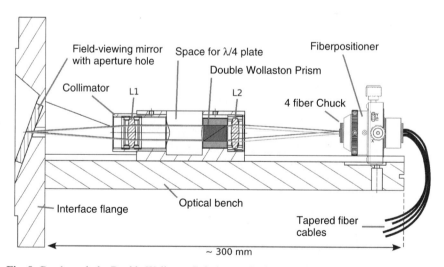

Fig. 8 Cut through the Double Wollaston Polarimeter. In the central parallel beam two Wollaston prisms, each covering about half the beam and separated by a thin opaque plate, split the incoming light into four images that are polarized at staggered angles (0°, 45°, 90°, and 135°) and arranged approximately on the corners of a square

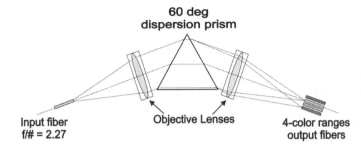

Fig. 9 Schematic of the prism spectrograph for simultaneous photometry in 4 colors

tapered fibers are mounted in a chuck. For further detail refer to the work of [10]. The system has been verified in the lab with polarized and unpolarized light and was used at Skinakas observatory during November 2005 and during a long campaign from July to September, 2006. From the 2005 commissioning a verification on polarimetry standard stars and on the phase dependent polarization of the Crab pulsar was obtained. The data from the 2006 observations are not yet fully analyzed. Further work is required to fully understand the cross-calibration of the four output channels.

1.3.3 Prism Spectroscopy

A prism spectrograph with 4 output channels has been added as an option to the OPTIMA system. A schematic of the spectrometer is shown in Fig. 9. The spectrometer fits in a small box and can be mounted inside the container with the photon counters. The input to the spectrometer is through a fiber from the telescope focal plane. The output spectrum spreads over about 1.1 mm for the wavelength range 500 to 900 nm. The output pick-up of 4 spectral bands consists of four 320 μm fibers placed next to each other and mounted on a 3 axis micrometer stage. The position of the output fiber bundle in the spectrum then determines the spectral boundaries picked up by each "color" fiber. The overall efficiency of $\sim$5–10% of this spectrometer is not very high, since the optics are of low quality and the fill factor of the pick-up fibers is small. It was primarily intended for use on rather bright source (e.g. Cyg X-1). The simultaneous fast timing of photons in several spectral bands could allow to investigate color variations during outbursts and timing correlations of photons with different energies.

2 Selected Measurements

2.1 The Crab Pulsar: Single Rotations and Polarization

The Crab pulsar is the most widely used target in high-energy astronomy for instrument calibration and timing verification. OPTIMA was similarly used between

1999 and 2006 to verify the instrument and to perform scientific measurements of the optical emission from this young pulsar. Figure 10 shows an overlay of the OPTIMA bundle apertures on a HST image of the Crab pulsar (R. Romani, private communication and [11]). The scale in Fig. 10 corresponds to the plate scale of the NOT telescope (7.3 "/mm, 137 μm/"). About 0.65" southeast of the pulsar a knot of optical emission can be noted ([3]). Under ground-level seeing conditions this knot cannot be separated from the pulsar and it is clear that the measured intensity in the central fiber contains this light as well as the much brighter pulsar emissions. Figure 11 shows a trace of of the Crab light curve resolving single rotations with a time resolution of 1 ms. Folding the barycentric arrival times of the Crab with the rotational ephemerides derived from radio observations of Jodrell Bank [7] full agreement of the OPTIMA lightcurve with that derived from the high-speed photometer aboard the Hubble Space Telescope ([12]) was demonstrated ([22]). This confirms that OPTIMA introduces no detectable timing noise or other non-linear intensity responses to the optical signal of the pulsar. The stability of the OPTIMA timing system is demonstrated by the constancy of the Crab light curve as measured in many epochs.

For verification and in a first application of the rotating polaroid polarimeter we observed the Crab nebula and pulsar in January 2002 at Calar Alto using the 3.5 m telescope. About 3 hours of exposure were achieved with the polarimeter ([6]).

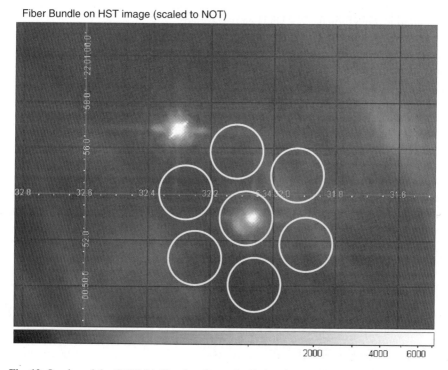

Fig. 10 Overlay of the OPTIMA fibre bundle on the Crab pulsar and its environment on a HST image. The scale (diameter of a fibre $\sim 2.3"$) corresponds to the focal plate scale of the 2.5 m NOT telescope where observations were performed in Nov. 2003

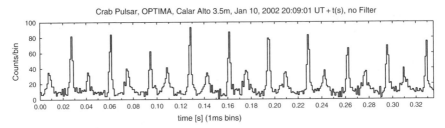

Fig. 11 Ten rotations of the Crab pulsar observed with 1 ms time resolution at the 3.5 m telescope on Calar Alto on 10 Jan 2002

The analysis requires the separation of the highly polarized emission of the Crab nebula from the radiation of the pulsar point source. This is done with the fibers in the hexagonal bundle, where the central fiber is on the pulsar and the surrounding ring of fibers, at a distance of ∼2", record the nebular environment. All fibers include of course the sky background. Figure 10 plotted for the NOT focal plane scale (7.3"/mm)which applies nearly also for the CAHA 3.5 m telescope scale of 5.9"/mm. The nebular emission is then interpolated at the pulsar position from the ring of fibers. The Crab pulsar is detected at all phases of rotation, i.e. also in the so-called off-pulse phase with an intensity of about 1.2% (CAHA result [6], [5]) or about 2% (NOT result [15]) compared to the intensity of the first peak. This level could partly be explained by the presence of the unresolved inner knot, although a quantitative modelling has not yet been performed.

Preliminary polarization characteristics of the Crab pulsar were derived from the measurement ([6]) and are shown in Fig. 12. The degree of polarization (uncorrected for polaroid efficiency) and the position angle (PA) of the E-vector are plotted with a resolution of 500 phase bins. The result agrees generally well with previous measurements [17], but shows details with much better definition and statistics. A measurement at the NOT in 2003 ([16] and [15]) accumulated about 14 hours of useful data. These results reveal even more details than all earlier measurements.

The interpretation of the optical polarization in terms of magnetospheric or pulsar wind emission models still involves an amount of uncertainty. The classical polarcap and outer gap emission models fail to reproduce the observed polarization characteristics. Recent magnetospheric models like the two-pole caustic model ([1],[2]) predict qualitatively the large swings in the PA and the phases of the observed minima of polarization. Another approach [13] places the optical emissions into the pulsar wind zone, and is also able to explain some features of the data.

2.2 Cataclysmic Variables: Eclipsing AM Her Binaries

The OPTIMA photometer was used to observe several short period binary systems. The selected targets were mostly of AM Her type and showed eclipses. As an example we present light curves for HU Aqr, which was observed repeatedly.

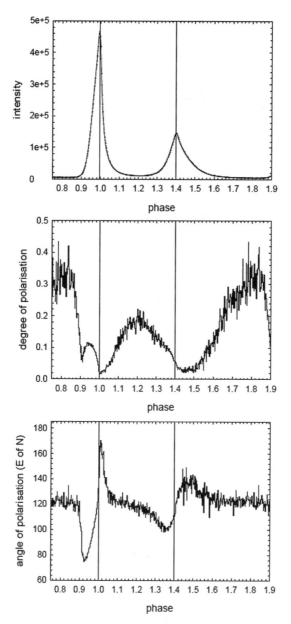

Fig. 12 Linear polarization of the Crab measured with the unfiltered polaroid: the pulsar's rotation is divided in 500 intervals of about 70 μs duration ([6]). The degree of polarization has not been corrected for the wavelength dependent efficiency of the polaroid filter, because the spectrum in the low intensity parts of the lightcurve (bridge and off-pulse phase) is not well known. For an updated result using an IR-blocking filter see [15]. The derived polarization in the bridge region is then typically 30% higher

HU Aqr is a close eclipsing binary system containing a highly magnetic white dwarf and a secondary star of type M4V. The orbital period is ∼125 minutes. With a range of observed magnitudes from ∼15 to 18 it is one of the brightest sources of this type. Very short timing signatures on the sub-second level were expected in this object, in particular at the eclipse entry and exit of the white dwarf. Figure 13 shows three sample lightcurves with 1 second resolution measured on July 5, 2000, September 21, 2001, and July 18, 2004, respectively. Two eclipses of the white dwarf are the dominant features of these light curves. The sky background in the vicinity of the source has been subtracted using the hexagonal bundle of fibres. The low count rate level in the eclipse is due to the secondary M4 star of magnitude 19.1^m.

The ingress and egress of the eclipse of the accretion spot on the white dwarf have a duration of about 7 seconds. Using the orbital velocity of the white dwarf of ∼200 km/s this corresponds to a spot size of about 1400 km. In 2000 sharp spikes of optical emission with clearly resolved time scales of 1–2 seconds and a brightness increase of up to a factor of 2 were detected. These features could be due to inhomogeneities in the mass accretion flow. The second observation in September 2001 showed an intensity of HU Aqr already at an level less than 25% of the 2000 values and the strong spikes in the light curve were not detected again. Later observations, like in 2004 and up to 2006, indicate that HU Aqr has faded even more and has not yet returned into a high intensity state.

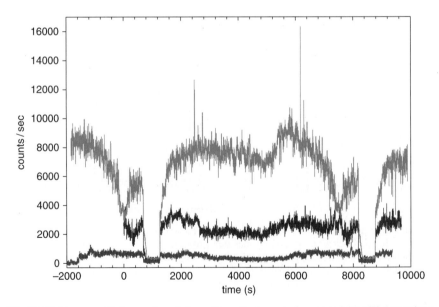

Fig. 13 Lightcurves with 1 second resolution of the eclipsing cataclysmic variable HU Aqr at three epochs: Jul 5, 2000 (upper curve), Sep 21, 2001 (middle curve), and Jul 18, 2004 (lowest curve). The observations were taken at Skinakas observatory, Crete, Greece

2.3 The Black Hole Binary Transient KV UMa: Correlation of X-ray and Optical Variability

From January to August 2000 the bright X-ray transient XTE J1118+48 (= KV UMa) was in an unusually long and intense state of outburst. This provided a unique opportunity for simultaneous X-ray and optical observations. KV UMa is a nearby binary system (~2 kpc) at high galactic latitude and contains a compact star with more than 6 solar masses. OPTIMA was used on the 1.3 m telescope at Skinakas observatory for a simultaneous exposure of 2.5 hours with RXTE over the nights of July 4–7, 2000 [4].

The variable emission from the black hole candidate was recorded with a timing accuracy of a few millisecond at X- and optical wavelengths. The X-ray to optical cross-correlation (Fig. 14), shows that the optical emission rises suddenly following an increase in X-ray output. The positive optical response lags the X-rays by typically 500 ms with a very fast onset on a timescale of 30 ms. Although this delayed optical emission is suggestive of a reprocessing scenario (light echo), the autocorrelation of the X-ray and optical time series shows that the latter has intrinsically a much faster timing structure. This argues strongly against reprocessing. It is therefore proposed that the optical light is separately generated as cyclo-synchrotron emission in a region about 20000 km from the black hole. The delay is then explained as a time of flight delay of disturbances in a relatively slow (~0.1 c) magnetically controlled outflow. A curious dimming of the optical light is also apparent 2–5 seconds before

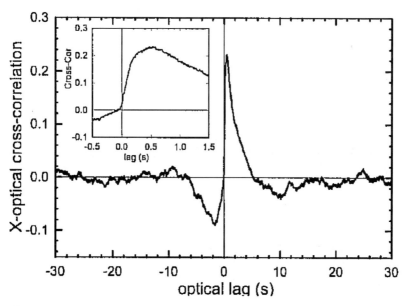

Fig. 14 The optical response correlated to X-ray variations of KV UMa (XTE J1118+48) shows time delayed emission and a preceding dip (pre-cognition dip)[4]

the X-ray maximum. More detailed analyses of the optical response components, especially the mysterious "pre-cognition dip" were presented by [19] and [8].

3 Conclusions

The scientific potential at the 'timing' frontier of astrophysics can be well demonstrated by the observations performed with the small photo-polarimeter instrument OPTIMA. The fast ($\sim$ μs) and single photon sensitive photometer is now augmented by options for polarimetry and coarse spectroscopy. In particular the observation of optical light from the vicinity of compact objects, like black holes, neutron stars, pulsars, and white dwarves, has provided data on radiation processes in the most extreme astrophysical environments. The polarization measured in the Crab pulsar provides important constraints on the distribution and spectrum of high-energy electrons in the magnetosphere of a young pulsar. The absolute timing accuracy of OPTIMA allows correlated observations at different wavelengths with a precision of $\sim$ms. The multi-wavelength correlations obtained between the X-ray and optical band for the black hole binary system KV UMa in July 2000 showed very fast correlations between the variations in both ranges and revealed new and unexpected phenomena of radiation from the vicinity (jet, inner accretion disk) of a black hole.

The ongoing developments of fast ($\sim$ ms) 2-dimensional detectors with high quantum efficiency over a wide spectral band (avalanche amplified pnCCDs, e-m (electron multiplication) CCDs) and the 3-dimensional (location and energy) cryogenic detectors (tunnel junctions or transition edge sensors) will lead to the 'instruments of choice' for many scientific investigations in HTRA. The realm of ultrafast measurements will however still be covered best with single photon sensitive, discrete detectors. In the future we hope to adapt OPTIMA to the new, large telescopes and make use of their high light collecting power to investigate fainter sources with good statistics.

The search for optical counterparts to high-energy sources (bursts and steady sources in the gamma-ray range from sub MeV to TeV), which we expect from the current and near-future ground based (TeV imaging Cherenkov telescopes) and satellite observatories (SWIFT, INTEGRAL, AGILE, GLAST), will be an exciting field for all HTRA instruments. We also hope to contribute to these investigations with our suitably advanced OPTIMA.

References

1. Dyks, J., Rudak, B.: *ApJ*, **598**, 1201, (2003)
2. Dyks, J., Harding, A.K., Rudak, B.: *ApJ*, **606**, 1125, (2004)
3. Hester, J.J., P.A. Scowen, R. Sankrit, et al.: *ApJ*, **448**, 240, (1995)
4. Kanbach, G., Straubmeier, C., Spruit, H.C., Belloni, T., *Nature*, **414**, 180 (2001)
5. Kanbach, G., Slowikowska, A, Kellner, S., and Steinle, H.: AIP Conf. Proc. **801**, 306, *'Astrophysical Sources of High Energy Particles and Radiation'*, eds. T. Bulik, B. Rudak, and G. Madejski (2005)

6. Kellner, S. *Einsatz und Weiterentwicklung von OPTIMA als hochzeitauflösendes Photo- und Polarimeter (= Use and development of OPTIMA as a high time resolution Phot-Polarimeter), Diploma Thesis (in German) at the Technical University Munich*, (2002) available at http://www.mpe.mpg.de/gamma/ instruments/optima/www/publications.html
7. Lyne, A. G., Pritchard, R. S. & Roberts, M. E.: University of Manchester, Nuffield Radio Astronomy Laboratories, Jodrell Bank, Macclesfield, Cheshire, UK,(http://www.jb.ac.uk/pulsar/crab.html) (1992)
8. Malzac J., Belloni T., Spruit H.C., Kanbach G.: *Astron. & Astrophys.*, **407**, 335 (2003)
9. D.D.E. Martin, P. Verhoeve , A. Peacock , et al: NIM-A **520**, 512 (2004)
10. Mühlegger, M. *Entwicklung und Einsatz eines Wollaston-Polarimeters für die Hochgeschindigkeitsastronomie mit OPTIMA-Burst (= Development and use of a Wollaston-polarimeter for high-time resolution astronomy with OPTIMA-Burst), Diploma Thesis (in German) at the Technical University Munich* (2006) available at http://www.mpe.mpg.de/gamma/instruments/optima/www/publications.html
11. Ng, C.-Y. and R. Romani: *ApJ*, **644**, 445, (2006)
12. Percival, J. W., Biggs, J. D., Dolan, et al.: *ApJ*, **407**, 276 (1993)
13. Petri, J., Kirk, J.G.: *ApJ*, **627**, L37, (2005)
14. R.M. Redfern: Vistas Astron. **34**, 201 (1991)
15. Slowikowska, A., 'Pulsar Characteristics across the Energy Spectrum', PhD Thesis at the Nicolaus Copernicus Astronomical Center, Dept. of Astrophysics, Torun 2006, available at http://www.mpe.mpg.de/gamma/instruments/optima/www/publications.html
16. Slowikowska, A., Kanbach, G., Stefanescu, A., *poster JD02-67 at XXVIth IAU GENERAL ASSEMBLY*, (2006)
17. Smith, F.G., Jones, D.H.P., Dick, J.S.B., Pike, C.D.: *MNRAS*, **233**, 305 (1988)
18. Sollerman, J., Lundquist, P., Lindler, D., et al.: *ApJ*, **537**, 861 (2000)
19. Spruit H.C. and Kanbach G.: *Astron.& Astrophys.*, **391**, 225 (2002)
20. Stefanescu, A. *Anpassung und Einsatz des OPTIMA Photometers zur Messung von GRB-Afterglow-Transienten (= Adaption and use of OPTIMA for the measurement of GRB-afterglow-transients), Diploma Thesis (in German) at the Technical University Munich* (2004) available at http://www.mpe.mpg.de/gamma/instruments/optima/www/optima-papers.html
21. Straubmeier, C., *OPTIMA - Entwicklung und erste astronomische Messungen eines optischen Hochgeschwindigkeitsphotometers (= OPTIMA - Development and first astronomical measurements with an optical high-speed photometer), Doctoral Thesis (in German) at the Technical University Munich* 2001, available at http://www.mpe.mpg.de/gamma/instruments/optima/www/publications.html
22. C. Straubmeier, G. Kanbach, F. Schrey, *Exp. Astron.* **11**, 157 (2001)

From QuantEYE to AquEYE—Instrumentation for Astrophysics on its Shortest Timescales

C. Barbieri, G. Naletto, F. Tamburini, T. Occhipinti, E. Giro and M. D'Onofrio

Abstract Current astronomical instrumentation essentially exploits only spatial or temporal coherence properties of the incoming photon stream. However, beyond this first-order coherence, and encoded in the arrival times of the individual photons, information lies about the details of emission mechanisms such as stimulated emission or of subsequent scattering. The Extremely Large Telescopes of the future could provide the high photon flux needed to export to the astronomical field the photonic techniques currently applied in the laboratory. These ideas were developed in a conceptual study of a focal plane instrument (QuantEYE) for the 100 m OverWhelmingly Large Telescope of the European Southern Observatory. QuantEYE would be a novel astronomical photometer capable to push the time tagging capabilities toward the pico-second region. We are now building a prototype of QuantEYE for the Asiago 182 cm telescope (AquEYE), to be followed by a larger instrument for existing 8–10 m telescopes. This paper expounds the adopted technological solutions and the first steps performed to develop such a prototype.

1 Introduction

Current astronomical instrumentation essentially exploits only spatial or temporal coherence properties of the incoming photon stream. However, beyond this first-order coherence, and encoded in the arrival times of the individual photons,

---

C. Barbieri · F. Tamburini · M. D'Onofrio
Department of Astronomy, University of Padova, Vicolo dell'Osservatorio 2, I-35122 Padova, Italy
e-mail: cesare.barbieri@unipd.it

G. Naletto · T. Occhipinti
Department of Information Engineering, University of Padova, Via Gradenigo 6/B, I-35131 Padova, Italy
e-mail: giampiero.naletto@unipd.it

E. Giro
INAF, National Institute of Astrophysics, Astronomical Observatory of Padova, Italy

information lies about the details of emission mechanisms such as stimulated emission or subsequent scattering. We recall on this respect the classic papers by Glauber [11, 12, 13] about second and higher order correlation functions. The concepts at the basis of the quantum mechanical theory of light, and useful for the astronomical applications can be found in detail in [8, 9, 10], so they won't be repeated here. We only recall that the amplitude of such functions increases with the square of the intensity (implying an enormously increased sensitivity of the future Extremely Large Telescopes, ELTs, over the existing 10 m class telescopes), and that in order to explore this new realm of information contained in the light coming from astrophysical sources, the time resolution and time tagging capabilities must be pushed towards the picosecond region. Indeed, by making recourse to Heisenberg's principle, a 0.1 nm wide filter would require a 10 picosecond time tagging capability. Silicon-based solid state devices capable to reach the nanosecond regime are already available, or are being developed in several laboratories. See for instance [1, 4, 15, 17].

Regarding our contributions, in the frame of the several studies performed in 2005 for the 100 m OWL telescope of ESO, we proposed a very high time resolution instrument for astronomical applications (QuantEYE), able to fully exploit these technological advancements of these very high speed and high quantum efficiency detectors [2, 3, 9].

For this conceptual study, we selected the Single Photon Avalanche Photodiodes (in the following SPAD) developed by S. Cova and collaborators at the Milano Polytechnic University, as further motivated in the following section.

QuantEYE would be capable to go into the domains of nanosecond variability and beyond, coping with GHz photon count-rates. QuantEYE would thus have the power to detect and examine statistics of photon arrival times (e.g. photon bunching, power spectra, autocorrelation functions). These capabilities would enable detailed studies of phenomena such as variability close to black holes, surface convection on white dwarfs, non-radial oscillation spectra in neutron stars, fine structure on neutron star surfaces, photon-gas bubbles in accretion flows, and possible free-electron lasers in the magnetic fields around magnetars. Photon-correlation spectroscopy would enable spectral resolutions exceeding $\mathcal{R} = 100$ million (as probably required to resolve laser-line emission around sources such as Eta Carinae). QuantEYE could also be adapted to more conventional high-speed astrophysical problems, using even small telescopes. Furthermore, given two distant ELTs, QuantEYE would permit a modern realization of the Hanbury Brown–Twiss intensity interferometer [14].

A more complete description of the astrophysical problems that can be tackled by QuantEYE has been given by Dravins [10]. On the technical side, QuantEYE was designed taking into account the characteristics foreseen in 2005 for the 100 m OverWhelmingly Large (OWL) telescope of the European Southern Observatory (ESO). Although the final OWL design will differ from the adopted characteristics, the QuantEYE concept maintains its full scientific appeal, and can be easily adapted to different ELTs.

2 QuantEYE Basic Opto-Mechanical Design

The basic optical design of QuantEYE [16] has been driven by two main factors. The first was the OWL telescope characteristics and performance at the date: namely a 100 m aperture with an $f/6$ focal ratio, and 6 mirrors. The available focus was fully corrected for geometric aberrations and limited by seeing (in the absence of an adaptive optics system). On the basis of the experience with large telescopes in "normal" seeing conditions, we assumed that OWL would be able to concentrate a large fraction of the light coming from a point-like source at infinity within a 1 arcsec image over a fairly satisfactory percentage of the observing time. Regarding the spot size at the telescope focus, owing to the 600 m telescope focal length, 1 arcsec field of view gives a 3 mm spot diameter. The second factor was the limited selection of available very fast photon counting detectors. After examination of several detectors, from photomultipliers to Avalanche Photodiodes, the best selection for our study seemed at the time the above mentioned SPAD, with active area of 50 micrometers, peak quantum efficiency better than 45% over the green-yellow region, very good timing accuracy (better than 50 ps), very low dark count rate (of the order of 50 counts/s when cooled to approximately $-20°C$), dead time less than 70 nanoseconds, and acceptable afterpulsing (less than 1%) ([6]). Moreover, these detectors had the important property of being available within our limited resources.

These SPADs also have a drawback, namely that CCD-type arrays do not exist yet, because of optical cross-talking problems among contiguous cells when high speed detection is required. However, a "distributed array" could be realized in several ways, by putting together a number of active cells separated by dead zones. After this choice, the instrument optical design essentially followed as a consequence. In brief, the basic optical design of QuantEYE foresees a first collimation of the beam after the OWL focus, and a following subdivision of the system pupil into N^2 sub-pupils (see Fig. 1), each of them focused on a single SPAD, thus giving a total of N^2 SPADs. In this way, since the "distributed array" detector is essentially sampling the telescope pupil, a system of N^2 parallel smaller telescopes is realized, each one acting as a fast photometer (i.e. with no imaging capability). In this case we have considered $N = 10$, because by putting together 100 SPADs, photon counting rates higher than 14 GHz (namely star Vega through a 0.1 nm wide filter) can be handled.

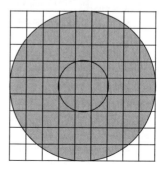

Fig. 1 The 10×10 pupil slicing concept. The non-illuminated detectors will be used for dark count statistics

In more detail, QuantEYE consists of a reversed Cassegrain telescope with 600 mm focal length and 100 mm diameter ($f/6$), which collimates the light beam after the focus of the OWL telescope (see Fig. 2). After collimation, the radiation beam has an annular shape of about 100 mm diameter where it is foreseen the possibility of inserting filters and/or polarizers, stacked in a suitable location on the collimator side (see Fig. 3). Several filters can thus be accommodated, from the very narrow ones needed in conjunction with polarizers for truly quantum optics applications, to broad band filters needed for more conventional high speed photometry, without degradation of their optical quality.

The filtered beam is then collected by a 10×10 lenslet array sampling the instrument pupil. Each lenslet is an ad-hoc compound system of two doublets (with SF11 and BaFN10 glasses) having a squared section of 10×10 mm^2 and giving a 10 mm total focal length ($f/1$). The doublet conceptual design adopts only spherical surfaces for easiness of fabrication.

The system pupil subdivision in 100 smaller subpupils allows each "channel" to obtain a very large 1/60 demagnification, that otherwise would be extremely difficult to reach, with very small losses. With this optical design, the 3 mm diameter (1 arcsec) focus of OWL is reduced on each channel to 50 μm, corresponding to the baseline SPAD active area. To direct the lens focus on the SPADs, an optical fiber link has been assumed. Figure 4 shows the plot of the encircled energy on the lenslet focal plane assuming a 1 arcsec extended source located at infinity. In this simulation, performed with the ZEMAX® software, the intensity of the source is assumed to be uniform over all its extension: the obtained result shows that about 90% of the source energy falls within the 50 μm diameter of the fiber core. Since in the real case the illumination distribution is of Gaussian type, we expect a still higher percentage of energy flux entering the fiber. There are other flux losses along the optical path from the telescope focus, mainly at the lenslet array, because of the foreseen lens lateral mounting. However, with an optimal mounting these losses can be rather small, less than 10% of the total incoming flux. Due to the SPAD quantum efficiency, the instrument working spectral range is 400–900 nm. The main optical head is fixed on the on-axis telescope focus. In addition, a second moving

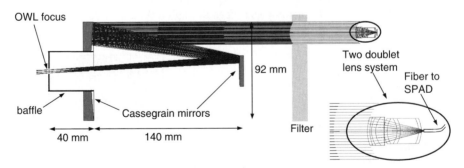

Fig. 2 The optical concept for QuantEYE. The collimator-lenslet system demagnifies 1/60 times (collimator focal length = 600 mm, lens focal length = 10 mm), giving a nominal spot size of 50 μm for a 1 arcsec source. For sake of simplicity, only one lenslet of the whole 10×10 array is shown

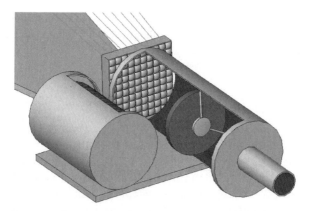

Fig. 3 3D view of the focal reducer/collimator. Light entering the baffle cone on the right is collimated by the telescope and is collected by the $N \times N$ lenslet array. Each lenslet is coupled to a single SPAD via optical fibers. The cylinder on the side of the telescope is the box accommodating the stack of filters and polarizers

detector head, able to span the whole 3 arcmin OWL scientific field of view, has been foreseen to observe a comparison star: the signal obtained from this second head will be used as a reference during a typical observation.

The proposed solution of sampling the telescope pupil and saving in 10×10 parallel channels the photon information, greatly reduces the limiting dead time problem of the detectors: in such a way, a fixed-area very high speed photometer with a tremendous dynamic range, from the 5th to the 30th stellar magnitude (a factor 10^{10} in brightness), can be obtained. In addition, it is likely that in the next few years the development of new detectors and new optical clocks can push the

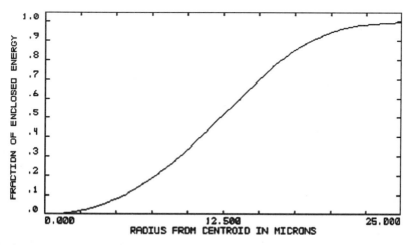

Fig. 4 Extended source encircled energy plot: percentage of energy falling within a circle of given radius for a uniformly illuminated 1 arcsec extended circular source. The 25 μm limit corresponds to the fiber acceptance diameter

time resolution into the picosecond region for several hours of continuous operation. Other non-imaging optical designs (e.g. enlarging the acceptance field to 1.2 arcsec, or using 3:1 tapered fibers, or entirely avoiding the optical fibers), have been studied and shown feasible. We refer for details to [8].

3 An Imaging Version for QuantEYE

Some consideration was given to the possibility of realizing an "imaging" detector in the absence of a CCD-type SPAD array, which seems far from being realized. In the following, we describe two designs we have considered, based on the availability of "granular" arrays of SPADs, namely a number of individual detectors integrated on the same silicon wafer or on a mosaic of wafers separated by a small dead area. The first design makes use of a 10×10 microlenses array (300×300 μm^2 section each) located at the OWL focus, acting as field lenses; this first array needs to be followed by a second 10×10 microlenses array, of the same section but different focal length, acting as relay lens. Figure 5 shows a schematic of this system: instead of "real" lenses, the so called "paraxial" ideal lenses (drawn as a straight line: they behave as ideal lenses of a given focal length, and do not introduce any aberration) have been simulated for simplicity. The field lenses, located on the OWL focus and collecting a 0.1×0.1 arcsec^2 field of view, have a 1.8 mm focal length; each of them sends the light on a corresponding relay lens, which has the same size and a focal length of 0.26 mm, focusing the beam to 0.3 mm distance. Owing to the suitable demagnification of the system, the spot size on the focus of this ideal system has a 50 μm diameter: this is schematically shown in Fig. 6, where only the extreme and the central fields are shown. At this point, light can be collected either by fibers bringing the light on the detectors, or by a granular array of suitably matched SPADs.

This design has two main problems: a) the availability of suitable microlenses array. In fact, these lenses can only be manufactured as an "array" (i.e. not as single lenses and then integrated to form the array): this implies that the lens shape control is rather poor, and that there are limitations in realizing the desired lens surface shape. Moreover, the large required spectral range implies the need of a "complex" lens (at least one doublet for the field lens, and probably two doublets for the relay lens: in addition, the latter lens has an extremely short focal length compared to its size, making very critical its shape), or of several interchangeable arrays optimized for shorter spectral ranges. b) the matching of the lens array foci with the detectors. In fact, unless a matrix SPAD for direct coupling is available, the fiber matching would be extremely critical because of the size of the system. This is due to the fact that the foci are separated by 300 μm, and the fiber cladding diameter is probably of only slightly smaller size.

The second proposed design is very similar to the previous one, but a further technological development of the detector towards a "true" SPAD array is foreseen: in this case, we have assumed the availability of a 30×30 matrix of SPADs, in which the 50 μm active area heads are center-to-center spaced by 100 μm. The latter distance has been assumed so small that having a "fiber" optical matching with the detector is practically not feasible, and a direct optical matching is necessary. In

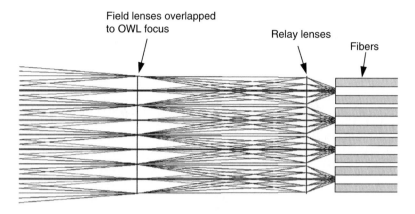

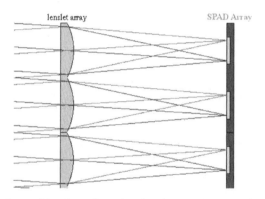

Fig. 5 Example of a possible scheme for an imaging detector. Upper panel: a number of "channels" of the two 10 × 10 lens arrays coupled to a fiber. We have used in the simulation the so-called "paraxial" lenses, here represented as straight lines. Lower panel: A possible solution to couple the light directly to a matrix of SPAD detectors

this case, an optical system of 30 × 30 microlenses array (100 × 100 μm^2 section each), acting as field lens, can be located on the OWL focus; then, a second 30 × 30 microlenses array (100 × 100 μm^2 section each), acting as relay lens, focuses the light on the SPAD active areas. This design would eliminate the problem of interfacing the lenses with the SPADs through the fibers, so simplifying the global design. Unfortunately, it suffers of the same limitations described for the previous case: that is not only the detector availability, but also the capability of providing the required lenslet array (that in this case would be still more critical). Finally, as a general comment, an imaging solution should also consider the need of inserting somewhere the very narrow filters required for quantum optics applications without degradation. This point has not been discussed here, but it is clear that this is fairly difficult to be realized with the above designs: an additional optical relay stage would have to be added between the telescope focus and the first microlenses array.

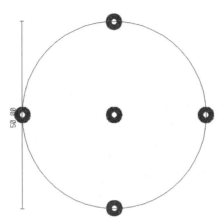

Fig. 6 Spot diagram of the focus of the ideal imaging system; only the extreme and the central fields are represented. The circle has a 50 µm diameter

4 Data Storage and Analysis

To preserve the photon time arrival information at a maximum rate of 1 GHz, Quant-EYE has a central storage unit with a minimum capacity of 50 Terabytes, connected to an a-posteriori analysis system with a high bandwidth transport channel. The arrival time of each photon is given as input to an asynchronous post-processor, which guarantees data integrity for the subsequent scientific investigation. The a-posteriori processing unit is a cluster of CPUs in which specific parallel algorithms will work together, optimizing the computations of the high order correlation functions between the time tagged photons. Furthermore, the capability of performing real time correlation functions among different detectors (on-line correlation unit) has been foreseen. A schematic of the overall electronics system is shown in Fig. 7. The number of detectors to be analyzed in real time depends on the computational power of reprogrammable logic circuits and processors, which is connected to the technological progress of CPUs, FPGAs (Field Programmable Gate Arrays) and ASIC (Application-Specific Integrated Circuits).

The electronics consists of four main elements (see Fig. 7):

1. The front-end system inside the primary and secondary optical heads. This system takes the signal coming from all the SPADs in parallel, and feeds several TDC (Time to Digital Converters) to have the precise time tags of each photons. Inside the primary and secondary heads there are microcontrollers for the data exchange with the Storing and Preprocessing unit and the on-line correlation unit. 2. A Start/Stop element connected to the precise and stable time reference. This element creates the necessary control signals for each TDC and controllers. The external timing reference system is positioned outside the QuantEYE "instrument envelope" and is connected to it via an optical link: this allows to share a 20 MHz reference timing given by an ultra-precise clock (e.g. a hydrogen maser). 3. A big Storing and Preprocessing unit able to handle all the data coming from the front-end system during a typical observation. It can also process the data for an efficient transmission

Instrumentation for Astrophysics on its Shortest Timescales

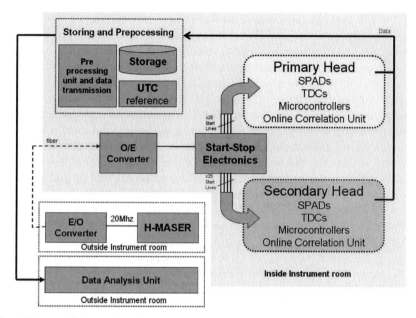

Fig. 7 The overall electronics scheme

toward the final analysis system, time tagging them through an external universal time reference (UTC) given by a GPS (or the Galileo Navigation Satellite System when available). 4. The final Data Analysis Unit that performs the a-posteriori calculations of the correlation functions and of other scientific tasks.

A final consideration is relative to the analysis of the huge, multidimensional database generated by QuantEYE. The necessary computational power is extremely large, and new algorithms have to be developed. In particular we think that it will be necessary to migrate to parallel architectures; for example using two or more dual/quad-cores processors and writing software that uses this computational power and parallelism. We have also a large interest in the fields of quantum information and quantum computation. Likely, in the near future, practical and easy to use quantum computers will become available. Then, quantum algorithms like the QFT (Quantum Fourier Transform) and the Grover's algorithm (see [22] and [21]), which allow to search in a highly efficient way an element inside a big database, will become of routine utilization.

5 A Small Prototype for QuantEYE

We are currently realizing a first scaled-down prototype of QuantEYE for the Asiago-Cima Ekar (Italy) 182 cm telescope, an instrument that we will call AquEYE (Asiago Quantum Eye). This telescope offers us a good availability of observing time and excellent environment, from the mechanical shop, to clean areas, to control rooms. A further advantage of this telescope is the existing AFOSC (Asiago Faint Object Spectrograph and Camera) focal reducer instrument ([7]), which can be easily adapted to our goals. AFOSC (see Fig. 8) is mounted on a flange which takes care of many observational needs, from pointing and guiding to field vision and rotation, etc. A simple way of realizing this small prototype is by dividing the telescope pupil in four parts only ($N = 2$). This can be easily obtained by mounting a simple pyramidal mirror at the exit of AFOSC. The beams reflected by the pyramid are independently sent along four perpendicular directions, and each of them can be imaged on a SPAD through a train of four commercial doublets (see Fig. 9). Obviously, in this case, the instrument specifications are much less stringent than in QuantEYE. In fact, the Asiago telescope focal length is only 16.1 m, and a 3 arcsec extended source (the average size of a point-like star, due to the limited seeing) gives a spot size at the telescope focus of about 0.23 mm. In addition, AFOSC introduces an almost 1/2 demagnification factor, bringing the size of the spot at the AFOSC output at about 130 µm. So, the lens train after the pyramid has to further demagnify the spot of only a factor 1/4 to have a final spot size of the order of 40 µm: it is because of this rather relaxed specification that it has been possible to design the system with only commercial lenses. As in QuantEYE, the system losses are negligible, and essentially limited at the edges of the pyramid, where the radiation beam is splitted. As detectors, we have selected and acquired the 50 µm SPADs produced by the MPD (Micro Photon Devices, Bolzano, Italy) company. According to the data sheet, their quality is as expected: quantum efficiency in the visible band is better than 45%, the dead time is around 70 nanoseconds, the time tagging capability is better than 50 picoseconds, and the afterpulsing probability is less than 1%. The internal thermoelectric cooling produces a dark count lower than 50 s^{-1} at $-20°$C. We plan to confirm these data, and characterize each detector using the facilities available at the Catania Observatory ([5]). An important simplification of AquEYE is the possibility of feeding the SPADs without the need of the optical fibers, which were motivated in QuantEYE by the difficulties of accommodating in "parallel" the relatively large detector boxes at the microlenses focal positions.

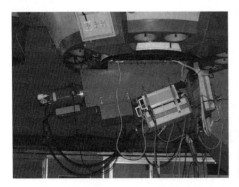

Fig. 8 Left panel: the side view of AFOSC mounted at the Cassegrain focus of the 182 cm telescope in Asiago. Right panel: the exit lens of AFOSC at the center of the mounting flange of AquEYE

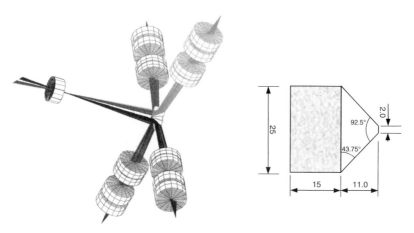

Fig. 9 Left panel: following the last lens of AFOSC shown at the left, a pyramid splits the light to four separate channels to the SPADs. Right panel: mechanical details of the pyramid

In this case, since the four beams are "independent", we can directly mount the SPADs on the lens foci. The optical performance of the designed system, obtained by a ZEMAX simulation, is excellent over the 50 μm size active area, from the blue (420 nm) to the red (720 nm), as shown in Figs. 10 and 11. Here both the extended source focal spot, obtained considering the central and four extreme fields of view, and the corresponding encircled energy are shown. It is evident the instrument is capable of focusing more than 90% of the incoming flux on the detector active area.

Furthermore, to simplify the construction of AquEYE, the second movable head foreseen in QuantEYE will not be built. The electronics scheme, based on available commercial products, is shown in Fig. 12. The core of the electronics system is

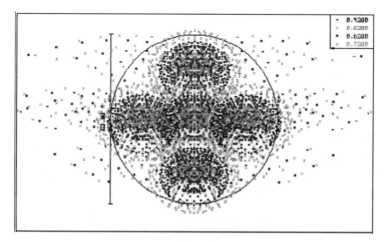

Fig. 10 The energy concentration between 420 and 720 nm, for five positions (at the center and at the edges) over the Field of View. The circle represents the sensitive area of the SPAD

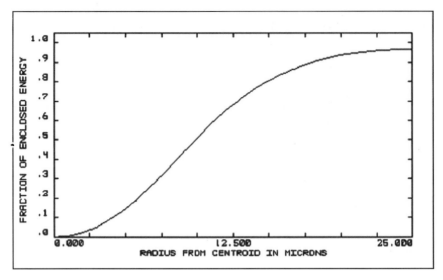

Fig. 11 Extended source encircled energy plot: percentage of energy falling within a circle of given radius for a uniformly illuminated 3 arcsec extended circular source. The 25 μm limit corresponds to the SPAD sensitive area

a CAEN (Costruzioni Apparecchiature Elettroniche Nucleari S.p.A.) TDC board. This board will take the TTL inputs coming from the four SPADs and will process the signals. Each time tag will be buffered inside the board and then put into a standard VME bus. The TDC will be able to tag each event with a time precision of 35 ps per channel. As the electronics system will be attached to the telescope, the

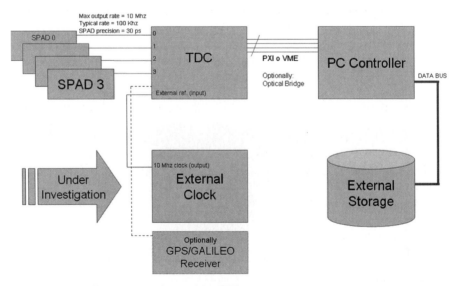

Fig. 12 The overall electronics scheme of AquEYE

CAEN board will transfer all the data time tags to an external personal computer able to save each tag to a mass storage. The external computer will perform the a-posteriori data analysis, saving the interesting scientific data in some removable support for further studies.

We now come to the crucial question of generating and maintaining a very accurate time for minutes and even hours. First of all, the start/stop commands will be tied to the UTC by means of the GPS (or the Galileo Navigation Satellite System GNSS if available) signal, so that the data can be referred to a common time scale adopted by all telescopes on the ground or in space (we recall the scientific interest of correlations with X-ray and Gamma ray timing). However, we have to overcome the jitter, that for both GPS and GNSS is of the order of ten nanoseconds or so. Taking into account that the available resources are fairly scanty, the "best" clock is still under investigation, with the help of experts by INRIM (the Italian National Institute of Metrological Research) in Turin and the Astronomical Observatory in Cagliari. Furthermore, we shall perform experiments to tie to a common reference, to better than 100 picoseconds, the clock in Asiago with a twin clock at the telescope Vega of Lubljana Observatory. This activity is seen as a practical demonstration of time distribution to very distant observers, in view of Very Long Baseline intensity interferometry quoted in [8].

6 Conclusions

AquEYE should be put in operation at Asiago toward the end of 2006. With this telescope, of modest size, we will not be able to achieve significant results on the quantum statistics of the photon streams from astrophysical sources. However, we plan to tackle with it several high time resolution astrophysical problems (occultations, cataclysmic variables, flickering and flare stars, exoplanet transits, and so on), taking advantage of the possibility of an a-posteriori integration of the counts. Aqueye will thus be able to: a) provide a way to test on a telescope the feasibility of quantum optics with future ELTs, and to elucidate some of the problems likely to be encountered; b) improve on the time resolution achieved by the pioneering observations made in the past with fast photometers (such as MANIA, see [18]), or more recently with micro-channel plates and APDs (e.g. TRIFFID [19, 20], and OPTIMA [15]). Even the use of Aqueye in Optical SETI projects could be considered; c) finally, we will test the feasibility of distributing the time and synchronizing two distant telescopes at the 50 picosecond level.

The subsequent goal is to define an upgraded version to be brought to existing 8–10 m telescopes, such as the VLT or the LBT. In particular, the two mirrors of LBT would allow to perform a modern version of the Hanbury Brown–Twiss Intensity Interferometer ([14]), with a 100-fold improved sensitivity thanks to the augmented quantum efficiency, the much higher electrical bandwidth, the higher collecting area, the much superior optical quality of the telescopes. Finally, consideration will be given to the possibility of mounting a quantum detector in the central pixel of the Cherenkov light collector MAGIC on the Roque, as a precursor for an ELT.

Acknowledgments The QuantEYE projects have been partly supported by ESO and the University of Padova. AquEYE enjoys the collaboration of researchers from INAF Observatories in Rome, Cagliari and Catania, from INRIM Torino and Lubljana University in Slovenia. The work presented in this paper has been partially supported by OPTICON/HTRA.

References

1. Aull BF, Loomis AH, Young DJ, Heinrichs RM, Felton BJ, Daniels PJ, Landers DJ (2002). In: Geiger-Mode Avalanche Photodiodes for Three-Dimensional Imaging. Lincoln Laboratory Journal Vol 13, Nr 2, p 335
2. Barbieri C, Da Deppo V, D'Onofrio M, Dravins D, Fornasier S, Fosbury RAE, Naletto G, Nilsson R, Occhipinti T, Tamburini F, Uthas H, Zampieri L (2006) QuantEYE, the Quantum Optics Instrument for OWL. In: Whitelock P, Leibundgut B, Dennefeld M (eds) The Scientific Requirements for Extremely Large Telescopes. IAU Symposium 232. Cambridge University Press, pp 506–507
3. Barbieri C, Dravins D, Occhipinti T, Tamburini F, Naletto G, Da Deppo V, Fornasier S, D'Onofrio M, Fosbury RAE, Nilsson R, Uthas H (2006) Astronomical applications of quantum optics for extremely large telescopes. Journal of Modern Optics. Special issue of on "Single-Photon: Sources, Detectors, Applications and Measurement Methods". In press
4. Belluso M, Mazzillo Cataldo M, Bonanno G, Billotta S, Scuderi S, Calì A, Timpanaro MC, Sanfilippo D, Fallica PG, Sciacca E, Lombardo S, Morabito A (2005) SPAD Array Detectors for Astrophysical Applications. Memories of the Italian Astronomical Society. In press
5. Bonanno G, Bruno P, Calì A, Cosentino R, di Benedetto R, Puleo M, Scuderi S (1996) Catania Astrophysical Observatory facility for UV CCD characterization. In: Siegmund OH, Gummin MA (eds) EUV, X-Ray, and Gamma-Ray Instrumentation for Astronomy VII. Proc. SPIE, vol 2808, pp 242–249
6. Cova S, Ghioni M, Lotito A, Rech I, Zappa, F (2004) Evolution and prospects for single-photon avalanche diodes and quenching circuits. Journal of Modern Optics, vol 51, nr 9, pp 1267–1288
7. Desidera S, Fantinel D, Giro E, Navasardyan H (2003) AFOSC User Manual, Version 1.2. Internal report
8. Dravins D, Barbieri C, Da Deppo V, Faria D, Fornasier S, Fosbury RAE, Lindegren L, Naletto G, Nilsson R, Occhipinti T, Tamburini F, Uthas H, Zampieri L (2005) QuantEYE. Quantum Optics Instrumentation for Astronomy. In: OWL Instrument Concept Study, ESO document OWL-CSR-ESO-00000-0162
9. Dravins D, Barbieri C, Fosbury RAE, Naletto G, Nilsson R, Occhipinti T, Tamburini F, Uthas H, Zampieri L (2006) Astronomical Quantum Optics with Extremely Large Telescopes. In: Whitelock P, Leibundgut B, Dennefeld M (eds) The Scientific Requirements for Extremely Large Telescopes. IAU Symp. 232, Cambridge University Press, pp 502–505
10. Dravins D (2006) Photonic Astronomy and Quantum Optics. This volume
11. Glauber RJ (1963) Photon Correlations. Phys. Rev. Letters vol 10, p 84
12. Glauber RJ (1963) The Quantum Theory of Optical Coherence. Phys. Rev. vol 130, p 2529
13. Glauber RJ (1963) Coherent and incoherent states of the radiation field. Phys. Rev. 131, pp 2766–2788
14. Hanbury Brown R (1974) The Intensity Interferometer. Taylor and Francis, New York.
15. Kranback G (2006) The OPTIMA photo-polarimeter: new developments and lessons learned. This conference
16. Naletto G, Barbieri C, Dravins D, Occhipinti T, Tamburini F, Da Deppo V, Fornasier S, D'Onofrio M, Fosbury RAE, Nilsson R, Uthas H, Zampieri L (2006) QuantEYE: A Quantum Optics Instrument for Extremely Large Telescopes. In: Ground-Based and Airborne Instrumentation For Astronomy, SPIE Proc. Vol 6269, p 62691W-1/9

17. Strueder L (2006) High Speed single photon imaging with AApnCCDs. This conference
18. Beskin G, Komarova V, Neizvestny S, Plokhotnichenko V, Popova M, Zhuravkov A (1997) The investigations of optical variability on time scales of $10^{-7} \div 10^2$ s: hardware, software, results. Exper. Astron. vol 7, pp 413–420
19. Redfern M et al (1992) First Scientific Results from TRIFFID. Gemini Newsletter, vol 38 p 1
20. http://www.ing.iac.es/PR/
21. Grover LK (1996) A fast quantum mechanical algorithm for database search. In: Proc. 28th Annual ACM Symposium on the Theory of Computing (STOC), p 212–219
22. Deutsch D, Jozsa R (1992) Rapid solution of problems by quantum computation. In: Proc. R. Soc. London A439, p 553–558

Fast Spectroscopy and Imaging with the FORS2 HIT Mode

Kieran O'Brien

Abstract The HIgh-Time resolution (HIT) mode of FORS2 has 3 sub-modes that allow for imaging and spectroscopy over a range of timescales from milliseconds up to seconds. It is the only high time resolution spectroscopy mode available on an 8 m class telescope. In imaging mode, it can be used to measure the pulse of pulsars and spinning white dwarfs in a variety of high throughput broad- and narrowband filters. In spectroscopy mode it can take up to 10 spectra per second using a novel "shift-and-wait" clocking pattern for the CCD. It takes advantage of the user-designed masks which can be inserted into FORS2 to allow any two targets within the 6.8' × 6.8' field of view of FORS2 to be selected. A number of integration, or more precisely 'wait', times are available, which together with the high throughput GRISMs can observe the entire optical spectrum on a range of timescales.

1 General Description

FORS is the FOcal Reducer and low dispersion Spectrograph for the Very Large Telescope (VLT) of the European Southern Observatory (ESO). It is designed as an all-dioptric instrument for the wavelength range from 330 nm to 1100 nm. It is capable of imaging as well as low to medium resolution spectroscopy, with a wide range of filters and grisms (see Tables 2.2, 2.4 and 2.5 of the FORS user manual[1] for a comprehensive list). Single and multiple object spectroscopy options are available using long slits, a set of 19 moveable jaws (referred to as MOS mode) and a magazine capable of holding up to 10 laser-cut Invar masks (referred to as MXU mode).

FORS-2 saw first light in 1999 and, together with its non-identical twin FORS-1, has accounted for more refereed publications than any other instrument at the VLT. Following an upgrade in April 2002, FORS-2 has been equipped with a mosaic of two 2k × 4k MIT CCDs (pixel size of 15 × 15 μm) with a pixel scale of

---

Kieran O'Brien
European Southern Observatory, Santiago, Chile
e-mail: kobrien@eso.org

[1] http://www.eso.org/instruments/fors/doc/

0.125"/pixel, although it is operated with binning 2 × 2 as standard delivering a final pixel scale of 0.25"/pixel. The HIT mode of FORS-2 was available from the beginning of operations, although it was subsequently unavailable for over a year following the upgrade of the CCD camera.

The general principle of the HIT mode is to increase the duty cycle of imaging and spectroscopic observations by reducing the exposed region of the CCD and shifting the charge from this small region to an unexposed 'storage' region of the CCD. By shifting the charge in this manner a number of times, it is possible to store a time series of images or spectra on the CCD without needing to read-out the CCD after each exposure. The shutter remains open throughout the observation and is only closed once a predefined number of shifts have occurred. The CCD is then read-out using the standard low noise read-out mode used for all spectroscopic observations. Further details of the different clock patterns used can be found in Sects. 2.1 & 2.3.

The HIT mode is similar in concept to the (now decommissioned) Low Smear Drift (LSD) mode used by ISIS on the William Herschel Telescope on La Palma [1]. This drift mode was limited by the size of the memory on the DMS (16 Mb) and allowed read-out of several spatial pixels in a given exposure. Another previous drift mode used the RGO spectrograph on the Anglo-Australian Telescope at Siding Spring, New South Wales [2]. This mode was similar to the LSD mode on ISIS, again using a physical buffer which limits the number of spectra that can be obtained in one exposure. In addition to these drift modes, a continuous readout mode was available as a visitor instrument on LRIS for a short time. This mode did not have the limitations of the buffer size and allowed a continuous read-out of the CCD. However, it did not allow for a comparison star and suffered from several instrumental effects that were never overcome (see [3]).

2 HIT Modes

There are three modes available to users that offer very different characteristics and are suitable for a wide variety of applications.

- HIT-I: Imaging mode
- HIT-OS: One-shift spectroscopy mode
- HIT-MS: Multiple-shift spectroscopy mode

2.1 HIT-I

The HIT-I mode is the only available imaging mode. It uses the moveable slitlets of the MOS unit located in the top section of the instrument to form a pseudo-longslit by aligning them along the same column or columns. The slit width can be set to a value in the range 0.2–30" and is placed so that the projection of the slit falls on the row nearest to the serial register in the unvignetted region of the field of view.

The basic scheme of the HIT-I clock pattern is shown in the top panel of Fig. 1. The imaging mode is available with five variations on this clock pattern. In each the charge is shifted across the unvignetted region of the chip (3280 unbinned pixels) in the direction away from the serial register, but the speed at which the charge is moved is varied to simulate a longer or shorter exposure time. The transfer from $t = 1$–3280 takes place in either 1, 4, 16, 64 or 256 seconds, which identifies the clock pattern (i.e. HIT-OS1-1sec ... HIT-OS5-256sec). This means each row is shifted every 0.61, 2.4, 9.8, 39.0 or 156 milliseconds. The speed of the slowest mode is determined by technical limitations. It is similar to the shortest exposure time for full frame imaging available with FORS-2 (0.25 seconds). Once the final shift has

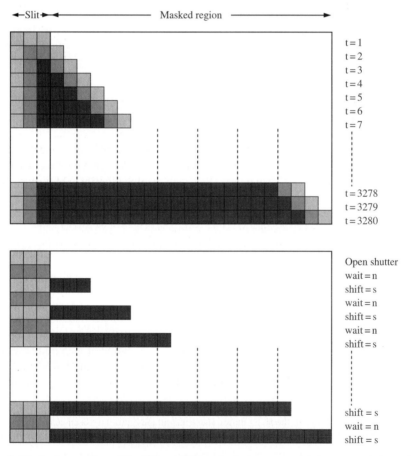

Fig. 1 The basic operation of the different HIT mode clock patterns: **(a)** the top panel shows the clock pattern for the HIT-I and HIT-OS patterns, where each row is shifted individually until the first row shifted reaches the end of the unvignetted region; **(b)** the bottom panel shows the clock pattern for the HIT-MS mode, where a number of rows are grouped together and shifted out from the exposed region into the masked region where they wait for a time while the pixels under the slit are exposed to light from the source. 's' refers to the size in unbinned pixels of the slit, whilst 'n' is the number of seconds to wait between shifts

taken place, the shutter is closed and the frame is read-out in the normal manner, using the low read-out noise mode of FORS, used for spectroscopic observations (100khz, 2 × 2, high).

The time resolution of the resulting lightcurve depends on the slit width, as each pixel is exposed as long as it remains in the unmasked region of the CCD. As can be seen in Fig. 1 there are also some edge effects if the image of the slit is projected onto more than one pixel.

All of the filters available in the FORS filter-set can be used, which includes standard Bessel filters, as well as Gunn and an number of narrow-band filters centred on common emission lines (Hα, HeI, OII). In addition it is possible for users to supply their own filters (details of this process can be found on the FORS webpages). The slit created by the MOS, which covers the entire 6.8' field, can be rotated on the sky to include a comparison star for relative photometry.

The limiting magnitude for the HIT-I mode is difficult to quantify as there are several parameters that have a large influence on the limit. In addition, many of these also have direct influence on other characteristics of the final data-set. Here I will summarize the most important of these:

- **Clock pattern:** The clock pattern determines the amount of time the pixels are exposed to light and can be thought of in the same way as the exposure time in traditional imaging.
- **Slit width:** Increasing the slit width will let more light onto the chip and therefore allow fainter objects to be observed. However, it will also increase the projected area of the slit on the CCD and mean that each pixel is illuminated for more than one clock-cycle, thus reducing the time resolution of the resulting lightcurve by convolving the true lightcurve with a filter with a width equal to the number of pixels exposed (and hence the number of time steps).
- **Filter:** Obviously, the filter transmission has a direct influence on the limiting magnitude.
- **Atmospheric conditions:** Atmospheric conditions such as seeing and transparency will lead to larger slit losses.

The magnitude limits in Table 1 correspond to a S/N of 5 per time step using the most popular mode, HIT-OS4-16sec. They have been calculated for a slit-width of 0.25" (equivalent to 1 pixel), which results in a time resolution of 9.75 milliseconds. They are calculated assuming a dark sky, clear conditions, a seeing FWHM of 0.8" and an airmass of 1.2, and have been determined for a point source of zero colour (A0V star).

Table 1 Magnitude limits for the HIT-I mode. These magnitude limits have been determined using the HIT-OS4-16sec mode with a slit-width of 0.25", assuming dark sky and typical conditions for Paranal

Filter	Magnitude limit
U	12.6
B	16.5
V	17.1
R	17.3
I	16.7

Obviously, fainter targets can be observed by increasing the slit width and/or changing the readout mode. However, this increase is at the expense of the time resolution (due to increasing the number of pixels the slit is projected onto) and the photometric accuracy (due to the increased chance to image motion effects within the slit, see Sect. 3.3).

2.1.1 HIT-I Observations of the Crab Pulsar

As a test of the timing accuracy we performed observations of the Crab pulsar. An example of the resulting dataset can be seen in Fig. 2, which clearly shows the quality of the data that is produced by the HIT-I mode. In Fig. 2, the position of the slit is clearly marked on the left hand-side of the image together with a further marker indicating the direction the charge is shifted across the CCD. A section of the image has been enlarged to show the trace of the pulsar (bottom) together with that of a nearby bright comparison star to show what can be expected from this mode. The "major peak" can be clearly seen, as well as the fainter "minor peak". The minor peak occurs slightly before $\phi_{pulse} = 0.5$ in the 33 millisecond pulse period (note that since the charge is shifted from left to right, the first images will appear on the right-hand side of the final image. These images were taken with the HIT-OS1-1sec mode with a slit width of 5″ under average conditions (an airmass of 1.5, FWHM of the seeing of 0.8″, dark sky) through the R_SPECIAL filter.

Whilst the HIT-I mode of FORS2 is relatively unique in offering the capability to perform imaging on a sub-second timescale using a CCD detector, it remains inferior to more dedicated HTRA instruments in a number of ways. For instance, ULTRACAM[2] has the advantage of having three colours simultaneously allowing colour changes to be determined, as well as a much larger field-of-view, which allows for a range of comparison stars to be selected, leading to much better deter-

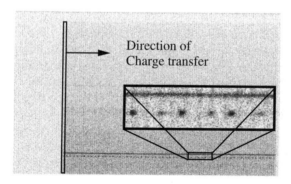

Fig. 2 HIT-I image of the Crab pulsar. The position of the slit is shown on the left hand-side of the field, which represents the exposed region of the CCD. Charge is shifted in the direction shown. The enlarged region shows both the crab pulsar (lower), including the major and minor peaks, as well as a nearby comparison star (upper). Time runs in the opposite direction to the charge transfer

---

[2] See the contribution elsewhere in these proceedings for a description of this instrument.

mined photometry. SALTICAM again has a large field-of-view from with to select comparison stars for relative photometry, but is limited in the amount of time it can observe a given target for. OPTIMA[2] and other similar photon-counting devices are able to observe a range of different phenomena on even faster timescales, but suffer from low throughput. In addition, the HIT mode is permanently mounted on an 8.2 m telescope (the VLT) and is available in service mode, for observations on very short notice, with a wide range of filters. The VLT allows for much longer observations than are possible with SALT and can react very quickly to transient phenomena over a large fraction of the sky.

2.2 HIT-OS

The HIT-OS mode was the first high time resolution spectroscopy mode available at the VLT. It uses masks cut with the laser cutting machine of the VIMOS instrument. As the user is limited to single target spectroscopy, there are a number of pre-cut masks for the HIT-OS mode with square apertures of varying sizes (0.5–5.0″). It is only possible to put a single target in the aperture cut in the mask as the charge would otherwise be shifted under the dispersed light of the subsequent apertures.

Due to the fact that the standard dispersion direction is perpendicular to the serial register which is incompatible with the direction in which the charge is shifted in the HIT modes, only 2 grisms, which have been rotated by 90°, are available for the HIT-OS mode. The characteristics of these are shown in Table 2.

The HIT-OS mode uses the same clock patterns as the HIT-I mode and is therefore limited in the number of targets that can be observed due to the relatively high limiting magnitude, shown in Table 2. The limiting magnitudes have been calculated using the HIT-OS5-256sec clock pattern, which results in a time resolution of 0.6*(slitwidth in pixels). However, if you want to reach this limit and ensure some degree of photometry, it is necessary to use a mask with a large aperture in order to minimize the slit losses. This photometric stability comes at the cost of the spectral resolution (which will be limited by the seeing profile parallel to the slit), the spectral stability (due to image motion parallel to the dispersion direction), the temporal resolution (which will be limited by the seeing profile perpendicular to the slit) and the short time-scale photometric stability (which will be limited by image motion perpendicular to the slit). However, if you choose a narrow, short slit (i.e. a small aperture) then you will only be able to measure equivalent width and line profile changes. The stability issues and the relatively bright limiting magnitude greatly limit the applicability of the mode, which is why we have introduced a second spectroscopic mode, the HIT-MS mode which overcomes many of these problems.

Table 2 Details of the grisms available for the HIT-OS and HIT-MS modes, together with the limiting magnitude for the HIT-OS mode

Grism name	Central wavelength(nm)	Dispersion (nm/pixel)	Limiting magnitude
600B	445	0.075	15.8
300I	857	0.162	15.9

2.3 HIT-MS

The HIT-MS mode can be thought of as " 'normal' spectroscopy with large gains in duty cycle" as it allows the user to operate in a pseudo-MOS mode, storing multiple exposures of the same target (or targets) on the CCD between successive read-outs of the CCD.

2.3.1 Multiple Shift Clock Pattern

The HIT-MS mode allows users to operate the CCD with a "shift-and-wait" clock pattern, as shown in the bottom panel of Fig. 1. In contrast to the HIT-I and HIT-OS modes, where the charge is constantly being shifted from one row to the next and only the rate at which it is shifted is changed, the HIT-MS clock pattern shifts a pre-defined number of rows very fast ($\sim$2.5 microseconds per line) and then integrates for a pre-defined 'wait' time before the sequence is repeated. This continues until the maximum number of line-shifts has occurred and the CCD is read-out using the standard low read-out noise spectroscopic mode (100kHz,2 $\times$ 2,high).

2.3.2 Mask Design

The HIT-MS mode again uses masks inserted into the MXU of FORS2 and is thus limited to FORS2. However, in contrast to the HIT-OS mode, the user designs the masks, as it is possible to have a number of slitlets and hence targets. There is no real limit to the number of slitlets, as long as the number of rows shifted each time is equal to the number of rows between the projection of the top of the uppermost slit to the bottom of the lowermost slit. In practice, it is usually necessary to have just two slits (one for the target and one for a nearby comparison star to be used to correct for slit-losses, small instrumental wavelength shifts and seeing variations). With just two slitlets it is possible to rotate the instrument so that the projection of the bottom of the uppermost slit falls on the row above the top of the lowermost slit, thus minimizing the number of rows that are shifted each time.

A sample acquisition image is shown in Fig. 3 which highlights all of the important features. Towards the top of the image are the alignment stars (*circles*) which the software (based on a number of MIDAS routines) uses to determine the small-scale offsets of the mask with respect to their nominal positions. Once these offsets have been calculated they are applied to the telescope before the mask is inserted in order to ensure the optimal alignment on sky. The position of the target (*small square*) is shown, which, in this case, is offset from the centre of the field in order to maximise the spectral coverage of the resulting spectrum. As dispersion is in the y-direction it is important to ensure that the gap in the CCD covers a relatively uninteresting part of the spectrum. The position of the comparison star (*triangle*) which has been orientated so that it is 5″ away from the target in the x-direction, ensuring that the images from the slit do not overlap, as described above.

One consequence of the offset between the target and comparison star slits is that the central wavelength of the resulting spectra will not be the same (due to

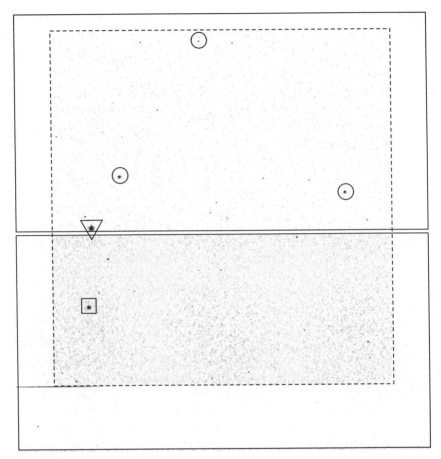

Fig. 3 A typical acquisition image showing the relative orientations of the upper and lower CCDs (*solid rectangles*), the field of view of the instrument (*dashed square*), the alignment stars (*circles*), the comparison star (*triangle*) and the target (*small square*)

their offset in the y-direction, the direction of the dispersion). This will make the correction for slit losses slightly more complicated, as the wavelength coverage is not the same for both targets, as would be the case in long-slit spectroscopy. Up to 10 masks can be inserted into the instrument during daytime in order to allow for either a number of targets, or a range of slit-widths to counter the problem of changing atmospheric conditions during the night. In contrast to the HIT-OS mode, narrow slit-widths are possible without the complete loss of photometric accuracy due to the presence of a comparison star, whose purpose it is to correct for such effects. This removes the problems caused by motion across the slit when the star under-fills the slit. In addition, accurate sky extraction is possible from regions around both the target and the comparison stars.

Both the target and comparison stars are moved to the left-hand side of the field to allow far as many shifts as possible before the CCD needs to be readout.

Fast Spectroscopy and Imaging with the FORS2 HIT Mode

Table 3 The duty cycles for the currently offered HIT-MS clock patterns. The overheads per read-out include read-out of the CCD (~10 seconds) and set-up time (~30 seconds). These are not significantly improved by windowing of the CCD

Clock pattern	Wait time (secs)	Cycle time (secs)	HIT duty cycle (%)	'Normal' duty cycle (%)	Gain factor
HIT-MS1-01sec	0.1	44.1	9.3	0.2	46
HIT-MS2-02sec	0.2	48.2	17	0.5	34
HIT-MS3-05sec	0.5	60.5	34	1.2	28
HIT-MS4-1sec	1.0	81.0	51	2.4	21
HIT-MS5-2sec	2.0	122	67	4.8	14
HIT-MS6-5sec	5.0	245	84	11.1	7.6
HIT-MS7-10sec	10.0	450	91	20.0	4.5
HIT-MS8-20sec	20.0	860	95	33.3	2.9

2.3.3 Duty Cycle Gains

The number of spectra that can be stored on the CCD before it is read out depends on the number of pixels shifted per 'shift-and-wait' cycle. As described in the previous section, the user would typically define two slitlets, one each for a target and a nearby comparison star. Assuming each slitlet is $5''$ long, which is long enough to sample the sky region around the star under typical conditions, then the number of rows shifted each step would be 80 (2 slitlets $* 5'' * 8$ pixels/$''$). The maximum rows available is 3280, meaning a maximum of 41 shifts before the CCD needs to be read out.

Table 3 shows the duty cycles for 'normal' and HIT mode spectroscopy together with the currently offered MS clock patterns. As can be seen from this table, for short integration times, the HIT mode is almost fifty times more efficient than using the traditional 'single exposure + readout' pattern. Even for wait times as long as 20 seconds, the HIT mode remains 3 times more efficient. In addition to these gains in efficiency, the HIT mode does not suffer from long deadtimes between exposures, making it possible to probe the variability on timescales of the order of the exposure (or 'wait') time, which is not possible with other (more traditional) modes of operation.

3 Characteristics of the HIT Sub-Modes

3.1 Timing Accuracy

The time signal of the observatory clock is distributed in two ways; firstly, via Network Time Protocol (NTP) for systems needing an internal accuracy of ~ 10 milliseconds and secondly, via dedicated timing (TIM) boards that deliver an internal accuracy of microseconds. These times are synchronised to a rubidium clock that is in turn synchronised annually via a GPS receiver. This system is expected to give an absolute accuracy on the timestamp of around 1 millisecond or better. The timestamp is placed on each frame and represents the start time of the exposure (in fact it is the time when the shutter has just begun to open, which is when the shifting starts). In order to calculate the time for a given pixel it is also necessary to know how many shifts have taken place.

In order to determine the timing accuracy of the HIT mode, we illuminate the CCD using a small bundle of LEDs that are set to trigger on the 1 pulse-per-second signal from the TIM board and last for 200 milliseconds. We then record a number of frames with the HIT-OS1-1sec mode and determine the position of the leading edge of the pulse. If we assume that our accuracy is better than one second, then any residuals from a integer value allow us to determine the accuracy of the time-stamps relative to the observatory clock and, in turn, UTC. The results of three such runs are shown in Fig. 4. As can be seen, the absolute timing accuracy (as given by the mean offset) and the relative timing accuracy (i.e. one stamp relative to another, as given by the square root of the variance) are $\sim$ 50 & 7.7 milliseconds respectively. This means that there is an instrumental delay of 50 milliseconds that can safely be removed as an offset for a given set of observations (and can be calibrated before an observing run). However, the 7.7 millisecond relative delay is the 'jitter' between frames that cannot be removed. This means that between two frames (or, equivalently, lightcurves) there is a random offset of $\sim$11 milliseconds.

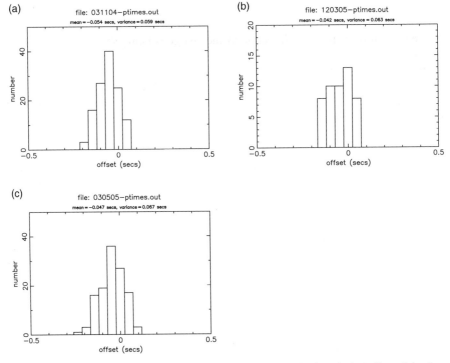

Fig. 4 Histograms of the timing marks used to determine the absolute timing offset of the data and the variance in that value. The data were taken on: (**a**) 3rd November 2004 (top left); (**b**) 12th March 2005 (top right); (**c**) 3rd May 2005 (bottom left)

3.2 CCD Parameters & Noise Model for the HIT Mode

The FORS2 detector consists of two 2k × 4k MIT CCDs (15 μm pixel size). The MIT CCDs were optimised to provide much higher response in the red wavelength range beyond 800 nm, with impressively low fringe amplitudes. However, this was at the expense of the response below 400 nm, which makes them sub-optimal for many HTRA applications. The standard CCD parameters are summarised in Table 4. As stated previously, the CCD is operated with 2 × 2 binning as standard.

The noise model for the HIT mode is the same as that for the standard spectroscopic readout mode. The only additional source of noise is simply due to the longer dark time for the exposure. In the case of the longest mode (HIT-MS8-20sec) the charge remains on the chip for 1000 seconds, so will typically incur a penalty of ~ 1 e^-/pix.

3.3 Image Motion

Image motion within the slit is a major source of uncertainty for the photometric accuracy of the HIT-I and HIT-OS modes, but not necessarily for the HIT-MS mode, as it is possible to use narrow slit widths. If the direction of motion is along the slit, then the variability is simply seen as image motion. However, if the image motion is across the slit, then the effect is seen as a series of maxima (at times when the image motion is in the same direction as the charge is being shifted) and minima (when the motion is in the opposite direction) in the lightcurve. This can be seen in the example below where a standard star was placed in a 5″ wide slit using the HIT-OS1-1sec mode.

In order to study the effects of the image motion on the lightcurves, we have fit the profile of the star at each time step using a minimum chi-squared goodness of fit approach. The model profile used was a Gaussian with the peak, centroid and width as free parameters. The best fit values for each of these can be seen in the top, lower-middle and lower panels of Fig. 5.

As the centroid has been derived from the 1-D profile of the star along the slit, it is fair to assume that this is also a plausible model for the motion in the direction **across** the slit, which is otherwise impossible to determine. Likewise, it is also reasonable to assume that the seeing varies symmetrically, so that the variability of the FWHM of the seeing (or more accurately the measured image

Table 4 CCD parameters for the '100kHz, 2 × 2, high' readout mode of the FORS2 detector; source FORS user manual and ODT pages

Parameter	Unit	Chip 1	Chip 2
read-out noise	ADU	2.7	3.0
gain	e^-/ADU	0.7	0.7
Dark current @-120°C	e^-/pix/hour	~ 3	~ 3
linearity	% RMS	0.1	0.23
readout time	seconds	41	

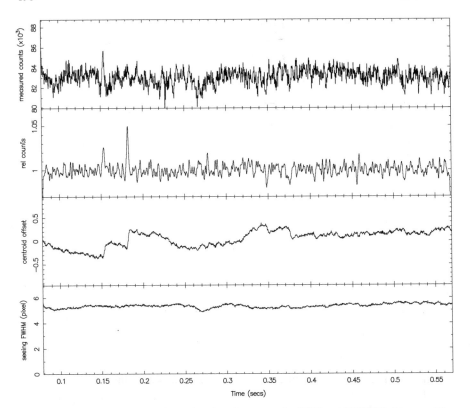

Fig. 5 The effects of image motion and seeing changes on the HIT-I and HIT-OS lightcurves. The upper panel shows the flux lightcurve extracted by fitting the seeing profile from an observation of a photometric standard star. The upper-middle panel shows the model lightcurve using the parameters from the fit (as described in the text). The lower-middle and bottom panels show the position of the centroid (in pixels relative to the first point) and the FWHM of the measured seeing profile respectively. The image scale is 0.25 "/pixel

quality) is qualitatively the same in both directions. For these reasons, we have used these parameters to investigate the effects of image motion and seeing on the observed flux from a non-variable source, e.g. a standard star. The model lightcurve is created from the parameters described above by projecting the 'star' through a 'slit' and calculating the expected flux at each time-interval/shift. The resulting lightcurve for a 5″ slit width (the same as in the real observations) is shown in the upper-middle panel of Fig. 5. There are a number of interesting features in this lightcurve;

- $t = 0.155$ seconds: This feature is present in both the observed and the model lightcurves and seems to be caused by the rapid image motion within the slit. This indicates that the motion took place both along and across the slit.
- $t = 0.18$ seconds: This feature is only present in the model lightcurve and again seems to be due to the image motion. However, as it is not present in the observed lightcurve, we can infer that the motion was only along the slit.

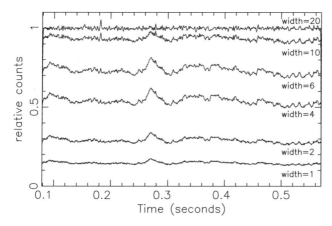

Fig. 6 The effect of the slitwidth on the model lightcurves. The lightcurves have been calculated using a range of slitwidths from 1–20 binned pixels, which corresponds to 0.25″–5″

- $t = 0.27$ seconds: The dip in the observed lightcurve is associated with a slight improvement in the 1-D seeing. However, such an improvement should not have any effect on the lightcurve, due to the large slit width. We can infer from this that either the seeing degraded significantly in the direction across the slit, leading to slit losses, or the transparency changed briefly.

The effects of the slit width can be seen in Fig. 6, where the parameters from the fitting mentioned above are again used to simulate a lightcurve. This time the slitwidth in pixels is varied from 1 to 20 pixels (0.25″ to 5″). When the slitwidth is 20 pixels no major deviations from unity are seen, indicating no slit losses. However, even with a slitwidth of 10 pixels some losses are seen except in the case where the seeing is at its best. This trend continues until the slitwidth becomes comparable to the seeing when it again becomes rather insensitive to the seeing (and image motion), but the average value is by this time far from unity, indicating a large slit-loss.

4 FO Aqr—a Worked Example

FO Aqr belongs to the sub-class of Cataclysmic Variables (CVs) called Intermediate Polars (IPs). IPs contain a magnetic white dwarf which is accreting material from a low-mass companion. FO Aqr has a 291 minute orbital period, which is clearly seen in photometry and spectroscopy of the system. In addition to this there is a clear 20.9 minute 'pulsation' period that is caused by the spin of the white dwarf, that has been spun up to these speeds by accretion torques. The inner accretion disk is truncated by the magnetic field of the spinning white dwarf which causes a modulation of the optical (and X-ray) flux on the spin and spin-orbital beat frequencies. This range of phenomena made FO Aqr an ideal target for highlighting the capabilities of fast spectroscopy on a complex but relatively well understood system (see e.g. [4, 5]).

We observed FO Aqr between UT07:42-10:37, 07 August 2005 using the HIT-MS mode with a 'wait' time of 2.5 seconds. This resulted in a total of 3224 spectra of FO Aqr. We observed a simultaneous comparison star which was used to correct for slit losses and slight wavelength drifts due to high airmass of the source at the time of observation. We used the 600B grism and covered the range 380–675 nm. A slit width of $1''$ was used resulting in a resolution of 780 at the central wavelength of 465 nm.

4.1 Data Reduction

The steps needed to reduce HIT-MS mode data are similar to those needed for any other spectroscopic observation. Each image can be thought of as a MOS image, with identical pairs of spectra taken during each wait time. A simple recipe for the data reduction based loosely around the steps used in PAMELA and MOLLY reduction software[3] would be;

- **Bias subtract**—The CCD is read-out in a completely standard way once the charge transfer steps have been completed and the shutter closed, so standard bias frames can be used.
- **Flatfield**—The flatfield frame for the HIT-MS mode is different to other modes, as the effects of charge traps, etc. are spread over a number of rows due to the charge transfer steps. Special screen flatfield frames are taken during the daytime with the telescope at zenith.
- **Target location**—The centroid of the spatial profile of the target and comparison star are determined by fitting a Gaussian profile to the collapsed 1-D spatial profile
- **Trace dispersion**—the tilt of the spectrum (i.e. the displacement of the centroid of the spatial profile along the columns) is traced to ensure that the sky spectrum produced is the correct one. The tilt on the spectrum can be large between the two extremes of the spectrum ($\sim$ 20 pixels) and must be taken into account.
- **Sky spectrum**—a 2-D image of the sky is created from regions sufficiently far away from the stellar profiles. This process also takes into account the tilt of the spectrum.
- **Spectral extraction**—The spectra of the target and the comparison star are extracted using either normal or optimal extraction routines that use information from the spectral profile, the trace and sky images.
- **Wavelength calibration**—The spectra are calibrated in wavelength using arc lamp spectra taken during the daytime with the telescope at zenith. Further calibration can be performed using night sky lines and/or any features in the spectrum of the comparison star.

---

[3] Use of the MOLLY and PAMELA software developed by T. R. Marsh is gratefully acknowledged.

- **Slit losses**—The relative slit losses, due to the changing conditions, can be compensated for by fitting a low order polynomial to the comparison star spectrum and comparing these to the flux level in the average spectrum. In addition, if a spectrum of the comparison has been taken through a wide slit in photometric conditions, it is possible to calibrate the spectra on an absolute scale. As there will almost certainly be a discrepancy between the wavelength coverage of the target and the comparison star, it will be necessary to extrapolate the polynomial fit to cover the entire spectral range of the target star. As this is a smoothly varying function over the possible wavelength ranges, this should still give a satisfactory correction.
- **Flux calibration**—If a spectro-photometric standard star was observed during the night, it is also possible to extract its spectrum using the same calibration steps and determine the absolute flux scale. As in the previous step, the standard star may not have the same wavelength range as the target star and the calibration will need to be extrapolated.

As can be seen in Fig. 7, there are a number of different types of variability that can be seen in the trailed spectrograms of FO Aqr. In the Balmer series of Hydrogen, clear S-waves can be seen in the phase-binned data which are $180°$ out of phase. In the He II ($\lambda 468.6$ nm) and He I ($\lambda 447.2$ nm) trailed spectrograms a number of knots can be seen, as well as transient higher velocity features, possibly originating from the same locations as those described by [5]. The bowen blend line profile is interesting and complicated by the multiple components that comprise the blend. I have chosen $\lambda 464.0$ nm as the central wavelength (one of the NII lines), although it is clear that other lines of the blend might be more suitable (e.g. $\lambda 464.7$ nm CIII). Further analysis is needed and is beyond the scope of this work (see e.g. [6]). The primary goal of showing such data is to show the excellent quality that can be obtained.

5 Future Upgrades

The HIT mode is now fully operational, and it is possible to imagine a number of different ways it could be improved upon. However, it is only possible to make wholesale changes to the instrument on the request of the ESO community, so it is impossible to tell at this stage which ones are likely to be put into effect. Possible upgrades include;

- A new high-throughput, high dispersion grism more capable of resolving the velocity profiles in Interacting Binaries.
- An improved time stamping system that has a direct connection to GPS in order to give an independent measure of the arrival time of the photons and remove uncertainty in the accuracy of the current time stamps, which are vital to many of the scientific applications.
- An EMCCD detector for photon counting spectroscopy and/or continuous readout. This would enable the user to perform fast spectroscopy without the need

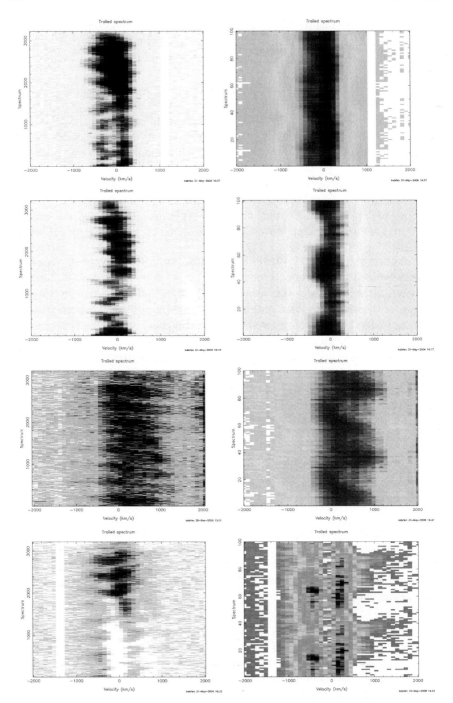

Fig. 7 Trailed spectrograms of (from top to bottom) Hβ (486.1 nm), He II (468.6 nm), the bowen blend (464.0 nm) and HeI (447.2 nm). The left-hand panel shows the un-binned trailed spectrogram and the right-hand panel shows the same data folded on the spin period of the white dwarf (the complete spin cycle is repeated for clarity)

to stop and read-out the CCD every few tens of frames. This would increase the duty cycle even closer to 100% and ensure evenly sampled lightcurves.
- The addition of imaging and/or spectro-polarimetry, which would require a hardware upgrade of FORS2.

6 Conclusion

The HIT mode of FORS2 is in fact three separate modes, one for imaging and two for spectroscopy. In imaging it is capable of taking images of variable objects with a time resolution in the range 0.6–156 milliseconds through any of the filters of the FORS filter-set. The images are taken through a pseudo-longslit formed by the moveable jaws of the MOS unit and the charge is continuously shifted from the exposed region and stored in the masked region. Once this region is full the CCD is read-out and the process begins again. The only deadtime that occurs is during the read-out of the CCD.

For spectroscopy there are two further modes. The first uses the same clocking scheme as in the imaging mode and hence offers the same time resolution. In addition to this is a second mode that operates a 'shift-and-wait' scheme, whereby the image of the slit is shifted very rapidly into the masked region of the CCD, thus exposing a new region under the slit. This region remains exposed for a defined amount of time before it too is shifted into the masked region. This process again continues until the masked region is full and the CCD is read out in the usual way.

The HIT mode is the only mode of its kind available on an 8 m class telescope and as such is a unique capability. I have summarised the issues and potential limitations of the mode as well as show some of the promise of the mode. I hope that in the future the HIT mode will be used in a number of applications and will help to show the strengths of the high time resolution spectroscopy with large telescope both in the 8 m and ELT eras.

Acknowledgments The author would like to thank the many people that have helped and continue to help with the implementation of the HIT mode on Paranal. These include Thomas Szeifert, Nicolas Haddad, Pedro Baksai, Mario Kiekebusch, Claudio Cumani and Karl-Heinz Mantel as well as many other past and present members of the FORS Instrument Operation Team. In addition, I would like to thank Jason Spyromilio and Andreas Kaufer for allowing me to experiment with the instrument.

References

1. D. Ives, T. Bennett & F. Gribbin, 1996, *New CCD Readout Mode (Low Smear Drift Mode) with Absolute Time Stamping*, La Palma Technical note #106
2. R. A. Stathakis & H. M. Johnston, 2002, *The RGO Spectrograph Manual*, http://www.aao.gov.au/AAO/local/www/ras/rgo/rgo.html

3. K. O'Brien, 2000, *X-ray and optical variability of X-ray binaries*, PhD thesis, University of St. Andrews
4. C. Hellier, K. Mason & M. Cropper, 1990, MNRAS, 242, 250
5. T. Marsh, 1996, NewAst, 1, 97
6. C. Hellier, 1999, ApJ, 519, 324

An Ultra-High-Speed Stokes Polarimeter for Astronomy

R. Michael Redfern and Patrick P. Collins

Abstract Optical polarization is a valuable diagnostic tool in astrophysics, frequently enabling asymmetries in source regions, magnetic field configurations, and magnetic field strengths to be investigated. In the context of high time resolution astrophysics polarimetry is particularly valuable in understanding the nature of the optical pulsations from pulsars, and whilst the averaged (over many cycles) linear polarization of the Crab nebula pulsar was determined many years ago, no attempts have been made to determine the Stokes vector of single, individual, pulse cycles. This may be very important information in the investigation into the nature of the enhanced optical pulses associated with random giant radio pulses. In the case of the Crab nebula this requires an instrumental response time of no more than 100 μs, and, possibly, the capability of tagging individual photons with nanosecond resolution. This imposes certain constraints on the design of the polarimeter, which rules out all of the popular designs. We describe the design of a novel polarimeter which will be able to meet these constraints. This is based upon a design by Compain and Drevillon [1] in which their design is modified, by the use of a different glass and prism angles, to work over a wide bandwidth of 400–800 nm and with a high polarimetric efficiency.

1 Introduction

A complete description of a light beam requires that the intensity and the polarization state be specified as a function of wavelength and of time. A convenient way of doing so is by means of the Stokes vector **S,** which has four components: I, Q, U and V

$$S = (I, Q, U, V)^T = (\langle I_x + I_y \rangle, \langle I_x - I_y \rangle, \langle I_{+45} - I_{-45} \rangle, \langle I_R - I_L \rangle)^T, \quad (1)$$

R. Michael Redfern · Patrick P. Collins
Physics Department, National University of Ireland, Galway, Ireland
e-mails: mike.redfern@nuigalway.ie, p.collins1@nuigalway.ie

where T denotes transposition, I_x, I_y, I_{+45}, and I_{-45} are the intensities of the linear polarization components of the input beam in the respective directions, and I_R and I_L are the intensities of the right and left hand circularly polarized components, respectively.

A quarter wave plate of uniaxial material — in which the optical path length for light plane polarized along one of its two orthogonal axes is greater than the other by $\lambda/4$—converts linear to circular polarization and vice versa, and is commonly used in Stokes polarimeters. Half wave plates introduce an optical path difference, defined in the same way, of $\lambda/2$. Since linearly polarized light can be split into orthogonal components by uniaxial prisms, which can then be separately determined, this effect enables the circular component to be measured. Descriptions of the physics involved, and detailed discussions of astronomical polarimetry and modern astronomical polarimeter design can be found in the literature [2], [3], [4], and [5]. It is not therefore proposed to spend much time in this article in describing the basics—rather, discussion will be particularly concerned with those aspects which are unique to high time resolution instruments.

It is true to say that attention has been largely focused upon characterizing the linear component of polarization. This is because it arises from many physical processes in the source (synchrotron emission, for example, in an oriented magnetic field component normal to the line of sight), or in transmission (by scattering, or by transmission through oriented dust grains in the interstellar medium)—all of which implies an asymmetry, which can be detected and modeled from the linear polarization.

Circular polarization arises from emission occurring in magnetic fields. A thorough discussion of mechanisms that produce linear and circular polarization may be found in, for example, Angel [6]. Circularly polarized bound-bound emission may be observed because of the Zeeman effect, and can be used to measure magnetic field strengths. In addition, mechanisms exist whereby free-free thermal emission in an intense magnetic field can also become circularly polarized to a high degree [7]. The observed degree of circular polarization (Stokes V) is generally small, but in magnetic white dwarf stars very high degrees, up to >20% may be observed [8], [9], and references therein. Free-free emission is thought to provide the main explanation, although bound-bound and bound-free mechanisms must also be involved. A similar mechanism might also account for the circular polarization observed in X-Ray pulsars [10]. X-Ray pulsars are thought to be magnetic neutron stars with fields up to 10^{12} G, so that accreting matter is constrained to the poles producing intense thermal bremmstrahlung—this model has been suggested to account for Cen X-3 and Her X-1.

Observations of the short-time variability of the Stokes vector, or, indeed, in just the degree and angle of linear polarization, has been largely neglected up to now. In only a few cases is the short-term variability considered to be of consequence. Chief amongst these are (a) short period accreting systems, such as magnetic cataclysmic variable stars (mCVs)—which vary on a time scale of seconds and minutes and display a high degree of linear and circular polarization—and (b) optical pulsars, which have been observed to display a high degree of linear polarization and can vary on a time scale of microseconds—circular polarization has not been measured, but can be presumed to vary on similar timescales. Both of these subjects are dealt with in detail in the next section. There are other sources where a rapidly varying Stokes vector can be observed, and is important to the understanding the physics of the source, for example in the Sun, but the design requirements are so different—the Sun is an intense extended source—that only mCVs and pulsars will be considered.

2 Sources of Polarized Radiation at High Time Resolution

2.1 Magnetic Cataclysmic Variable Stars

There are a considerable number of recent references in the literature to the polarimetry of magnetic cataclysmic variable stars—[8], [9] and references therein. Magnetic cataclysmic variable stars are one of the few types of object for which rapid variability of the Stokes vector can be observed.

Magnetic cataclysmic stars (mCVs, also called polars and AM Her systems) are accreting binary systems with a white dwarf primary and a late-type main sequence secondary, in which accretion onto the primary from the inner Lagrangian point, L_1 is dominated by a large magnetic field in the 10–200 MG range so that the accretion stream traces the geometry of the magnetic field. Reviews of the properties of polars can be found in Cropper [11] and Warner [12]. Accreting material, near to the white dwarf surface, is fully ionized in the presence of a strong magnetic field, producing high degrees of linear and circular polarization. Since the optical flux largely comes from this region, where the accretion stream strikes the surface, polarimetry is a crucial diagnostic tool in studying these objects.

Stokes parameters are very diagnostic of the state of the accreting region because they depend upon the angle by which the region is observed as well as on its physical properties. If the accretion region is small, the orbital behaviour of the linear polarization position angle can be described by a simple equation that depends upon the inclination of the system, ι, and the colatitude of the axis of the magnetic field, β. Besides the quantities mentioned above, the phase interval during which no cyclotron emission is observed also constrains ι and β. Estimates of the magnetic field as well as of some other plasma properties can be obtained through the modeling of flux and polarization variations with orbital phase [13], and references therein, on accreting column cyclotron models [9].

AM Her systems show orbital periods ranging from few thousand seconds to a few tens of thousands of seconds (AM Her itself has a period of $\sim$11,000 s), and accretion streams are thought to be of the order of 1% of the primary radius. One might therefore expect Stokes variability to be manifest on a time scale of seconds, unless there are magneto-hydrodynamic instabilities in the accretion column itself. In high inclination systems, the white dwarf itself, the magnetic and ballistic accretion streams and the accretion hotspot are eclipsed by the secondary, producing structured photometric eclipse profiles—offering an opportunity to investigate these spatial structures and the binary parameters. Polarimetric changes due to eclipse and/or viewing angle changes offer further insights into mode of accretion onto the white dwarf [14], [15]. Various authors have attempted to "inverse map" periodic fluctuations in the Stokes vector to generate images of the intensity and distribution on the surface of the white dwarf star.

The "Stokes imaging" method [15] allows objective mapping of the cyclotron emission regions in mCVs in terms of their location, shape, and size, see, for recent examples [8] and [9]. Stokes imaging uses broadband emission, 350 nm–900 nm in the case of Potter et al., this being the unfiltered response bandwidth of the GaAs photomultiplier used in the University of Capetown Polarimeter (UCTPol), [16]. This was necessary to achieve sufficiently high signal to noise with small telescopes (SAAO 1.9 m telescope [8]), and 1.6 m Perkin-Elmer Telescope at the *Observatorio do Pico dos Dias,* Brazil [9]. Both authors report very high levels of linear and circular polarization from AM Her stars, and are able to produce credible tomographic maps. Figure 1 below (from [9]) shows the very high degree of polarization observed at certain phases of the *ROSAT* polar candidate 1RXS J161008.0+035222.

A different method—"Zeeman tomography" [17],[18],[19]—inverse maps synthetic Zeeman line profiles obtained from modeling various binary parameters along

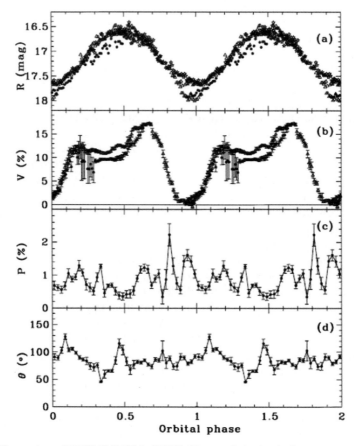

Fig. 1 Observations of 1RXS J161008.0+035222. Linear polarization is shown as percentage polarization and angle, rather than as Stokes Q and U [9]

with accretion stream geometry and magnetic field strength. The authors were able to obtain impressively detailed tomographic maps, but are much more limited in their ability to track short timescale changes because of the much higher fluxes demanded by the high resolution spectro-polarimetry which they performed. In fact, they were only able to do so using the ESO FORS1 spectro-polarimeter on the 8.2 m VLT.

2.2 Optical Pulsars—the Crab Nebula Pulsar

There are only 5 known optical pulsars, nevertheless they represent a most intriguing problem for ultra-high speed polarimetry—see Table 1, below, from Shearer (this volume)

The Crab nebula pulsar stands out, because of its young age and closeness, as by far the brightest and therefore by far the best candidate for detailed polarimetric analysis. Polarization of the optical emission was discovered by Wampler, Scargle & Miller [20]. Even for the Crab pulsar, the photon flux is so low that phase averaging techniques over many pulsar cycles are necessary to obtain an acceptable signal to noise, and the literature also contains measurements of Stokes I, Q and U, but not V. A prototypical observation of Stokes I, Q, and U was made using the 2.5 m Isaac Newton Telescope in 1985 [21]. Because in this observation the authors were able to measure the background and instrumental polarization they were able to report that polarized emission can be detected throughout the whole pulse cycle, and that the plane of linear polarization makes two complete revolutions per cycle—which they interpret as being due to detectable emission from both the north and south hemispheres of the pulsar magnetosphere. Figure 2, from Smith et al. [21] shows this effect.

Subsequent observations [22], [23] show a similar picture, extending into the UV. Details and limitations of the polarimeters that were used will be discussed in the next section, which deals with design requirements for an *ultra-high speed* polarimeter, but all of these observations made use of a phase averaging technique and report the linear polarization averaged over many, perhaps millions, of cycles.

An important property of the Crab pulsar is that it exhibits the giant radio pulse (GRP) phenomenon, in which there is a sporadic and random emission of intense pulses on average 1000 times larger than normal pulses. This has been known since the discovery of the Crab pulsar in 1968 [24], and was, indeed, the reason why the

Table 1 The five known optical pulsars

Pulsar	B Magnitude	Period(ms)	B Photons per rotation	
			8.2 m VLT	42 m E-ELT
Crab	17	33	3,300	85,000
PSR 0540-69	23	50	17	450
Vela	24	89	12	310
PSR 0656+14	25.5	385	13	340
Geminga	26	237	5	130

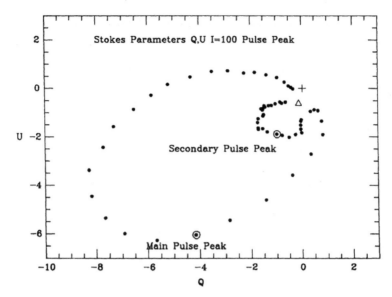

Fig. 2 Stokes parameters Q, U plotted as a vector diagram for the Crab nebula pulsar [21]

discovery was made in the first place. GRP come in short episodes, about 5 to 20 minutes in duration, and appear extremely prominent during such phases [25]. The pulse structure varies as a function of frequency, with varying power in various components, namely the main pulse (MP), interpulse (IP), precursor to the main pulse (P), a low frequency component (LFC) and two broad high frequency components (HFC1, HFC2). At 8.3 GHz the histogram of the peak strengths can be described by a power law spectrum with slope $\sim$–3.34 [25], and this is roughly consistent over GRPs coincident with all of the normal components. GRPs are very strongly polarized—almost 100% linear polarization has been derived by Karastergiou et al. [26], although Moffet & Hankins [27] found that GRP coincident with the IP were polarized at 50% and HFCs at 70–80%—still strongly polarized. It is not clear, however, what is the relationship between GRPs and non-thermal high-energy emission. Some observations suggest that the Crab exhibits GRP at phases where no high energy emission is known [25], but other observations contradict this. There is a correlation between X-Ray and radio pulses for Vela [28], and Shearer et al. [29] detected a correlation between GRP emission and optical emission. They found that optical pulses coincident with GRP were $\sim$3% brighter on average—roughly an energy excess similar to that of a typical GRP.

Recently, Hankins et al. [30] reported on observations of GRP from the Crab made with extremely high time resolution at very high radio frequencies (5.5 and 8.6 GHz) and with high bandwidth (0.5 and 1 GHz). They found that even over short timescales of a few minutes the arrival time of GRP jitters by several hundred microseconds in phase (that is to say, relative to a fixed point on the pulsar surface), but that pulse phases were confined to either the MP or the IP. Most of the pulses showed one or more noisy bursts separated by fractions of a microsecond. It was clear that there was structure at the limit of the receiver bandwidth (Δt=2 ns) and that the emissions were strongly circularly polarized, either right or left handed, and the handedness varied randomly on nanosecond timescales. Figure 3, below, from Hankins et al., illustrates the intensity, polarization and time structure of subpulses. In the bottom panel, illustrating the polarization structure, the RMS noise is 18 Jy.

The relationship between GRP and the associated optical pulses is not at all clear, but what is clear is that polarization is extremely important to the understanding of the GRP phenomenon in the radio region, and it is also likely to be important to an understanding of the associated optical phenomenon—even though no plausible physical model for the connection exists at present. Furthermore, whilst the averaged GRP is, on average, a few thousand times larger than a normal pulse, the nano-pulses may be thousands of times larger again, making them briefly the brightest objects in the sky. This suggests the intriguing possibility that the extra 3% observed by Shearer et al. may itself have nano-pulse structure. The requirements to investigate optical counterparts to GRP and possible nano-pulse structure therefore preclude the sort of phase-averaged polarization measurements which have been made up to now and require time resolution of the order of microseconds for individual pulses and photometry with an absolute time resolution of nanoseconds for correlation with radio nano-pulse structure, if that is possible.

A distinction must of course be made between radio and optical polarimetry. Radio receivers are able to amplify the incoming EM-waves for two orthogonal directions, and the amplitudes are recorded directly. In the optical the photon energies are large and, with photon counting devices it is obviously not possible to measure the polarization of a single photon, so that polarization measurements rely upon

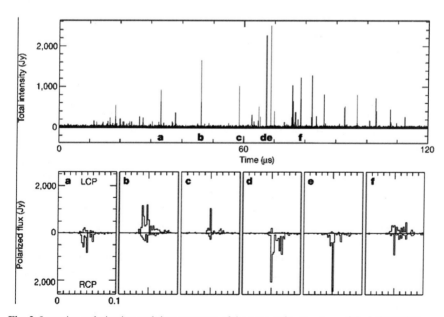

Fig. 3 Intensity, polarization and time structure of the nano-pulse structure of Crab GRP [30]

the detection of several photons, and measurements are subject to Poisson fluctuations. This is the main reason why this undertaking is so challenging. Polarization measurements of single GRP nano-pulses are of course entirely impossible in the visible region, although the average Stokes vector on longer timescales within a whole GRP could be determined with reasonable precision—as discussed in Sect. 6, below—given a sufficiently large telescope.

3 Requirements for an Ultra-High-Speed Stokes Polarimeter

The design requirements discussed below form the major drivers for the Galway Astronomical Stokes Polarimeter.

3.1 Speed

In the previous section reference was made to a requirement, at the very least, to resolve structure within an individual cycle of the Crab nebula pulsar whose period is ∼33 ms. Within the pulse structure both the main pulse and interpulse have structure at least down to a few hundred microseconds [31], [32]. This entirely precludes the designs of typical polarimeters—such as the Durham electronographic polarimeter, (DEP) [33], UCTPol [16], ESO FORS1 on the VLT, etc.—in which a rotating phase retarder acts in front of an analyser (usually a Wollaston or Foster prism), so that the two resulting polarized beams can measure Stokes I, U and Q (a half wave plate) or Stokes I, U, Q and V (quarter wave plate). In the case of imaging polarimeters like DEP or FORS1, half of the field must be obscured in strips corresponding to the position of the second image formed by the analysing prism. In order to obtain a full field the telescope would then need to be slewed by one grid spacing so as to image the previously obscured half field. Of course, high speed sources are likely to be point sources, but even so the requirement to image only half the field would make it difficult to get good background measurements in cases, like the Crab nebula, where the sky background is substantial. In the case of UCTPol a quarter wave plate is rotated in front of the analysing prism, and the resulting modulated light is detected by (non-imaging) photomultiplier tubes.

The main requirement is to measure the Stokes vector within a single, randomly chosen, Crab main pulse. There must be sufficient, intrinsic, time resolution to determine the Stokes vector within a period of one hundred microseconds. The ability to make this measurement will, of course, also depend upon having a telescope of large aperture and detectors of high detection efficiency to reduce Poisson fluctuations. This is discussed in Sect. 6, below.

A secondary requirement is the ability to look for overall photon arrival time correlations with GRP nano-pulses selected over a whole observing run by simultaneous radio observations. This imposes a stronger requirement on detector time resolution, unfortunately ruling out the use of anything other than true photon-counting detectors.

The speed requirement rules out any polarimeter design in which the components of the Stokes 4-vector are measured with mechanically moving or rotating components.

3.2 Throughput

The two classes of object identified in Sect. 2 as being promising candidates for high time resolution polarimetric observations are extremely faint—white dwarfs because they are intrinsically faint, and optical pulsars because they are rare events and therefore likely to be both distant and old. In the case of optical pulsars, Table 1 gives the numbers of photons to be expected with a perfectly efficient instrument on a large telescope, and this corresponds to no more than 40 photons per 100 microseconds on a (currently available) large telescope. It is therefore clear that any polarimeter for high time resolution astrophysics must have an extremely high throughput. This rules out designs involving linear dichroic filters, for example, or, indeed, any design in which all of the photons are not detected and used, and dictates the use of high quantum efficiency detectors. High throughput also implies high polarimetric efficiency as well as the maximum detection of photons. This will be addressed in Sect. 4, below.

The design for a high speed Stokes polarimeter must not waste photons, should have a minimum of optical surfaces, and must be designed to maximize polarimetric efficiency.

3.3 Optical Bandwidth

The primary difficulty in making high time resolution polarimetric observations is lack of photons. Even the brightest sources likely to show rapid variability are still relatively faint. This means that a high speed polarimeter must operate over a wide bandwidth and have wide bandwidth detectors.

A polarimeter for high time resolution astrophysics must have as wide a bandwidth as possible, consistent with maintaining efficient measurement of the Stokes 4-vector. The aim, discussed in the next section is to achieve a bandwidth of at least 400–800 nm.

3.4 Imaging

Shearer et al. [34] showed that aperture photometry is most sensitively performed with a 2-d imaging photometer, and with photon-counting, recorded, post-processed data. This is particularly true in cases where there is a strong nebular background—as for the Crab nebula pulsar. The problem is that where the usual fixed aperture is defined by a diaphragm, or by the entrance pupil of an optical-fibre-coupled photon counting device (as in the OPTIMA [23] or TRIFFID [29] high speed photometers), the aperture must be made very large to take into account varying atmospheric conditions, and also to be able to encompass guiding errors and telescope wobble. In order to exclude the nebular background by having an optimally small aperture to maximise signal-to-noise, which nevertheless can track small centroiding errors, the aperture needs to be adaptive. Such an adaptive aperture can most easily be defined by post-exposure data selection from a photon-counting 2-d imaging field, and very large (>25) signal-to-noise ratio improvements can be achieved compared to conventional single-pixel photometry.

Indeed, if the object of interest is faint, and can only be detected after post-processing of the data, there is a danger that a fixed aperture of small size (required to maximise signal-to-noise) may not be centered on (and may not even include) the object at all if the astrometry is at all uncertain. This danger is obviated by data selection from a relatively large 2-d field, with the required astrometry being performed off-line rather than in real-time.

The field size will depend upon the target. Most mCVs are single, isolated stars, so that a field of 15" would be sufficient to encompass seeing and pointing and tracking errors, in all but the very worst seeing conditions. This would also be sufficient for the Crab nebula pulsar, where there is a very convenient reference star, of similar magnitude to the pulsar itself, $\sim$5" away (Trimble 21). Of course, initial acquisition may be very difficult with a field as small as 15", and so a desirable additional feature would be an acquisition/guiding channel with a larger field of view.

The design for a sensitive, high-speed Stokes polarimeter must be based upon a system in which a 2-d image can be post-processed by data selection to define an optimum aperture, to track centroiding errors, or to search for the detection of a faint object.

4 Design of the Galway Astronomical Stokes Polarimeter (GASP)

The need for fast and rugged polarimeters—with no moving parts or modulators—has prompted the development of many polarimeter designs (see, for example, [35],[1], [36], and references therein). As in (1), the polarization state of a beam is defined by the Stokes vector **S**,

$$S = (I, Q, U, V)^T = (\langle I_x + I_y\rangle, \langle I_x - I_y\rangle, \langle I_{+45} - I_{-45}\rangle, \langle I_R - I_L\rangle)^T,$$

where T denotes transposition, I_x, I_y, I_{+45}, and I_{-45} are the intensities of the linear polarization components of the input beam in the respective directions, and I_R and I_L are the intensities of the right and left hand circularly polarized components, respectively.

In order to determine these four components the beam must be divided into four components with intensities i_{1-4} (forming the output column vector of intensities, **I** = $(i_1, i_2, i_3, i_4)^T$), which must be linearly dependent in some way upon all of the components of **S**. The division can be done by many methods, which fall into two classes—these are called (by analogy with interferometers), Division Of Amplitude Polarimeters (DOAP) and Division Of Wavefront Polarimeters (DOWP), along with separation into orthogonal linearly polarized beams, and with quarter wave phase retardation (to turn circular into linear polarization) in two of the beams. **S** may then be derived, in certain circumstances, from **I**. The relationship between the input and output vectors is given by the system matrix of the polarimeter, which is a real 4×4 characteristic matrix **A**:

$$S = A^{-1}I \tag{2}$$

The components of **A** are determined by calibration. If **A** can be inverted, then **S** can be determined from **I**.

The design of GASP is based upon a DOAP concept described by Compain & Drevillon [1]. Figure 4, shows the principle of the design, which is based upon a modification of the Fresnel rhomb, which was the first broadband quarter wave retarder, described by Fresnel in 1817, (see Fig. 5). Reflections at A and B introduce a phase delay to the component of the light polarized parallel to the surface, which is dependent upon the refractive index of the glass. For glasses of reasonable refractive index this can be 45° at each reflection—producing an effective quarter wave retardation overall (see Fig. 6, below). The modification utilizes the reflection of light on uncoated dielectric surfaces at oblique angles to create a beam splitter, thus the modified rhomb or DOAP prism acts as a beam splitter and a broadband

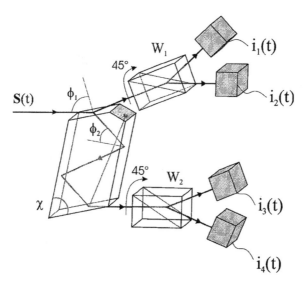

Fig. 4 Principle of a visible-near-infrared DOAP with an uncoated dielectric prism. The light beam is separated first into two by a prism, and then into four by two Wollaston prisms (W1 and W2) orientated at 45° to the plane of incidence. The internal beam has a quarter wave uniaxial phase delay created by two internal reflections in the rhomb-type prism. S can be determined, in principle, from the ratios i_1-i_4. A small fraction (∼20%) of the energy is inevitably reflected at the second partial transmission and is absorbed by a blackened vertex of the prism. From [1]

Fig. 5 A Fresnel rhomb

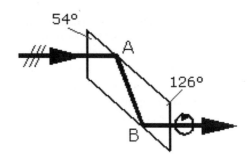

quarter wave plate, which we refer to as a retarding beam splitter (RBS). The result is a prism of rectangular cross-section and parallelogram top and bottom faces—like the original Fresnel rhomb, but with a grazing incidence beam incident on the top end of the prism. In this scheme the value of I+/−U is provided by Wollaston prism which separates beams i_1 and i_2. I+/−V is provided by the Fresnel rhomb and the Wollaston prism which separates beams i_3 and i_4. The modulation in Q is provided by the first grazing incidence reflection/refraction at the top face of the prism.

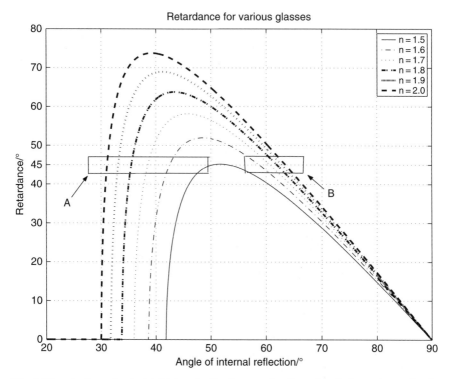

Fig. 6 The retardance due to total internal reflection as a function of angle for various refractive index glasses. Note that in region B the curves are more closely spaced, so that different refractive indices (over a range n=1.5 to n=2.0) require a smaller range of angles to produce 45° retardance, lessening the effect of dispersion

A critical requirement of the GASP is that it should be an effective polarimeter over a broad spectral range—this requires that precise quarter wave retardance be achieved over a broad wavelength range with realizable low-dispersion glasses. In addition, a requirement of the beam splitter design is for a prism angle so that the reflected and transmitted components are equal, while maintaining the necessary angle so that the internally reflected angle will give a 45° retardance as required. Figure 7 below shows the addition of these two requirements. We note that around n=1.59 the slope is approximately zero, therefore a low dispersion 1.59 index glass will have a large bandwidth and a retardance error <<1%. Commercially available glasses of high transparency can be found to meet this requirement.

In Fig. 4, it can be seen that there is an inevitable ~20% of the light reflected from the second face of the rhomb-type prism. In the original design one vertex of the prism was removed and the resulting surface roughened and blackened in an attempt to absorb this waste light, which might otherwise result in stray reflections. In the design of GASP an "extractor" prism was designed to produce transmission of 80% of the stray light into an external beam which could be used for finding, guiding, and/or photometry—but not to determine components of the original Stokes vector. The final waste light is absorbed by a blackened ground face as seen in Fig. 8.

A design of the GASP has been completed, as shown in Fig. 9. The performance of this design, both in the laboratory, and in practice, on a telescope, will be the

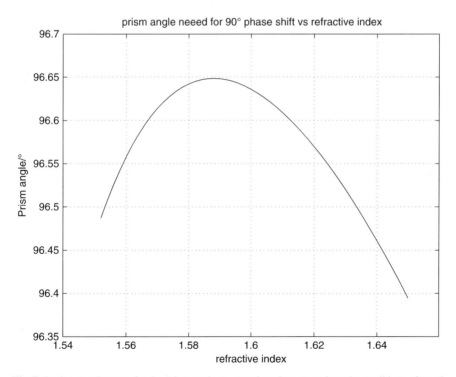

Fig. 7 Design requirement for the prism angle vs. wavelength to meet the twin conditions of equal transmitted and reflected beams and a precise quarter wavelength internal retardation

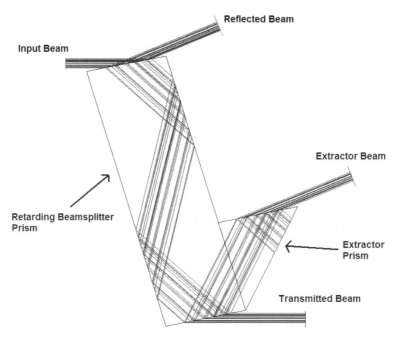

Fig. 8 Final layout of the retarding beam splitter and the extractor prism showing the input, reflected, transmitted and extracted beams. The input, collimated, beam is incident from the top left of the diagram and splits into two by refraction/reflection. The external, reflected, beam is split into two by a Wollaston prism at 45° (not shown) and re-imaged onto a detector (not shown). The internal (refracted) beam has two total internal reflections, which creates $\lambda/4$ retardance. Refraction/reflection at the bottom face creates an external (refracted) beam which is split into two by a Wollaston prism and re-imaged onto the same detector (not shown). The internal, reflected, beam passes into the "extractor" prism and is finally refracted/reflected at the top face to create an external beam which is of large field of view, which is re-imaged onto a separate detector (not shown) for finding/guiding and photometry. The internal, reflected, beam is absorbed by roughening/blackening of the bottom face of the extractor prism. The top and bottom faces of the rhomb-type prism and the top face of the extractor prism are parallel, so that the three external, collimated, beams are of the same geometry

subject of later articles. The important consideration of the degree of precision in the low photon flux range will be considered in a later section. The four primary beams are re-imaged onto a single detector at the same magnification, but the light path of the (waste light) extracted beam is not shown for clarity. It is a particular difficulty of this sort of detection scheme—in which the four components of **S** are determined by inversion from four images either on separate detectors or different parts of the same detector, rather than by time division onto a single detector—that the final performance of the detector will ultimately depend upon the relative stability of these detections. Time multiplexing is precluded by the science requirements. In GASP we use a single detector in order to reduce this difficulty as far as possible. Determination of the magnitude of these effects will be the subject of later articles after further laboratory and telescope trials.

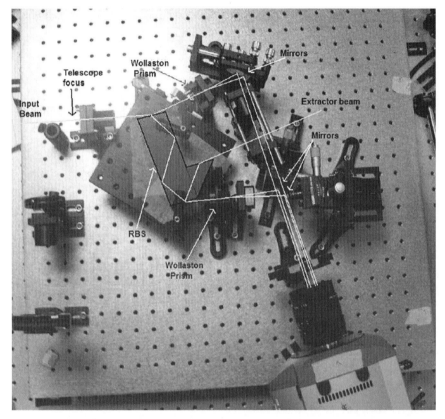

Fig. 9 Laboratory realization of the design for the Galway Astronomical Stokes Polarimeter. In this setup the four primary beams are imaged onto a single detector at the same magnification. The top reflecting surface of the RBS is conjugate to the telescope pupil

An unvignetted field of at least 15 arc seconds for each of the four "Stokes" beams is achieved for a 10 m telescope in this design. The extracted beam has an unvignetted of 45×15 arcseconds, and a partially vignetted field of 45×45 arcseconds.

For any polarimeter calibration is an essential and difficult process, and components to achieve this will be permanently incorporated into the GASP design. A beam of known polarization will be introduced into the collimated beam by a diagonal flat for calibration purposes. The calibration beam will generate four possible Stokes vectors, whose components are linearly independent of each other, namely: horizontal linear polarization: [1 1 0 0], vertical linear polarization: [1 −1 0 0], 45 degrees linear polarization: [1 0 1 0] and left circular polarization [1 0 0 1]. The exact states which will be generated in the side-arm are: [0.88 0.88 0 0], [0.94 −0.94 0 0], [0.91 −0.03 0.89 −0.16], and [0.91 −0.03 0.16 0.89] which takes into account the effect of the fold mirror in the calibration arm to produce the desired vectors previously specified. The four ideal states are then arranged into a 4×4 matrix (the polarization state generator [PSG] matrix), as are the corresponding intensity vectors [i_1 i_2 i_3 i_4] from the polarimeter. The system matrix **A** is the product of the intensity matrix and the inverse of the PSG matrix.

A general method for determining the system matrix in the laboratory is described in Compain, Poirier & Drevillon [37]—which they call the eigenvalue calibration method (ECM). Four known samples—air (nothing), a vertical polarizer, a horizontal polarizer, and a quarter wave plate orientated at 28 degrees—are introduced directly into the collimated beam, and the Mueller matrix determined for each of them. A comparison between the measured and specified Mueller matrices enables the system matrix to be calibrated. The calibration will vary slightly across the field because of the small angular changes occurring at the reflecting/refracting surfaces. Essentially, each part of the field will have its own calibration.

However, the system matrix (**A**) must, of course, include any polarization effects in the atmosphere and the telescope, which is more difficult to achieve because the instrumental polarization will have both components which have both fast and slow variability due to the varying zenith angle and the evolution of oxide and dust layers in the atmosphere. In particular, if GASP is to be used in an off-axis focus, such as a Nasmyth focus, this will have a considerable influence upon **A**, mainly by introducing unwanted degrees of linear polarization to the beam. However, in general it will still be possible to invert **A**, although the condition number will be lowered. Calibration sources will be built into GASP, and (hopefully) small corrections to the coefficients of **A** can be made by observations of known "linear polarization calibration" stars [38].

5 Detectors for GASP

Section 2 discussed the requirements for GASP, and we have concluded an optical design which will enable these requirements to be met. However the same requirements severely limit the choice of suitable detectors (a) for the four "Stokes" beams, and (b) for the extracted finder/guider/photometry beam. In particular, design requirements for a detector suitable to investigate the GRP phenomenon in the Crab pulsar become even more stringent:

(a) "Stokes" beams

- Photon counting with high Detector Quantum Efficiency (DQE) (i.e. displaying statistical fluctuations corresponding to a high detected number of photons)
- Absolute time resolution better than 100 μs (GRP pulse structure)
- Absolute time resolution ∼nanoseconds (GRP nano-pulse structure)
- Imaging, with at least 500 × 500 pixels, to accommodate four "Stokes" images
- High bandwidth

(b) Finder/Guider/Photometry beam

- Photon counting, with high DQE
- Imaging, with at least 250 × 250 pixels

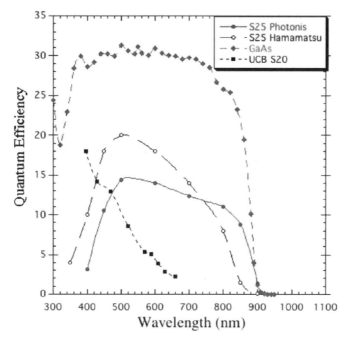

Fig. 10 Quantum efficiency of the image tube for GASP [Siegmund, private communication]

- High Bandwidth
- Absolute time resolution ~milliseconds to resolve the Crab light curve

For requirement (a) ("Stokes" beams) we have chosen an image tube based upon the use of a high gain channel plate and a delay-line anode array—as developed by Siegmund, et.al. and described elsewhere in this volume. It is intended to use a GaAs cathode, with the properties shown in Fig. 10.

The relevant properties, which are a good approximation to the GASP requirements are in Table 2.

For requirement (b) (finder/guider/photometer) we have chosen an E2V EMCCD (model CCD97—thinned, back illuminated) from Andor Instruments (model Andor iXon DV-887-BV) with the following properties (see Table 3).

Table 2 GASP Requirements

Specification	Value
Area	512×512 pixels
Resolution of GPS time tagged photons	25 ns
Maximum overall count rate	$> 5 \times 10^6$ cps
Maximum DQE	30%
Bandwidth	400–850 nm

Table 3 Properties of Andor iXon DV-887-BV

Specification	Value
Area	512 × 512 pixels
Max frame rate	450 fps
Peak QE	92.5% @ 575 nm

6 Error Analysis & The Sensitivity of Gasp

As shown in (2), the unknown Stokes vector can be found from the inverse of the system matrix **A**. The system has been designed so that the system matrix is well conditioned and non-singular. This has the effect of maximizing the polarimetric efficiency. From the design the system matrix of the polarimeter is:

$$A = \begin{bmatrix} 0.1884 & -0.0858 & 0.1677 & 0.0000 \\ 0.1884 & -0.0858 & -0.1677 & 0.0000 \\ 0.1884 & 0.0964 & 0.0000 & 0.1618 \\ 0.1884 & 0.0964 & 0.0000 & -0.1618 \end{bmatrix} \qquad (3)$$

This takes into account (a) the effects of extinction in the prism, which are small because the glass was chosen to be very transparent over the specified 400–800 nm bandwidth, (b) the effects of dispersion over the same wavelength range, again small because a low dispersion glass was chosen, and (c) the effects of angular spread of the entrance beam—which is small for point sources. No account has been taken of gain variations between the four channels. All of these effects will be dealt with in a later article.

The variance of the kth (k=1,4) element of the input Stokes vector is given by

$$\sigma_{S_k}^2 = \sigma_I^2 \sum_{i,k=1}^{4} \left(A_{ki}^{-1} \right)^2 \qquad (4)$$

where σ_I^2 is the variance due to Poisson counting statistics (equals the number of counts, n). From the variances of the Stokes vector we can see how these errors propagate through the equations for finding the polarization angle (ψ), the ellipticity (χ), the degree of polarization (DOP), the degree of circular polarization (DOC), and the degree of linear polarization (DOL).

$$\begin{aligned} \chi &= 1/2 \arctan\left(V/\sqrt{Q^2 + U^2}\right) \\ \psi &= 1/2 \arctan(U/Q) \\ DOP &= \sqrt{Q^2 + U^2 + V^2}/I \\ DOC &= \sqrt{V^2}/I \\ DOL &= \sqrt{Q^2 + U^2}/I \end{aligned} \qquad (5)$$

In general, a function of several variables $Z = f(x, y, \ldots)$ has a variance

$$\sigma_z^2 = \left(\frac{\partial z}{\partial x}\right)^2 \sigma_x^2 + \left(\frac{\partial z}{\partial y}\right)^2 \sigma_y^2 + \ldots$$

and thus the variances of the components of (5) are

$$\sigma_\chi^2 = \frac{\sigma_V^2 Q^4 + 2\sigma_V^2 Q^2 U^2 + \sigma_V^2 U^4 + \sigma_Q^2 V^2 Q^2 + \sigma_U^2 V^2 U^2}{4\left(Q^2 + U^2\right)\left(Q^2 + U^2 + V^2\right)^2} \tag{6}$$

$$\sigma_\psi^2 = \frac{\sigma_Q^2 + \sigma_U^2}{4\left(Q^2 + U^2\right)^2} \tag{7}$$

$$\sigma_{DOL}^2 = \frac{\sigma_Q^2 I^2 Q^2 + \sigma_U^2 I^2 U^2 + \sigma_I^2 Q^4 + 2\sigma_I^2 U^2 Q^2 + \sigma_I^2 U^4}{I^4 \left(Q^2 + U^2\right)} \tag{8}$$

$$\sigma_{DOC}^2 = \frac{\sigma_V^2 I^2 + \sigma_I^2 V^2}{I^4} \tag{9}$$

$$\sigma_{DOP}^2 = \frac{\sigma_Q^2 I^2 Q^2 + \sigma_U^2 I^2 U^2 + \sigma_V^2 V^2 I^2 + \sigma_I^2 Q^4}{\left(Q^2 + U^2 + V^2\right) I^4} +$$
$$\frac{2\sigma_I^2 U^2 Q^2 2 + \sigma_I^2 V^2 Q^2 + \sigma_I^2 U^4 + 2\sigma_I^2 U^2 V^2 + \sigma_I^2 V^4}{\left(Q^2 + U^2 + V^2\right) I^4}, \tag{10}$$

where $\sigma_I^2, \sigma_Q^2, \sigma_U^2, \sigma_V^2$ and I, Q, U, V are the variances and parameters of the Stokes vector respecitvely

In certain circumstances the variances will asymptotically approach infinity—for example, the variance in the angle of linear polarization will become undefined as U and Q approach zero and as V/I approaches $+1$ or -1 (100% right or left hand circularly polarized light). We assume that these situations can never arise in practice, and a maximum of 50% polarization will be assumed in determining maximum errors. It can be seen from (6)–(10) that differing Stokes vectors will generate different errors for the same degree of polarization and counts. For this reason, each DOP and Counts data points on Fig. 11 through 14 is the maximum error on the Poincaré sphere. Hence these errors are a worst case scenario and further development is being pursued to a better way of representing these errors.

Figures 11 through 14 show the errors arising from the propagation of counting errors for small numbers of detected photons through the inversion of **A**, and do not take into account detector noise counts—which in very low light situations could dominate these errors. Figure 11 shows the errors in the ellipticity angle (circular angle) measured in radians. In sensible circumstances—more than a few hundreds

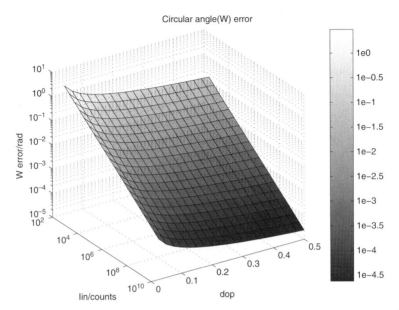

Fig. 11 Error in the ellipticity angle of polarization as a function of intensity (counts from 10^2 to 10^{10} within a particular time bin) and degree of polarization (10% to 50%). An error of 1° (.0175 radians) will occur for ∼100 counts for a 10% polarized beam

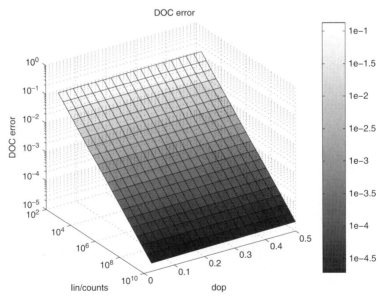

Fig. 12 Error in the degree of circular polarization as a function of intensity (counts from 10^2 to 10^{10} within a particular time bin) and degree of polarization (10% to 50%). An error of 1% will occur for ∼10^4 detected counts—purely due to the effects of counting statistics

An Ultra-High-Speed Stokes Polarimeter for Astronomy

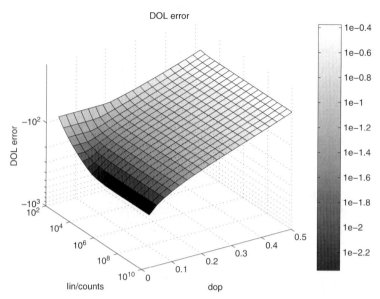

Fig. 13 Error in the degree of linear polarization as a function of intensity (counts from 10^2 to 10^{10} within a particular time bin) and degree of polarization (10% to 50%). An error of 1% will occur for $\sim 10^4$ counts at 10% degree of polarization

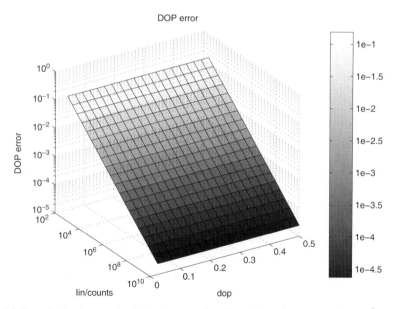

Fig. 14 Error in the degree of polarization as a function of intensity (counts from 10^2 to 10^{10} within a particular time bin) and degree of polarization (10% to 50%). An error of 1% will occur for $\sim 10^4$ counts

of photons and a moderate degree of linear polarization (say 50%, as they approach 100% linear polarization, the ellipticity angle becomes undefined)—errors in this parameter are small. In all of these figures the x axis is the degree of polarization (dop), the y axis shows the number of counts in (Iin/counts) and the z axis is the relevant error for each figure. The colourbar scale is the representation of the errors in each figure.

7 Conclusions

We have demonstrated the need to develop an unusual ultra-high-speed Stokes polarimeter in order to investigate various magnetic phenomena in collapsed objects—such as magnetic CVs and particularly optical pulsars. Whilst "conventional" polarimeters have sufficient speed and sensitivity to perform detailed observations of CVs on time scales of minutes to seconds, no conventional polarimeter can investigate random phenomena occurring during particular rotation cycles of optical pulsars (see Table 1).

The GASP design has sufficient sensitivity and precision to enable a detailed study to be made of the Crab pulsar, at least, using today's large telescopes, but it is, of course, subject to many other sources of error and inefficiency, which we have attempted to minimize, but which will affect the performance. Measurements of the performance of GASP, in the laboratory and on a telescope will be the subject of later articles. If the theoretical analysis is correct, it will come into its own with the next generation of extremely large telescopes, because GASP has a design bandwidth covering the B, V, R & I bands, a high throughput and a high DQE detector. A prototype has been demonstrated in the laboratory, and will be used on the sky in summer 2007.

Acknowledgments The authors would like to thank:

Science Foundation Ireland, who supported the development of GASP under grant 05/RFP/PHY 0045,

The Applied Optics Group in NUI Galway for continued help and support,

Antonello de Martino, for advice and support and for the loan of equipment including the original "Compain & Drevillon" RBS,

PC would like to thank Science Foundation Ireland and the National University of Ireland, Galway for financial support.

References

1. Compain E, Drevillon B (1998) Broadband division-of-amplitude polarimeter with uncoated prisms. Applied Optics 37:5938–5944
2. Tinbergen J (1996) Astronomical Polarimetry. Cambridge University Press, Cambridge, UK, pp 174
3. Di Serego Alighieri S (1997 Polarimetry with large telescopes. In: Jose M. Rodriguez Espinosa JM, Herrero A, Sánchez F (eds) Instrumentation for Large Telescopes. Cambridge University Press, Cambridge, pp 287–329

4. Leroy J-L (2000) Polarization of light and astronomical observation. Gordon & Breach Science, Amsterdam
5. Keller CU (2002) Instrumentation for astrophysical spectropolarimetry. In: Trujillo-Bueno J, Moreno-Insertis F, Sanchez F (eds) Astrophysical spectropolarimetry. Proceedings of the XII Canary Islands Winter School of Astrophysics, Cambridge University Press, Cambridge, UK, pp 303–354
6. Angel JRP (1974) Mechanisms that produce linear and circular polarization. In: Gehrels T (ed) Planets, Stars, and Nebulae: Studied with Photopolarimetry. Proceedings of IAU Colloq. 23, held in Tucson, AZ, November, 1972. University of Arizona Press, pp 54–63
7. Kemp JC (1970) Circular polarization of thermal radiation in a magnetic field. Astrophysical Journal 162:169
8. Potter SB, O'Donoghue D, Romero-Colmenero E, Buckley DAH, Woudt PA, Warner B (2006) Photometric and polarimetric observations of the eclipsing polar SDSS J205017. 84−053626.8. Monthly Notices of the Royal Astronomical Society 371:727–731
9. Rodrigues CV, Jablonski FJ, D'Amico F, Cieslinski D, Steiner JE, Diaz MP, Hickel GR (2006) Optical polarimetry and infrared photometry of two AM Her binaries: 1RXS J161008.0+035222 and 1RXS J231603.9–052713. Monthly Notices of the Royal Astronomical Society 369:1972–1982
10. Lamb FK, Pethic CJ, Pines D (1973) A Model for compact X-Ray sources: accretion by rotating magnetic stars. Astrophysical Journal 184:271–290
11. Cropper M (1990) The Polars. Space Science Reviews 54:195–295
12. Warner B (1995) Cataclysmic Variable Stars. Cambridge University Press, Cambridge, New York
13. Wickramasinghe DT, Meggitt SMA (1985) The polarization properties of magnetic accretion columns. III - A grid of uniform temperature and shock front models. Monthly Notices of the Royal Astronomical Society 214:605–618
14. Bailey JA, Ferrario L, Wickramasinghe DT, Buckley DAH, Hough JH (1995) Polarimetry and photometry of the new AM Herculis system RE J1844–741. Monthly Notices Of The Royal Astronomical Society 272:579–584.
15. Potter SB, Hakala PJ, Cropper M (1998) Stokes imaging' of the accretion region in magnetic cataclysmic variables - I. Conception and realization. Monthly Notices of the Royal Astronomical Society 297:1261–1268
16. Cropper M (1985) Simultaneous linear and circular polarimetry of EF ERI. Monthly Notices Of The Royal Astronomical Society 212:709–721
17. Euchner F, Beuermann K, Reinsch K, Jordan S, Hessman FV, Gansicke BT (2003) Zeeman tomography of magnetic white dwarfs: General method and application to EF Eridani. In: de Martino D, Silvotti R, Solheim J-E, Kalytis, R (eds) White Dwarfs. Kluwer Academic Publishers, p. 195
18. Euchner F, Reinsch K, Jordan S, Beuermann K, Gansicke BT (2005) Zeeman tomography of magnetic white dwarfs. II. The quadrupole-dominated magnetic field of HE 1045–0908. Astronomy and Astrophysics 442:651–660
19. Euchner F, Jordan S, Beuermann K, Reinsch K, Gansicke BT (2006) Zeeman tomography of magnetic white dwarfs. III. The 70-80 Megagauss magnetic field of PG 1015+014. Astronomy and Astrophysics 451:671–681
20. Wampler EJ, Scargle JD, Miller JS (1969) Optical observations of the Crab nebula pulsar. Astrophysical Journal 157:L1
21. Smith FG, Jones DHP, Dick JSB, Pike CD (1988) The optical polarization of the Crab pulsar. Monthly Notices of the Royal Astronomical Society 233:305–319
22. Graham-Smith F, Dolan JF, Boyd PT, Biggs JD, Lyne AG, Percival JW (1996) The ultraviolet polarization of the Crab pulsar. Monthly Notices Of The Royal Astronomical Society 282:1354–1358
23. Kanbach G, Slowikowska A, Kellner S, Steinle H (2005) New optical polarization measurements of the Crab pulsar. In: Astrophysical Sources Of High Energy Particles And Radiation, AIP Conference Proceedings 801:306–311

24. Staelin DH, Reifenstein EC (1968) Pulsating radio sources near the Crab nebula. Science 162:1481–1483
25. Slowikoswska A, Jessner A, Klein B, Kanbach G (2005) Polarization characteristics of the Crab pulsar's giant radio pulses at HFCs phases. In: Astrophysical sources of high energy particles and radiation. AIP Conference Proceedings 801:324–329
26. Karasterigiou A, Johnston S (2004) An investigation of the absolute circular polarization in radio pulsars. Monthly Notices of the Royal Astronomical Society 352:689–698
27. Moffet DA, Hankins TH (1999) Polarimetric properties of the Crab pulsar between 1.4 and 8.4 GHz. Astrophysical Journal 522:1046–1052
28. Donovan J, Lommen, A, Arzoumanian Z, Harding A, Strickman M, Gwinn C, Dodson R, McCulloch P, Moffett D (2004) Correlations Between X-ray and Radio Pulses in Vela. In: Camilo F, Gaensler BM (eds) Young Neutron Stars and Their Environments. Astronomical Society of the Pacific, San Francisco, p 335
29. Shearer A, Stappers B, O'Connor P, Golden A, Strom R, Redfern M, Ryan O (2003) Enhanced optical emission during Crab giant radio pulses. Science 301:493–495
30. Hankins TH, Kern JS, Weatherall JC, Eilek JA (2003) Nanosecond radio bursts from strong plasma turbulence in the Crab pulsar. Nature 422:141–143
31. Golden A, Shearer A, Redfern RM, Beskin GM, Neizvestny SI, Neustroev VV, Plokhotnichenko VL, Cullum M (2000) High speed phase-resolved 2-d UBV photometry of the Crab pulsar. Astronomy and Astrophysics 363:617–628
32. Karpov S, Beskin G, Biryukov A, Debur V, Plothotnichenko V, Redfern M, Shearer A (2006) Short time scale pulse stability of the Crab pulsar in the optical band. In: Page D, Turolla R, Zane S (eds) Isolated Neutron Stars: from the Interior to the Surface. ApSS, London
33. Scarrott SM, Warren-Smith RF, pallister WS, Axon DJ, Bingham RG (1983) Electronographic polarimetry - the Durham polarimeter. Monthly Notices of the Royal Astronomical Society 204:1163–1177
34. Shearer A, Redfern M, Pedersen H, Rowold T, O'Kane P, Butler R, O'Byrne C, Cullum M (1994) The position of PSR 0540–69. Astrophysical Journal 423:L51–L53
35. A M El-Saba AM, Azzam RMA, Abushagar MAG (1996) Parallel-slab division-of-amplitude photopolarimeter. Optics Letters 21:1709–1711
36. Azzam RMA, De A (2003) Optimal Beam Splitters for the Division of Amplitude Photopolarimeter. JOSA A 20:955–958
37. Compain E, Poirier S, Drevillon B (1999) General and self-consistent method for the calibration of polarization modulators, polarimeters, and Mueller-matrix ellipsometers. Applied Optics 38:3490–3502
38. Fossatti L, Bagnulo S, Mason E, Landi Degl'Innocento E (2006) Standard stars for linear polarization. In: Sterken C (ed) The Future of Photometric, Spectrophotometric, and Polarimetric Standardization. Astronomical Society of the Pacific, San Francisco
39. Povel HP (2001) Ground-based Instrumentation for Solar Magnetic Field Studies, with Special Emphasis on the Zurich Imaging Polarimeters ZIMPOL-I and II. In: Mathys G, Solanki SK, Wickramasinghe DT (eds) Magnetic Fields Across the Hertzsprung-Russell Diagram. Astronomical Society of the Pacific, San Francisco, p 543

Use of an Extremely Large Telescope for HTRA

Oliver Ryan and Mike Redfern

Abstract Investigative studies have designed concept instruments for a proposed 42 m European Extremely Large Telescope (E-ELT) facility suite. An ELT will aid, at its most basic level, detection of the faintest stellar sources and instrinsic features, and detailed quantitative analysis of relatively bright sources. The development of such a telescope thus plays to the strengths of high time resolution (HTR) detectors which provide single photon detection capabilities and/or extremely high refresh rates—they can maximize detection of faint sources through photon counting and they can cope with the high photon flux from relatively bright objects building up extremely high S/N. Specific considerations and fast detectors then need to be adopted into the ELT design to accommodate high time resolution astrophysics (HTRA) efficiently. No one instrument or detector can cover all HTRA timescales (nanoseconds to seconds and 'slower'), and a dedicated facility HTRA instrument on the ELT is unlikely as yet, but chosen facility instruments can include HTRA options in their packages. Currently there are numerous HTRA options in existing telescope facilities around the world as well as many visiting HTRA instruments. This demonstrates an active HTRA community with worldwide support and interest in HTRA observation capabilities, which underscores the need for HTR capabilities to be designed into the ELT.

Keywords: Extremely Large Telescope · HTRA · high time resolution · design · fast detectors · instrumentation

1 Introduction

A High Time Resolution Instrument (HiTRI) 'small study' was undertaken to examine the astronomical community's requirements and need for such an instrument on an ELT. This was undertaken as part of a larger program of small studies to

Oliver Ryan · Mike Redfern
Dept. of Physics, National University of Ireland, Galway, Ireland
e-mail: oliver.ij.ryan@nuigalway.ie

produce concept ideas for a first light instrument suite for a 42 m ELT [1]. The study looks at the current status of HTRA detectors and also examines the constraints the instrument could put on the design of the European Extremely Large Telescope. This chapter reproduces the study's main points:

- To identify the facilities and characteristics required of a 42 m ELT to accommodate HTRA observations,
- To identify the available detectors for HTRA and hence a HTRA-specific instrument,
- To identify the technological limitations and shortcomings in detectors and data acquisition techniques that must be addressed, and to highlight technology advancements necessary, to develop a HTRA-specific instrument,
- To make a case for including a HTRA-specific instrument to the ELT suite of instruments,
- To make a case for including HTRA-friendly detectors into non-HTRA specific instruments in the ELT instrument suite.

At the most superficial level, a 42 m ELT is a 27.6 times improvement in photon collecting ability over the current 8 m class of telescopes. This advent of the ELT now extends the time domain to fainter as well as more distant sources, and also allows the application of high time resolution astrophysics studies down to microseconds with reasonable signal to noise, making HTRA a more readily available area.

There are a number of areas of observational interest where an improved understanding will be gained, and which generally share a common theme of compact objects, high energy densities, and extremely high magnetic fields. These are briefly mentioned in Sect. 2 but are more specifically dealt with in other chapters of this volume.

The ELT will need some basic and necessary characteristics and provisions in order to cater to HTRA science as this is a specialised area of detection and observation. This is discussed in Sect. 3.

As well as 'tuning' the telescope for HTRA, detectors with specific characteristics are needed to accomplish the observations. They have to be, in essence, fast large area detectors or arrays, and with high quantum efficiency (QE). Suitable detector types are discussed in Sect. 4.

2 Optical HTRA Science

A brief introduction is given to the wide variety of science pursued in the limited parameter space of high time resolution. This is extensively shown through the other chapters of this book, cf. Jeffery [2], Littlefair [3], Marsh [4], Shahbaz [5], Shearer [6] and others. Specific details are not entered into here.

A short list of fascinating HTRA science targets is given in Table 1, which in itself shows the variety of topics and targets that are of interest to HTR astrophysicists. Footnotes in the table briefly explain the fast and/or stochastic physical processes and object characteristics that require HTR observations to increase our understanding of them.

Dedicated instrumentation is key to HTRA. An example of an instrument designed with HTR observations in mind is ULTRACAM (this volume [8]). ULTRACAM is a leading HTRA instrument in high demand (~25 papers have been based in whole or in part on ULTRACAM data to date) which in itself indicates a high level of interest existing in HTRA. Important, though, is the wide variety of targets astronomers have been observing with this instrument—Table 2 (private communication from V. Dhillon) shows a percentage breakdown of ULTRACAM usage since its commissioning in May 2002.

Tables 1 and 2 are a simple way to highlight the beginning of the breadth of science possible in HTRA. Observations of these targets are photon-starved due to their instrinsic faintness and the HTR required to pick out their optical variabilities—the ELT by virtue of its photon gathering ability alone will hugely help the HTRA field. Some more specific details on the benefits of the ELT to HTRA are given in section 2.1.

2.1 HTRA Science on an ELT

Information in this section has been summarised from [6] (this volume) and [9]. Simply put, an ELT with HTR capabilities will allow:

- the study of stellar pulsations through sub-second observations,
- a more accurate determination of binary system eclipses giving stellar radii of both objects; in-depth studies of their flickering and flare phenomenoa on all timescales,
- the study of the accretion disk—compact object interactions of CVs (cataclysmic variables) containing white dwarfs, neutron stars, black holes on sub-second timescales,
- detailed examinations for optical spectral breaks in the optical and IR emissions of AXPs (anomalous X-ray pulsars, with pulse periods of 6–12 s),
- direct optical probing of the plasma surrounding pulsars (only fourteen of over 2000 radio pulsars have been observed at optical wavelengths) giving further information on pulsar emission mechanisms,
- searches for optical correlations to nanosecond structure routinely observed in giant radio pulses from pulsars,
- new detections of unknown young pulsars at optical wavelengths, and a sampling of pulsar polulations in other galaxies,
- galaxy surveys out to 20 MPc for young super nova remnants (SNRs) to quantify the fraction of supernovae that produce pulsars, AXPs, magnetars, black holes, no remnant, or other condensed object,
- optical observations of active galactic nuclei (AGNs) which show variability over nine magnitudes in timescale, giving constraints to other multiwavelength models of synchrontron emission.

Table 1 Examples of Targets for HTRA observations*

Name	Additional Information	Magnitude
	Low Mass X-Ray Binaries[1]	
Aql X-1	= V1333 Aql	$m_V \sim 18.2$
Cen X-4	= V822 Cen	$m_V \sim 15.9$
Cir X-1		$m_V \sim 21.4$
	Black Hole Candidates[2]	
J1650-4957		$m_V \sim 17\text{--}24$
KV UMa	= J1118+4802	$m_V \sim 12.5\text{--}19.0$
V381 Nor	= J1550-5628	$m_V \sim 16.0\text{--}22.2$
	High Mass X-Ray Binaries[3]	
Cyg X-1	spec.type O9.7Iab	$m_V = 8.9$
LMC X-1	spec.type O8III	$m_V = 14.5$
SMC X-3	spec.type O9 IVe	$m_V = 14.9$
	Pulsars[4]	
Crab pulsar	P = 33 ms	$m_V = 16.5$
Vela pulsar	P = 89 ms	$m_V = 23.6$
PSR J2051-0827	P = 4.5 ms	$m_R = 23$
	Magnetars[5]	
XTE J1810-197	P = 5.5 s	$m_K = 21$
1E 1048.1-5937	P = 6.5 s	$m_I = 26.2$
1E 2259+586	P = 7.0 s	$m_K = 21.5$
	Neutron Stars[6]	
Geminga	isolated neutron star, P=0.24 s	$m_V = 25.5$
RX J1856.5-3754	isolated neutron star	$m_V = 25.7$
	Cataclysmic Variables[7]	
CP Com		$m_B \sim 16.4$
DP Leo		$m_V \sim 17.5$
GG Leo		$m_V \sim 16.5$
	Magnetic Cataclysmic Variables[8]	
EU UMa		$m_V \sim 17.0$
HU Aqr		$m_B \sim 15.3$
HY Eri		$m_V \sim 17.5$
	Hot Emission-Line Stars[9]	
Cen X-3	high-mass X-ray binary with kHz fluctuations	$m_V = 13.3$
LMC X-4	high-mass X-ray binary; spec.type O7 IV	$m_V = 14.0$
MWC 349A	circumstellar recombination-line lasers	$m_R = 10.1$
	Active Galactic Nuclei[10]	
Cen A		$m_V \sim 7.0$
Mrk 421		$m_V \sim 13.5$
Mrk 501		$m_V \sim 13.8$
	White Dwarf Surface Structure and Dynamics[11]	
Procyon B		$m_V = 10.7$
Sirius B		$m_V = 8.4$
40 Eri B		$m_V = 9.5$

*Adapted with permission of D. Dravins from the Quanteye report [7]. See references therein.
[1]These systems typically contain neutron stars, and several have optical counterparts for HTRA investigation. Studies can be conducted on magnetohydrodynamic gas-flow instabilities in the accretion flow onto the neutron star surface.

Table 2 Statistics of ULTRACAM use, 2005 (courtesy of V. Dhillon)

Target	Percentage time
Cataclysmic variables, accreting white dwarfs	22%
Black hole X-ray binaries	19%
Subdwarf B stars, asteroseismology	15%
Kuiper belt objects	11%
Eclipsing white dwarf/red dwarf binaries	10%
Pulsars	5%
Ultra-compact binaries	4%
Flare stars	4%
Extrasolar planets	3%
Isolated white dwarfs	2%
Isolated brown dwarfs	2%
GRBs	1%
AGN	1%
Titan/Pluto	1%

HTRA is ultimately concerned with source variability (at all energy levels from radio through optical to X-ray and gamma ray) at their own demonstrated characteristic timescales which can vary from subseconds to days and weeks. The ELT will permit examination of the finer timing detail in variable output objects, as well as allowing the discovery of new sources of variability which are too faint and/or too distant for the current 8–10 m class of telescopes to see. Table 3 (from [6] and [9]) gives an indication of the increased time resolution the ELT will allow over

---

[2] These are a subgroup of LMXRBs, and some are also microquasars, i.e. sources of relativistic jets. The behaviour of matter and radiation close to the innermost stable orbit (and beyond), close to the black-hole horizon can be studied.

[3] These also include systems with black-hole candidates. Of interest is temporal flickering of polarization in line emission from the accretion disk.

[4] New studies are trying to find optical counterparts to radio nanopulses, studying their emission physics. Alfvén waves in neutron-star magnetospheres are also of interest. Only five pulsars have been observed optically. Numerous millisecond and X-ray pulsars have been found in radio or X-rays; very plausibly several of these will become observable optically with the ELT.

[5] To be studied here are electrical discharges close to the magnetic poles and free-electron laser emission in the extremely strong magnetic fields.

[6] At present, the only known isolated neutron stars are optically very faint. However, from statistical arguments, they should be numerous in our Galaxy, and quite possibly some nearby and relatively brighter ones will be found from X-ray imaging in the near future, offering more suitable targets for time-resolved studies, and optical identification. Of interest are small localized spots across neutron star surfaces, magnetic-field inhomogeneities, and acoustic spectra of non-radial oscillations.

[7] Under study is the magnetic dynamo in the differentially rotating accretion disk.

[8] These are also called polars, and AM Her-type objects. Example of astrophysical problem: Plasma instabilities inside the accretion column hitting the magnetic poles of the white dwarf.

[9] Examples of astrophysical problem: Spectral width of laser emission line components, and their photon statistics; wave propagation in the stellar wind; photon bubbles emerging through the gas in extremely luminous stars close to (or even above) the Eddington luminosity.

[10] Example of astrophysical problem: Rapid fluctuations in mass flow and radiation emerging from close to the central black-hole engine.

[11] Example of astrophysical problem: Timescales and dynamics of white-dwarf surface convection along the temperature sequence (with expected millisecond scales for the convective features).

Table 3 Science timescales showing current and future timescale possibilities, from [6, 9], reproduced with permission

Object	Science	Timescale Now	Timescale ELT era
	Stellar flares and pulsations	Seconds/ minutes	10–100ms
White Dwarfs	Stellar	1 μs–1 ms	1 μs–1ms
Neutron Stars	Surface Oscillations	–	0.1 μs
Close Binary	Tomography	100ms++	10ms+
Systems	Eclipse in/egress	10ms+	<1ms
accretion &	Disk flickering	10ms	<1ms
turbulence	Correlations (e.g. X & optical)	50ms	<1ms
Pulsars	Magnetospheric	1 μs–	ns (?)
	Thermal	100ms	ms
AGN		Minutes	Seconds

current facilities, permitting studies to reach sub-millisecond, sub-microsecond and possibly nanosecond time resolutions.

Read further through [6] and the other chapters of this volume for more information.

2.2 Current HTRA Instrumentation

Table 4 displays the wide variety of current HTRA instrumentation in development or in use around the world, each generally tailored to accommodate different timescales from microseconds to milliseconds and seconds. The footnotes tell of the science that has been pursued with them—studies of eclipsing CVs, polars, globular clusters, pulsars, neutron star accreters, X-ray binaries, optically violent variable jets, and more, through linear/circular/imaging polarimetry and fast photometry. Much more is available through the given references. Refer also to Table 1 of Dhillon's chapter, this volume [8], where further instruments and modes for high-speed astrophysics with CCDs are listed. All these different instruments from various scientific collaborations demonstrate the wide interest astronomers have in knowing the fine details of time in their observations. What is lacking are sufficient photons, which an ELT will provide, to permit these instruments to be used to their furthest limits.

Table 4 HTRA instruments, facility and visitor, existing and planned

Instrument	Detector	Photometry[@]	Polarimetry[*]	Spectroscopy
Quanteye[1]	100 SPAD	ps–ns	No	No
Aqueye[2]	4 SPAD	ns–μs	AFOSC	AFOSC
GASP[3]	GaAs Image Tube	ns–μs	Full Stokes	possibly
Salticam[4]	2x1 CCD	100 ms–secs	No	UBVRI
RSS[5]	3x1 CCD	50 ms–1.6 s	L, C, SP, FS	VPH, filters
ULTRACAM[6]	3 CCD	0.237s–10s	No	3 colour
LuckyCam[7]	L3CCD	> 40 frames/s	No	filters
TRIFFID[8]	3 APD, L3CCD	1 μs	No	3 colour
OPTIMA[9]	8 APD	1 μs	L	No
MPPP[10]	PSD	1 μs	Full Stokes	4 colour
FUSP[11]	PSD	1 μs	L, IP, SP	4 colour
IMPOL[12]	CCD	12 s frame rate	Full Stokes	No
ZIMPOL[13]	CCD	34 ms frame rate	Full Stokes	No
LRIS (Keck)[14]	CCD	72ms	L, C, IP+SP	Grism
FORS2 (VLT)[15]	CCD	2.3ms–2.3s	No, FORS1	Grism, VPH
FOCAS (Subaru)[16]	CCD	0.1s	L, C, SP	Grism, VPH
S-CAM3[17]	STJ	5 μs	No	Energy resolving, R 8–13
UCTPol[18]	photomultiplier tube	1 ms	L, C	UBVRI
AcqCam[19]	CCD	6–60 s	No	UBVRI
ISIS[20]	CCD	0.2–15 s	IP, SP	dichroics, blaze
Argos[21]	CCD	1 s	No	No
TES array[22]	TES	30μs	IP, LP, SP	Energy resolving, R $\sim$20

[@]Timing resolutions used during observations—instruments may be capable of higher resolutions
[*]L—linear polarimetry, C—circular polarimetry, IP—imaging polarimetry, SP—spectropolarimetry, FS—Full Stokes
[1]A proposed instrument from OWL studies (Oct. 2005). Quantum optics, X-ray binaries, AGNs, magnetic CVs, white dwarfs [7]
[2]Under construction for end 2006, a prototype for Quanteye to be mounted under AFOSC at the Asiago Telescope—AFOSC provides the polarimetric and spectroscopic abilities [10]
[3]Galway Stokes Polarimeter. Under construction for end 2006 [11]
[4]A first light instrument on the 11 m SALT [12, 13], Nov. 2005
[5]Robert Stobie Spectrograph, a first light instrument on the 11 m SALT [14, 13], Nov. 2005
[6]Eclipsing CVs, polars, pulsating subdwarf B stars, black hole binaries [15, 16, 8]. First commissioned May 2002 on WHT. Plan to possibly use EMCCDs when further developed
[7]Very low mass close binaries [17, 18, 19], in use since 2000
[8]Optical pulsars, globular clusters, AXPs, magnetars [20, 21, 22, 23, 24]. First commissioning in 1988
[9]Pulsars, eclipsing polars, neutron star accreter, X-ray binaries [25, 26, 27, 28], first light in 2000.
[10]Multicolor Panoramic Photometer Polarimeter, optical GRBs, pulsars, X-ray binaries [29], 2003
[11]Fast Universal Spectropolarimeter, fast response on robotic telescope [30], 2004–2005
[12]Molecular clouds, Bok globules, magnetic field effects on ISM [31, 32, 33], IMPOL's first use in 1997
[13]Solar photosphere, CHaracterising Exoplanets by Opto-infrared polarization and Spectroscopy (CHEOPS) [34, 35, 36, 37], ZIMPOL-I first design in 1992
[14]Spectral evolution of AeAquarii flares, viewing CVs on different timescales [38, 39, 40, 41, 42], commissioned 1993
[15]White dwarfs, AGNs, optically violent variables, jets [43, 44, 45]. FORS1 has polarimetric capabilities, first light in 1998
[16]QPOs, accreting magnetic white dwarfs, mass transfer magnetic CVs [46], first light in 2000
[17]AM Her systems, CVs, accretion disks, binaries, dwarf novae [47, 48, 49], first demonstrated 1996

Table 5 gives the photon rates expected from a one arcsec point source on a 42 m telescope in the V band. Note the fourth column with footnote 'c'—this lists the photons expected per second per metre-squared at the focal plane of the telescope. A typical HTR science target is the Crab pulsar (m_v ~16), observations of which (particularly polarimetric) are photon-starved on existing telescopes. With the ELT, an astounding 2.59×10^6 photons s^{-1} m^{-2} are incident on a detector in the focal plane, which will allow HTR astrophysicists to push current detectors to their limits and probe the object (and all others) in extremely fine detail. Please refer to section 4.7 which uses Table 5 to further show the increases in limiting magnitudes detectable by particular detectors on a 42 m versus a 10 m telescope.

The timescale and wavelength ranges that can be covered by current HTRA instrumentation on the ELT are shown in Fig. 1 and Table 3. A HTRA-specific instrument could fit into the optical slot covering a range of time resolutions from sub-microseconds to milliseconds and greater. Refer also to Fig. 2 in Sect. 4.7 which has the limiting magnitudes for a selection of HTRA detectors on the ELT. These figures show the breadth of HTRA coverage in timescale, magnitude, and wavelength.

3 HTRA Requirements of the ELT

HTRA observations are a specialised field for which the ELT (and its first light suite of instruments) should be ready. The telescope needs specific facilities and characteristics to make HTRA observations useful. These are described here, in no particular order.

3.1 ELT Polarization

Oblique reflections can change incoming light's polarization state (a 45° reflection can introduce linear polarization of a few percent at visible wavelengths) so polarimetry is best carried out at the unfolded Cassegrain focus [65]. Adaptive Optics (AO) introduces oblique reflections but this can be reduced by the use of compensating mirrors and retarders, and adaptive secondary mirrors can be used to improve spatial resolution without introducing additional reflections. Angled reflections need to be designed to minimize polarization.

---

[18] University of Cape Town Photometer/Polarimeter [50] (first light 1985). Stokes imaging and doppler mapping of magnetic CVs [51]

[19] A & G Acquisition Camera on Gemini South in Chile [52]. Fast photometry of X-ray transients [53], 2002

[20] ISIS on the 4.2 m WHT offers a rapid spectroscopy mode on the TEK4 CCD in its red arm [54], first light 1989

[21] Argos is a prime focus CCD photometer optimized for the 2.1 m telescope at McDonald Observatory [55], built 2004. ZZ Ceti stars, blue variables, pulsating white dwarfs

[22] TES array—Crab pulsar, polar ST LMi [56, 57, 58], first astronomical application in 1999

Table 5 Photon rates from a point source (1 arcsec assumed) collected by a 42 m telescope in V band[*], and integrating times for three detectors for S/N of 10

m_V mag	Flux[a] ph s^{-1} m^{-2}	Flux[b] ph s^{-1} m^{-2}	Flux[c] ph s^{-1} m^{-2}	Flux[d] ph s^{-1} m^{-2}	Exp. time, s 100%[e]	Exp. time, s GaAs[f]	Exp. time, s SPAD array[g]	Exp. time, s L3CCD[h]
0	8.794E+09	8.033E+12	6.507E+12	4.555E+12	3.293E-12	3.450E-11	2.278E-11	1.339E-11
1	3.501E+09	3.198E+12	2.590E+12	1.813E+12	8.272E-12	8.666E-11	5.723E-11	3.364E-11
2	1.394E+09	1.273E+12	1.031E+12	7.219E+11	2.078E-11	2.177E-10	1.438E-10	8.450E-11
3	5.549E+08	5.068E+11	4.105E+11	2.874E+11	5.220E-11	5.468E-10	3.611E-10	2.123E-10
4	2.209E+08	2.018E+11	1.634E+11	1.144E+11	1.311E-10	1.374E-09	9.070E-10	5.332E-10
5	8.794E+07	8.033E+10	6.507E+10	4.555E+10	3.293E-10	3.450E-09	2.278E-09	1.339E-09
6	3.501E+07	3.198E+10	2.590E+10	1.813E+10	8.272E-10	8.666E-09	5.723E-09	3.364E-09
7	1.394E+07	1.273E+10	1.031E+10	7.219E+09	2.078E-09	2.177E-08	1.438E-08	8.450E-09
8	5.549E+06	5.068E+09	4.105E+09	2.874E+09	5.220E-09	5.468E-08	3.611E-08	2.123E-08
9	2.209E+06	2.018E+09	1.634E+09	1.144E+09	1.311E-08	1.374E-07	9.070E-08	5.332E-08
10	8.794E+05	8.033E+08	6.507E+08	4.555E+08	3.293E-08	3.450E-07	2.278E-07	1.339E-07
11	3.501E+05	3.198E+08	2.590E+08	1.813E+08	8.273E-08	8.667E-07	5.723E-07	3.364E-07
12	1.394E+05	1.273E+08	1.031E+08	7.219E+07	2.078E-07	2.177E-06	1.438E-06	8.451E-07
13	5.549E+04	5.068E+07	4.105E+07	2.874E+07	5.221E-07	5.470E-06	3.612E-06	2.123E-06
14	2.209E+04	2.018E+07	1.634E+07	1.144E+07	1.312E-06	1.375E-05	9.079E-06	5.337E-06
15	8.794E+03	8.033E+06	6.507E+06	4.555E+06	3.301E-06	3.458E-05	2.284E-05	1.342E-05
16	3.501E+03	3.198E+06	2.590E+06	1.813E+06	8.320E-06	8.717E-05	5.757E-05	3.384E-05
17	1.394E+03	1.273E+06	1.031E+06	7.219E+05	2.108E-05	2.209E-04	1.459E-04	8.574E-05
18	5.549E+02	5.068E+05	4.105E+05	2.874E+05	5.411E-05	5.668E-04	3.748E-04	2.201E-04
19	2.209E+02	2.018E+05	1.634E+05	1.144E+05	1.432E-04	1.500E-03	9.935E-04	5.824E-04
20	8.794E+01	8.033E+04	6.507E+04	4.555E+04	4.055E-04	4.248E-03	2.824E-03	1.650E-03
21	3.501E+01	3.198E+04	2.590E+04	1.813E+04	1.308E-03	1.370E-02	9.166E-03	5.324E-03
22	1.394E+01	1.273E+04	1.031E+04	7.219E+03	5.109E-03	5.353E-02	3.610E-02	2.081E-02
23	5.549E+00	5.068E+03	4.105E+03	2.874E+03	2.435E-02	2.551E-01	1.732E-01	9.918E-02

(continued)

Table 5 (continued)

m_v mag	Flux[a] ph s^{-1} m^{-2}	Flux[b] ph s^{-1} m^{-2}	Flux[c] ph s^{-1} m^{-2}	Flux[d] ph s^{-1} m^{-2}	Exp. time, s 100%[e]	Exp. time, s GaAs[f]	Exp. time, s SPAD array[g]	Exp. time, s L3CCD[h]
24	2.209E+00	2.018E+03	1.634E+03	1.144E+03	1.338E−01	1.402E+00	9.555E−01	5.451E−01
25	8.794E−01	8.033E+02	6.507E+02	4.555E+02	7.944E−01	8.322E+00	5.684E+00	3.236E+00
26	3.501E−01	3.198E+02	2.590E+02	1.813E+02	4.887E+00	5.120E+01	3.500E+01	1.991E+01
27	1.394E−01	1.273E+02	1.031E+02	7.219E+01	3.052E+01	3.197E+02	2.187E+02	1.243E+02
28	5.549E−02	5.068E+01	4.105E+01	2.874E+01	1.918E+02	2.009E+03	1.374E+03	7.813E+02
29	2.209E−02	2.018E+01	1.634E+01	1.144E+01	1.208E+03	1.266E+04	8.657E+03	4.922E+03
30	8.794E−03	8.033E+00	6.507E+00	4.555E+00	7.618E+03	7.980E+04	5.459E+04	3.103E+04

*Using flux at $m_v(0) = 3460$ Jy, $\lambda_c = 550$ nm, $d\lambda/\lambda = 0.16$, 1 Jy $= 1.51 \times 10^7$ photons s^{-1} m^{-2} $(d\lambda/\lambda)^{-1}$
[a]Top of the atmosphere, with new moon sky background of 21.8 mag s^{-1} arcsec^{-1}
[b]Collected by 42 m telescope with 20% obscuration with atmospheric extinction of 0.21 mag
[c]Present at focal plane with 81% telescope throughput(four reflections of 95%)
[d]Present at detector with 70% detector optics thoughput
[e]Exposure time for a 100% efficient detector, zero dark count, zero read noise, to reach S/N ratio of 10
[f]Exposure time for a 35% efficient GaAs image tube with 55 μm pixels, 0.0006 dark counts per sec per pixel, zero read noise [59], to reach S/N ratio of 10
[g]Exposure time for a 53% efficient SPAD array, 50 μm pixels, 50 dark counts per pixel per sec, zero read noise [60, 61], to reach S/N ratio of 10
[h]Exposure time for a 90% efficient L3CCD, 30 μm pixels, dark count 0.003 counts per pixel per sec, 0.9 e$^-$ rms read noise [62, 63], to reach S/N ratio of 10

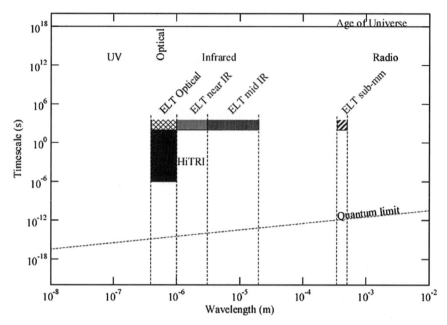

Fig. 1 The wavelength range and timescales to be covered by HiTRI and other instruments on the ELT, and the ELT proposed range of observations. Courtesy of A. Vick after the diagram of D. Dravins [64]

To minimize polarization effects, therefore, a minimum-mirror ELT design (Ritchey-Chretien, Cassegrain) should be considered.

3.2 ELT Timing

HTRA observations need absolute timing for their data, whereas other instruments may only need relative timing. The ELT needs accurate GPS time stamping to achieve absolute timing. A realistic time resolution would be to the useable sub-millisecond scale (as evidenced by science reviews in this book—see other chapters in this volume), with further niches possible to microsecond and nanosecond scales. The observatory needs telescope-wide accurate, synchronised timing, from an onsite clock (e.g. see chapter 4.2 of [7]), with links to GPS and/or the Galileo network (European satellite navigation system) [66]. Timing facilities are of the utmost importance for HTRA. Precise timing, not only available but actually implemented in the various ELT (non-HTRA) instruments would allow an easily accessible HTRA option.

3.3 Data Handling

Any HTRA instrument will have to have a standardised data output format to adhere to ESO protocols and archiving, taking into account evolution of standards for the

ELT, and new generations of detector controllers. This may be difficult depending on the detector chosen as it may be non-CCD based, or photon counting. A fast CCD-type detector in a (non-)HTRA instrument could conform easily, but timestamps, increased data storage and handling capacity needs to be addressed to collect the larger amounts of data as time resolution increases, and high-speed observations become more frequent.

From Table 5 a flux of 1.8×10^6 photons per sec detected for a 16th magnitude object (e.g. the Crab pulsar) needs to be time-tagged. Each photon could be represented by up to 32 (or 64 bits) of binary data (relative instantaneous time, two-dimensional position, UTC stamp, etc.) generating a data rate of 7.2 MBytes (14.4 MBytes) per second. In five hours this leads to 129.6 GBytes (259.2 GBytes) of data to be stored, per night of observation. On-site fast storage facilites are required, with fast off-site access to download the data elsewhere.

3.4 ELT Visitor Focus

A visitor focus/auxiliary port is crucial to accommodate new unexpected science (that will most certainly occur with the ELT), test instrumental developments and allow dedicated niche instruments proven on 4–8 m telescopes to pursue ELT science. Similarly, space in the focal plane of non-HTRA instruments could be given to allow observations with a fast detector.

3.5 HTRA with Non-HTRA Specific ELT Instruments

The ELT provides an opportunity to accommodate HTRA through its acquisition camera, a test/commissioning camera or FOSC (Faint Object Spectrograph Camera). FOSCs have been designed to be highly efficient with high throughputs for faint object work, and so are ideally suited for low flux HTRA studies [67, 68]. They also allow dispersive and polarimetric optics to be easily inserted and changed in front of a camera (e.g. L3CCD). Instead of a 'standard' CCD, a fast detector, L3CCD or similar should be installed as the acquisition camera, e.g. ESO is already developing L3CCDs as wave front sensors for AO [69]. Not only will this provide its primary function for acquisition, but it will allow simultaneous high speed observations to be taken, e.g. SALTICAM on the 11 m SALT [12, 13], and AcqCam on the 8 m Gemini South telescope in Chile [52, 53].

One of the first most important steps to be taken to enable HTRA observations in non-HTRA specific instruments, is to reduce the readout deadtime between detector exposures. Not only will this increase the efficiency of an observation but will save hours of telescope time. This not only benefits the HTRA community, but all telescope users. Minimising detector deadtime, getting the data out as quickly as possible, is a benefit to everyone, increasing telescope usage.

The other instruments under consideration within the previously mentioned small studies package can be used for HTRA, but only if this optional mode of observation is considered at an early stage in their design. The basic necessity is a fast detector capable of sub-second frame rates, and the ability to store the extra volume of frames when used at these HTRA rates.

- MIDIR: The MIDIR (a mid-infrared combined imager and spectrograph) will use AQUARIUS 1k × 1k Si:As BB detectors (private communication with B. Brandl). Their maximum frame rate is 150 Hz, giving in principle a time resolution of 6.7 ms, excellent for HTRA science. It should be technically feasible to store individual frames at that rate, with the ELT perhaps allowing sufficient S/N in an individual frame. The standard readout mode has 8 channels or 32 channels, which are "vertical strips". It may be possible to further increase the time resolution by scanning the object across the detector (i.e., across the strips).
- MOMSI, HiSPEC: MOMSI (a multi-object multi-field spectrograph and imager) and HISPEC (high resolution visual spectrograph) are planning to use detectors (Focal Plane Arrays) similar to those already in use in WFCAM on UKIRT, in VISTA (the Visible and Infrared Survey Telescope for Astronomy), and the JWST NIRCAM (private communication with C. Evans). WFCAM has four 2048 × 2048 Rockwell Hawaii-2 HgCdTe detectors. VISTA uses sixteen 2048 × 2048 Raytheon VIRGO HgCdTe detectors. The Rockwell Hawaii-2 FPA can have 8 Hz full frame readout (125 ms), or 16 ms readout of a 64 × 64 subarray using 32 outputs [70]. The Raytheon VIRGO FPA can have a maximum frame rate of 1.456 Hz (685.75 ms) using 16 outputs [71]. If either is used in MOMSI there is the possibility of HTRA observations at a very usable and useful subsecond time resolution.
- SCELT, sub-mm imager: Transition Edge Sensors (TES) will be used as they have been for SCUBA-2 [72, 73]. These have high capability for HTRA as discussed in Romani's chapter [56].
- EPICS, exoplanets imaging camera and spectrograph: The Differential Polarimetric Imager channel is based on the CHEOPS/ZIMPOL concept which uses a CCD with a frame readout rate of 34 ms (signal modulation occurs at 1–50 kHz) [36]. This frame rate allows HTRA observations to be performed.

3.6 ELT Tracking and Guiding

The diffraction limit of the 42 m ELT is to be ∼2.4 milliarcsec at 500 nm, ∼6 milliarcsec at 1000 nm. Blind pointing, tracking and guiding needs to be able to operate accurately and precisely to this level for up to 10 hours of continuous observation of time-tagged data, thereby allowing a continuous sequence of fast observations.

3.7 Derotator

A mechanical derotator in the interface between the telescope and HTRA instrument is preferable, as an optical detrotator introduces prohibitive polarization effects.

3.8 Atmospheric Dispersion Corrector (ADC)

Most of the ELT suite of instruments are likely to have an ADC incorporated into their internal optical train. Once these instruments also have a fast detector, HTRA is possible. The visitor focus should have an ADC built-in, and not rely on visiting instruments to have it—a HTRA specific instrument for the optical absolutely needs an ADC to allow broadband, high speed, maximum throughput, imaging and spectroscopic/polarimetric observations.

3.9 ELT Location

The ELT site is not as critical for optical HTRA observations as it is for spectroscopy and other instruments. HTRA targets exist and will be found in both hemispheres. Synergy with ALMA and JWST is necessary to allow HTRA simultaneous and follow-up observations, as mentioned in Marsh's chapter [4].

The location of the ELT and its pointing capability will put limits on the declination of observable targets. From Chajnantor (longitude: W 67° 45' 18" latitude: −23° 1' 22"), for an object to be sighted above 30° altitude (60° zenith angle for the ELT) its declination must lie within the bounds of $\sim +37°$—$\sim -83°$ (calculations using Staralt [74]). For example the Crab pulsar (RA: 5h 34m 32s dec: +22° 00' 52") rises to a maximum of 45° altitude (1.41 airmass) above the horizon at Chajnantor; Geminga pulsar (RA: 6h 33m 54s dec: +17° 46' 13") rises to $\sim 50°$; Vela pulsar (RA: 8h 35m 21s dec: −45° 10' 36") rises to $\sim 68°$.

3.10 Prototypes

Prototype development for visitor focus usage at 8–10 m class telescopes is ongoing to assess performance, use and interest from other astronomers in HTRA instruments (e.g. the Galway Stokes Polarimeter, Aqueye , S-CAM3, etc.). Established visitor and facility instruments (notably ULTRACAM, LuckyCam, FORS, and others—see Table 4) are already noteworthy in their successes. Positive responses at the prototype level will lead to further development to take advantage of an ELT visitor focus for niche, experimental and unexpected HTRA science.

3.11 Adaptive Optics

An AO system has a complex control structure with many nested feedback loops. One might expect modulation of Strehl, at some level, by control loop characteristic frequencies. If, then, a small aperture is used to derive a high SNR signal for subsequent time domain analysis, these control loop frequencies will appear at some level throughout the temporal power spectrum of the light. Since one approach to the detection of periodic signals in HTRA is to look for specific frequencies (where a period is a-priori known) or excesses (where there is no known period), periodic modulation of the signal (by AO) will reduce the SNR in the frequency domain.

The details of these effects will need to be investigated when the electro-mechanical designs for AO systems are being modelled.

Diffraction limited imaging is preferable for HTRA imaging to maximize signal-to-noise and for working with small fields of view, though in the optical achieving diffraction limits will be difficult. It is easier in the NIR where a lot of HTRA is being accomplished.

4 HTRA Detectors

The current standard in detectors for high time resolution astrophysics must be surpassed in order to promote new science in this parameter space, especially towards ultra-high time resolutions. These current state-of-the-art detectors are described to show the choices available to this initial HiTRI-HTRA study.

4.1 Single Photon Avalanche Diode (SPAD) Arrays

Refer also to Phelan and Morrison's chapter of this volume [75]. With further development in the near future it would be advantageous to photon imaging if a 256×256 or bigger array of geiger mode avalanche photodiodes was produced. Zappa et al. [60, 61] have developed a circular SPAD (Single Photon counting Avalanche photoDiode) array with 60 elements, each element containing 4 pixels of sizes 20 μm, 35 μm, 50 μm and 75 μm—only one size is active, chosen for science use and dark counts. Dark counts at modest cooling of $-15\,°C$ are 15 cps, 40 cps, 160 cps and 1000 cps per pixel size respectively. QE ranges with bias voltage above breakdown voltage from 30% to 50% at 500 nm, with a QE of >10% over 400–800 nm. Low voltages are required <40 V. Photon timing resolution is ~100 ps at FWHM. Maximum photon rates are ~20 Mcps for the array, with a best timing window of 10 μs. Higher QE comes with increased depletion regions in the APDs but with a cost of higher operating voltages (up to 500 V), and higher manufacturing costs. Dark counts decrease with cooling but the afterpulsing probability increases. Dark counts increase with diode size. The fill factor is small <5% (so a lenslet array is required to increase the fill factor to 100%) to eliminate optical crosstalk (each APD diode emits photons when an avalanche is initiated, whether by thermal electron-hole pairs or detected photon).

4.2 Photocathode Imager

Refer to Siegmund's chapter [76]. The image tube detector is well understood, has been used for numerous space instrumentation (e.g. FUSE, GALEX, IMAGE, SOHO), and has been developed into GaAs photocathode imaging detectors [59, 77]. It allows photon counting imaging over a broad range of 400–800 nm with a timing resolution of $\sim$2 ns, and a flat QE of $\sim$35% (which is the combined photocathode QE of $\sim$50% and open area efficiency of the MCP of $\sim$70%). The readout method may have an associated read noise (e.g. a CCD [78]), but generally a pixelated, cross delay line or cross strip anode is used which does not contribute to the noise [77, 79, 80, 81, 82]. The subsequent non-detector based DAQ electronics limit the timestamping to a possible maximum 25 ns with an improved GPS stamping system (GPS information quoted from Sematron UK Ltd. [83]). The in-development 512 × 512 ($\sim$25 μm) pixels will allow narrow field imaging as well as giving leeway for wide field work. Unfortunately high voltages (few kV) are required to operate the image tube, and the gain and photocathode will deteriorate over time ($\sim$1.2x10^5 hours to 50% of anode output [84, 85, 86]). The maximum overall counting rate is generally $\sim$1 Mcps where the detector output becomes non-linear. The maximum per pixel rate is $\sim$100 cps when used with the latest 'hot' channel multiplier arrays. Photocathode burning and degradation will occur at greater rates.

4.3 Intensified CCD (ICCD)

The ICCD is an imaging, photon counting, intensified CCD developed for high speed photometry [87]. The ICCD is optimised for the UV by the use of a RbTe photocathode on the input of a Z stack microchannel plate (MCP) (40 mm diameter, 10 μm pores on 12 μm pitch, $\frac{L}{D}$ of 40), the P20 output phosphor of which is coupled onto a CCD through a 1:3.6 fibre taper. The CCD has 512 × 512 15 μm pixels and can be read out at a maximum speed of 220 frames per second (if only using 512 × 128 pixels). With a readout time of 50.6 ns per pixel, an overall time of 4.5 ms is given to read half the CCD.

As each CCD frame is read out (with its maximum intrinsic time resolution of 4.5 ms) it is analysed by high speed FPGA programming to identify valid events, calculate position centroids, and produce a time tagged spatial coordinates list. A 3 × 3 pixel window is used to scan the frame at the 50.6 ns pixel clock for events. The detection and centroiding of each individual event is done before subsequent events occur.

The spatial coordinates are saved in series with time tags which denote the end of each successive frame, and further offline processing gives the time to each event. A spatial resolution of 25 μm FWHM is given, set by the granularity of the MCP and the proximity focusing of the photocathode.

4.4 pnCCDs

Refer to Streuder's chapter [88]. The Halbleiterlabor (HLL) of MPG (Max Planck Gesellschaft) have developed a frame store pnCCD, available in a format of 264 × 568 pixels (264 × 264 image and storage areas with double-sided readout), each pixel being 51 × 51 μm^2. 512 × 1024 pixels is upcoming. The device has an image transfer time of 20 μs, has a 100% fill factor, and an internal quantum efficiency of <70% over a broad wavelength range of 300–950 nm. The readout noise is dependent on the frame rate: at $-50°C$, 1.8 e^- at 10–400 fps; 2.3 e^- at 400–1100 fps; 2.3 e^- at 2200–4400 fps with 4 pixel binning. The QE is > 80% over the 400–1100 nm bandwidth. For photon counting the pnCCD will be coupled to an avalanche amplifier in a proportional mode, with a new readout strategy to decrease read noise to 0.5 e^-.

4.5 L3CCDs

Refer to Smith's chapter of this volume [89]. Electron multiplication CCDs (low light level CCDs, or L3CCDs) are CCDs incorporating hundreds of multiplication stages before the readout amplifier allowing the amplification of a detected photon's charge above a noise threshold. The overall gain is given by $(1+p)^n$, where n is the number of multiplication stages and p is the probability of each stage converting one electron into two. The gain process is statistical so there is a wide dispersion in the number of electrons generated. The gain can be set so it overcomes the readout noise of the output amplifier, but overall DQE is reduced by a half (due to the multiplication noise term of $\sqrt{2}$). The gain can be set higher again for photon counting so that the readout noise of the output amplifier is negligible and the full detector quantum efficiency (DQE) is restored by setting a threshold level that determines all detections to be a photon. In this mode the input photon rates have to be low to avoid coincidence losses, and the maximum rate without significant, though correctable, non-linearity is 1 photon per pixel per 30 frames (for the 512 × 512 CCD65 from E2V) [62]. With high coincidence losses the DQE can reduce towards half its value again, though these losses are the same as for other intensifier based photon counting systems (e.g. ICCDs), with the exception that the L3CCD is a relatively low-cost, solid state device with higher quantum efficiency when back-illuminated and has immunity to light overload [62]. The maximum frame rate is 40 frames per sec for the full frame of the 512 × 512 L3CCD, but a CCD with multiple outputs can allow faster frame rates as can sytems that read out a smaller sub-array of the CCD, e.g. a sub-array of 128 × 128 pixels on the CCD65 allows a 500 Hz (2 ms) frame rate [62]. The L3CCD is cooled to $-140°C$, as per normal CCDs, to eliminate dark current for photon counting. Gain stability is achieved with excellent clock-high stability, and with control of clock waveform ringing repeatability—the gain at a specific voltage can vary significantly with temperature, but with sufficient gain and a suitable detection threshold this should not affect photon counting.

Also the dynamic range of the detector is reduced by a factor equal to the applied gain, but again in photon counting mode at low fluxes this should not be a problem.

ESO and JRA2 OPTICON have funded e2v technologies to develop a compact packaged Peltier cooled 24 µm square 240 × 240 pixels split frame transfer 8-output back-illuminated L3Vision CCD for Adaptive Optic Wave Front Sensor (AO WFS) applications [69]. The device is to achieve sub-electron read noise at frame rates from 25 Hz to 1,500 Hz and a dark current lower than 0.01 e-/pixel/frame. To obtain high frame rates, multi-output EMCCD gain registers and metal buttressing of row clock lines are used. Deep depletion silicon devices will have improved red response and an electronic shutter will aid Rayleigh and Pulsed Laser Guide Star applications.

4.6 Cryogenic Imaging Spectrophotometers

A large array of photon counting, energy resolving pixels answers the requirements of a high time resolution imager with inherent spectroscopic abilities. The pixels could be superconducting tunnel junctions, transition edge sensors or kinetic inductance detectors. Polarization optics are essentially the only extra optics required.

4.6.1 Superconducting Tunnel Junctions

The current S-Cam 3 [90, 91] is a 10 × 12 array of STJ tantalum pixels 33 µm × 33 µm with 4 µm interpixel gap (76% fill factor). The array is divided into 4 electrically separated subarrays of 10 × 3, each subarray sharing a common return wire to a base electrode, making it easier to bias all pixels simultaneously. All pixels are low leakage (~30 pA) and show uniformity across the array with 1 $\sigma = 3.3\%$. They are sensitive to wavelengths up to 1 mm so rejection of thermal infrared photons is crucial. An anti-reflection coating optimises the optical response of ~30% at 500 nm, with a bandwidth at 10% of maximum covering 330–745 nm. The STJ array operating temperature is 285 mK, obtained using liquid helium, which needs to be refilled every 24 hrs. GPS time tagging gives the photon time of arrival. The intrinsic detector wavelength resolution at 500 nm is 8–13. There is essentially no readout noise. The contribution from noise events is always negligible but an estimate for the S-Cam3 mode allowing the highest count rates and using the standard filtering criterium (aimed at maintaining all real photons) is 1.4×10^{-3} events/pixel/s (compared with source count rates of ~5000 ph/s/pix) (private communication from T. Oosterbroek).

Achieved system throughputs are 10^4 photons per sec per pixel and 5×10^5 photons per sec for the whole array. Wiring problems limit the maximum size of the STJ array as each pixel needs to be individually biased and read out. The Distributed Read-Out Imaging Detector (DROID) may provide an avenue to large arrays. The DROID absorbs photons in a superconducting absorber with STJs at either end (1-D linear structure) or with STJs at corners of a square absorber (2-D structure). The DROIDS then act similarly to a position sensitive diode but the photon energy can be reconstructed from the sum of the amplitudes of all STJ signals. The geometry is under investigation with 1-D DROIDS seen as the likely candidate for larger arrays. The spectral resolving power of DROIDs is ~10%–50% poorer depending on pixel size, being worse for larger sizes.

4.6.2 Transition Edge Sensors (TES)

Refer to Romani's chapter [56]. Transition edge sensors are another superconducting detector used primarily for X-rays and sub-mm wavelengths (as in SCUBA2 - this contains two focal plane units, each containing 4 arrays with 40×32 pixels giving a total of 10240 pixels) but with applications in the optical. An optical imaging TES array from Stanford University [58] contains 8×4 20 μm tungsten pixels held at a temperature of 50 mK. They detect single photon events above a threshold of 0.3 eV (4 μm wavelength), with an energy resolution of 0.15 eV (best) at FWHM 3 eV (spectral resolution of ~20, at 415 nm), and with a photon arrival-time resolution of <300 ns. The tungsten has an efficiency for absorbing photons ranging from ~50% in the optical/UV to ~10% in the near IR. The pixels are sensitive up to the saturation energy of 10–15 eV (125–83 nm). As the sensors are photon counting with energy resolution there is zero dark current and zero read noise. There is the issue of a noise distribution associated with the shunt resistor's Johnson noise, read electronics, etc. This noise distribution generally has an rms of ~0.1 eV and is quite close to Gaussian.

A reflecting metalized mask shields photons from hitting interpixel biasing wires (hitting these would change the photon energy) and directs the photons directly onto the TES surface, increasing the effective fill factor. The maximum count rate is ~30 kHz (the pixel has a recovery time constant of 2 μs). A superconducting quantum interference device (SQUID) is used to amplify the TES pixel current produced in response to an absorbed photon. While it is possible to fabricate large format arrays (up to 40×32) and create larger pixel arrays through butt coupling these arrays (as for SCUBA-2), reading out each individual pixel is a challenge and requires multiplexing of the SQUIDs. Without multiplexing constraints on refrigeration cooling power and wiring complexity would limit direct SQUID readout to only a few TES pixels.

Single pixel behaviour has been characterised for the optical array and a scaling has begun to operate multiple pixels simultaneously up to the use of the full array. There is a runtime of 10 hours for devices with a critical temperature of 100 mK before having to regenerate the magnet.

4.6.3 Kinetic Inductance Detector

This uses a different method to STJs and TESs to readout the bolometer which could allow large pixel counts, high sensitivity, Fano-limited resolution, and more easily multiplexed arrays [92]. They have been mentioned in passing in the SCELT study [93] where admission is made that they are not as developed as TESs and STJs. Primary use is for sub-millimetre to X-ray observations. Photons absorbed in a superconductor affect the surface reactance (dubbed kinetic inductance) such that this reactance is a function of the density of the quasiparticles produced, even at temperatures $\ll T_c$ (critical superconducting transition temperature). For readout, the superconductor can act as an inductive element in a resonant circuit. A change in the kinetic inductance changes the resonant frequency which can be monitored by measuring the transmission phase of a constant-frequency microwave probe signal transmitted through (or reflected from) the resonator (using a high electron mobility transistor—HEMT). Then, frequency multiplexing offers the prospect of large arrays with wireless monitoring, each superconductor circuit using a different resonant frequency.

4.7 Detector Limiting Magnitudes

Figure 2 shows V band limiting magnitude curves (for S/N of 10) for seven main HTRA detectors on a 42 m aperture telescope, based in part on Table 5. This table has the photon rates expected to be collected by the 42 m ELT, and the integration times necessary to reach a S/N ratio of 10 for three of the detectors (times were calculated using the Massey CCD equation found in [94]). Table 6 gives the assumed

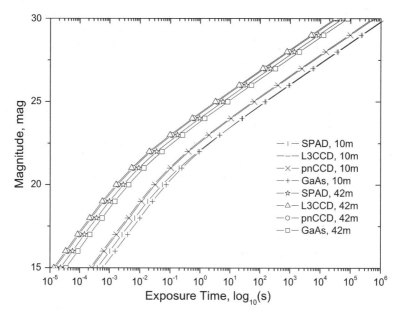

Fig. 2 Limiting magnitude graph for 42 m and 10 m telescope apertures. Exposure times (for S/N = 10) for the range of magnitudes were calculated using the Massey CCD equation found in [94]. See Tables 6 and 7, and main text

Table 6 Assumed ELT, detector, and sky parameters for a 1 arcsec object imaged onto 2 × 2 pixels in V band. See also Table 7, and [95]

Parameter	Value
V band flux at m=0	3640 Jy
V band dλ/λ	0.16
1 Jy	1.51×10^7 photons s^{-1}m^{-2}(dλ/λ)$^{-1}$
Night sky, new moon, V	21.8 mag per arcsec^2
Tel. aperture obscuration	0.2
Signal to Noise	10
Image Sampling	Nyquist (2 × 2 pixels)
Atmospheric extinction	0.21 mag
Telescope throughput	0.81
Detector optics throughput	0.7

telescope and sky characteristics, while Table 7 gives the detector characteristics used in the calculations.

From Fig. 2 the 42 m aperture allows ∼1.6 mag increase in the detection limit over a 10 m aperture. This means an observation to 25th mag for the GaAs or L3CCD is ∼17.7 times faster in time, and for the SPAD is 27.5 times faster. The spread in detector limits at 42 m is ∼0.6 mag implying that the detectors are quite close in terms of detection capabilities, though not all allow imaging (SPAD, TES, STJ, PerkinElmer's Slik APD, etc.), nor photon counting (pnCCD as yet). They all have fast frame rates or high time resolutions.

An eight hour integration is 2.88×10^4 seconds which, from Table 5, limits observations to targets of ∼29 or 30 mag for any of the three detectors listed.

Table 7 Detector parameters

Detector	Pixel Size μm	Dark Count ph. pix^{-1}s^{-1}	Read Noise e$^-$	QE	Op. temp °C
GaAs[a]	55	0.0006	0	0.35	−20
pnCCD[b]	51	0	2.3	0.8	−50
L3CCD[c]	20×30	0	1	0.9	−140
SPAD[d]	50	50	0	0.53	−20
Slik APD[e]	180	25	0	0.7	?
TES[f]	20	0	0	0.5	70 mK
STJ[g]	33	0	0	0.3	285 mK

[a]From [59]
[b]From [88, 96]
[c]From [62, 63]
[d]From [61]
[e]For SPCM-AQR-16, from [97]. Operating temperature not stated.
[f]From Sect. 4.6.2 and [56, 57]
[g]From Sect. 4.6.1 and [90, 91]

5 Conclusion

A 42 m Extremely Large Telescope (ELT) would be capable of allowing extremely detailed optical studies of objects to be conducted on millisecond timescales, but also on microsecond and possibly nanosecond timescales. There is already an amazing variety of science being pursued in HTRA on 8–10 m telescopes using polarimetry, photometry and spectroscopy to study exoplanet transits, eclipsing CVs, polars, globular clusters, pulsars, neutron star accreters, X-ray binaries, optically violent variable jets, and much more. While a dedicated HTRA instrument is unlikely to be adopted into the ELT first light suite of instruments, the ELT should and needs to be ready to make use of its photon-gathering abilities and accommodate high speed sub-millisecond observations—a time resolution expressed to be most useful by attendees at the Galway HTRA Workshop 2006 [98], and contributors to this book. This can be achieved with a fast detector acquisition camera (e.g. an L3CCD as is already in production as an ESO wave front sensor [69]), through HTRA options in non-HTRA specific instruments (high frame rate CCDs or focal plane arrays), or through HTRA instruments at the visitor focus. In the case of 'ELT unexpected science' we cannot afford not to have a fast detector capable of imaging with at least sub-second frame rates.

HTRA detectors are inherently fast detectors. Their application not only lies in capturing fast variability in astronomical objects, but in having a much higher saturation limit for any (non-)HTRA bright object due to the fast refresh rate. A fast detector can handle the increased photon rates expected from an ELT (the technical advantage of an increase in telescope size is in reduced scintillation on bright objects) and will also help to maximize usable telescope time through minimizing inter-frame deadtimes.

5.1 Recommendations

To accommodate HTRA observations, the ELT needs:

- minimized (possible with a minimum-mirror configuration) and calibrated/monitored polarization at all altitudes and rotation positions
- integrated and implemented absolute timing throughout the observatory
- a fast detector/acquisition camera that is HTRA compatible
- the first light instruments to contain a HTRA option in the form of a fast detector, or shared focal plane with a HTRA instrument
- decreased deadtime in all detectors used, which aids HTRA and all other observations by maximizing useful time spent on the telescope
- data storage and handling capacities for HTRA data rates (10^6 photons per second leading to hundreds of GBytes to be stored per night)
- a visitor focus, or FOSC, to bring HTRA-specific (and other specialised/experimental) instruments to the ELT
- the AO system to be modelled and investigated for effects in the data of HTRA

Through 'prototype' developments and specialised HTRA instruments (e.g. OPTIMA, ULTRACAM, GASP, etc.), and facility instruments (UCTPol, LRIS, FORS2, SALTICAM, AcqCam, etc.), all niches of HTRA are covered which no one instrument can easily do. A visitor focus then becomes all the more important. The popularity of these existing instruments (Table 4) and the amount of observing time already granted are indications of the useful science astronomers are actively pursuing. These instruments will lead the way to future HTRA instrument developments and hopefully a permanent HTRA fixture on an ELT.

References

1. I. M. Hook, ed., *The Science Case for the European Extremely Large Telescope: The next step in mankind's quest for the Universe*, ESO, 2005. http://www.eso.org/projects/e-elt/science.html.
2. D. Phelan, O. Ryan, and A. Shearer, eds., *High Time Resolution Astrophysics*, ch. Stellar Pulsation and High Time Resolution Astrophysics by C.S. Jeffery. Springer, 2007.
3. D. Phelan, O. Ryan, and A. Shearer, eds., *High Time Resolution Astrophysics*, ch. High Time Resolution Observations of CVs by S. Littlefair. Springer, 2007.
4. D. Phelan, O. Ryan, and A. Shearer, eds., *High Time Resolution Astrophysics*, ch. High-Speed Optical Spectroscopy by T. Marsh. Springer, 2007.
5. D. Phelan, O. Ryan, and A. Shearer, eds., *High Time Resolution Astrophysics*, ch. High-speed optical observations of X-ray binaries by T. Shahbaz. Springer, 2007.
6. D. Phelan, O. Ryan, and A. Shearer, eds., *High Time Resolution Astrophysics*, ch. High Time Resolution Astrophysics and Pulsars by A. Shearer. Springer, 2007.
7. D. Dravins, C. Barbieri, V. Da Deppo, D. Faria, S. Fornasier, R. Fosbury, L. Lindegren, G. Naletto, R. Nilsson, T. Occhipinti, F. Tamburini, H. Uthas, and L. Zampieri, "Quanteye. quantum optics instrumentation for astronomy.," tech. rep., ESO, 2005. OWL Instrument Concept Study. ESO document OWL-CSR-ESO-00000-0162,280 pp. (2005).
8. D. Phelan, O. Ryan, and A. Shearer, eds., *High Time Resolution Astrophysics*, ch. ULTRACAM: an ultra-fast, triple-beam CCD camera for high-speed astrophysics by V. Dhillon. Springer, 2007.
9. A. Shearer, "High time resolution astrophysics and extremely large telescopes," in *Emerging Optoelectronic Applications. Edited by Jabbour, Ghassan E.; Rantala, Juha T. Proceedings of the SPIE, Volume 5382.*, A. L. Ardeberg and T. Andersen, eds., pp. 748–754, jul 2004.
10. D. Phelan, O. Ryan, and A. Shearer, eds., *High Time Resolution Astrophysics*, ch. From Quanteye to Aqueye—Instrumentation for astrophysics on its shortest timescales by C. Barbieri. Springer, 2006.
11. D. Phelan, O. Ryan, and A. Shearer, eds., *High Time Resolution Astrophysics*, ch. An Ultra-High-Speed Stokes Polarimeter for Astronomy by M. Redfern and P. Collins. Springer, 2007.
12. D. O'Donoghue, "Salticam specifications." http://www.salt.ac.za/telescope/instrumentation/salticam/specifications%/.
13. D. A. H. Buckley, P. A. Charles, K. H. Nordsieck, and D. O'Donoghue, "The Southern African Large Telescope project," in *IAU Symposium*, P. Whitelock, M. Dennefeld, and B. Leibundgut, eds., pp. 1–12, 2006.
14. E. Burgh, "Robert Stobie Spectrograph Observer's Guide." http://www.sal.wisc.edu/%7eebb/pfis/observer/overview.html#optdesign.
15. S. Beard, A. Vick, D. Atkinson, V. Dhillon, T. Marsh, S. McLay, M. Stevenson, and C. Tierney, "The Ultracam camera control and data acquistion system," SPIE, (Waikoloa, Hawaii, USA), 22–28 August 2002. http://www.roe.ac.uk/atc/electronics/publishedpapers/ultracomspie.pdf.
16. V. Dhillon, "Ultracam website." http://www.shef.ac.uk/physics/people/vdhillon/ultracam/.

17. N. M. Law, C. D. Mackay, and J. E. Baldwin, "Lucky imaging: high angular resolution imaging in the visible from the ground," *Astronomy and Astrophysics* **446**, pp. 739–745, Feb. 2006.
18. C. Mackay, "Luckycam." http://www.ast.cam.ac.uk/ optics/Lucky_Web_Site/.
19. N. M. Law, C. D. Mackay, and J. E. Baldwin, "Lucky imaging: high angular resolution imaging in the visible from the ground," *Astronomy and Astrophysics* **446**, pp. 739–745, Feb. 2006.
20. R. M. Redfern, A. Shearer, R. Wouts, P. O'Kane, C. O'Byrne, P. D. Read, M. Carter, B. D. Jordan, and M. Cullum, "Crowded field photometry using post-exposure image sharpening techniques," in *IAU Colloq. 136: Stellar Photometry - Current Techniques and Future Developments*, p. 147, 1993.
21. A. Shearer, B. Stappers, P. O'Connor, A. Golden, R. Strom, M. Redfern, and O. Ryan, "Enhanced Optical Emission During Crab Giant Radio Pulses," *Science* **301**, pp. 493–495, July 2003.
22. R. F. Butler, A. Shearer, R. Redfern, M. Colhoun, P. O'Kane, A. Penny, P. Morris, W. Griffiths, and M. Cullum, "TRIFFID photometry of globular cluster cores - I. Photometric techniques and variable stars in M15," *Mon. Not. R. Astron. Soc.* **296**, pp. 379–391, 1998.
23. O. Ryan, *Development of a Photon Counting Imaging Polarimeter for Astronomy.* PhD thesis, National University of Ireland, Galway, 2004.
24. O. Ryan, M. Redfern, and A. Shearer, "High speed imaging photon-counting system," in *Proceedings of the SPIE: High Energy, Optical, and Infrared Detectors for Astronomy II*, D. Dorn and A. Holland, eds., **6276**, pp. 62760G–1–8, 2006.
25. C. Straubmeier, G. Kanbach, and F. Schrey, "OPTIMA: a photon counting high-speed photometer," *Experimental Astronomy* **11**, pp. 157–170, 2001.
26. G. Kanbach, A. Sᵃowikowska, S. Kellner, and H. Steinle, "New optical polarization measurements of the crab pulsar," in *American Institute of Physics Conference Series*, T. Bulik, B. Rudak, and G. Madejski, eds., Nov 2005.
27. G. Kanbach, "OPTIMA website." http://www.mpe.mpg.de/gamma/instruments/optima/www/optima-papers.html.
28. D. Phelan, O. Ryan, and A. Shearer, eds., *High Time Resolution Astrophysics*, ch. OPTIMA: A High Time Resolution Optical Photo-Polarimeter by G. Kanbach et al. Springer, 2007.
29. V. Plokhotnichenko, G. Beskin, V. Debur, A. Panferov, and I. Panferova, "The multicolor panoramic photometer-polarimeter with high time resolution based on the PSD," in *PSD6: Sixth International Conference on Position Sensitive Detectors, Nuclear Instruments and Methods* **A513**, pp. 167–171, 2003.
30. S. Karpov, D. Bad'in, G. Beskin, A. Biryukov, S. Bondar, G. Chuntonov, V. Debur, E. Ivanov, E. Katkova, V. Plokhotnichenko, A. Pozanenko, I. Zolotukhin (Russia), K. Hurley (USA), E. Palazzi, N. Masetti, E. Pian, L. Nicastro, C. Bartolini, A. Guarnieri, D. Nanny, A. Piccioni (Italy), N. Brosch, D. Eichler, (Israel), A. Shearer, A. Golden, M. Redfern (Ireland), J.-L. Atteia, and M. Boer (France), "FAVOR (FAst Variability Optical Registration) - two-telescope complex for detection and investigation of short optical transients," *Astronomische Nachrichten* **325**, pp. 677–677, Oct. 2004.
31. A. Ramaprakash, R. Gupta, A. Sen, and S. Tandon, "An imaging polarimeter (IMPOL) for multi-wavelength observations," *Astronomy and Astrophysics Supplement Series*, pp. 1–7, March 1998.
32. A. K. Sen, R. Gupta, A. N. Ramaprakash, and S. N. Tandon, "Imaging polarimetry of some selected dark clouds," *Astronomy and Astrophysics Supplement* **141**, pp. 175–183, Jan. 2000.
33. A. N. Ramaprakash, S. N. Tandon, and R. Gupta, "Imaging Polarimetry of Nearby Molecular Clouds," *LNP Vol. 506: IAU Colloq. 166: The Local Bubble and Beyond* **506**, pp. 243–246, 1998.
34. C. Keller, "Instrumentation for astrophysical spectropolarimetry," in *Astrophysical Spectropolarimetry*, J. Trujillo-Bueno, F. Moreno-Insertis, and F. Sanchez, eds., pp. 303–354, Cambridge University Press, 2002.
35. H. Povel, "Imaging stokes polarimetry with piezoelastic modulators and ccd image sensors," *Optical Engineering* **34**, pp. 1870–1878, July 1995.

36. H. M. Schmid, D. Gisler, F. Joos, H. P. Povel, J. O. Stenflo, M. Feldt, R. Lenzen, W. Brandner, J. Tinbergen, A. Quirrenbach, R. Stuik, R. Gratton, M. Turatto, and R. Neuhäuser, "ZIM-POL/CHEOPS: a Polarimetric Imager for the Direct Detection of Extra-solar Planets," in *Astronomical Society of the Pacific Conference Series*, A. Adamson, C. Aspin, and C. Davis, eds., pp. 89–+, Dec. 2005.
37. J. Stenflo, "Imaging polarimetry: Opportunities and limitations," SPIE, (Waikoloa, Hawaii, USA), 22–28 August 2002. http://www.astro.phys.ethz.ch/papers/stenflo/stenflo_p_nf.html.
38. W. Skidmore, K. O'Brien, K. Horne, R. Gomer, J. B. Oke, and K. J. Pearson, "High-speed Keck spectroscopy of flares and oscillations in AE Aquarii," *Monthly Notices of the Royal Astronomical Society* **338**, pp. 1057–1066, Feb. 2003.
39. W. Skidmore, K. Horne, K. Pearson, R. Gomer, K. O'Brien, and B. Oke, "Rapid Keck Spectroscopy of Cataclysmic Variables," in *Revista Mexicana de Astronomia y Astrofisica Conference Series*, G. Tovmassian and E. Sion, eds., pp. 155–157, July 2004.
40. K. O'Brien, "Rapid optical spectroscopy of X-ray binaries with Keck II," *ArXiv Astrophysics e-prints*, Oct. 2001.
41. K. O'Brien, K. Horne, R. H. Gomer, J. B. Oke, and M. van der Klis, "High-speed Keck II and RXTE spectroscopy of Cygnus X-2 - I. Three X-ray components revealed by correlated variability," *Monthly Notices of the Royal Astronomical Society* **350**, pp. 587–595, May 2004.
42. D. Steeghs, K. O'Brien, K. Horne, R. Gomer, and J. B. Oke, "Emission-line oscillations in the dwarf nova V2051 Ophiuchi," *Monthly Notices of the Royal Astronomical Society* **323**, pp. 484–496, May 2001.
43. ESO, "Focal reducer and low dispersion spectrograph." https://www.eso.org/instruments/fors1/.
44. C. Cumani and K.-H. Mantel, "Phase resolved high speed photometry and spectroscopy of pulsars," *Experimental Astronomy* **11**, pp. 145–150, Apr. 2001.
45. D. Phelan, O. Ryan, and A. Shearer, eds., *High Time Resolution Astrophysics*, ch. Fast spectroscopy and imaging with the FORS2 HIT mode by K. O'Brien. Springer, 2007.
46. NAOJ, "Faint object camera and spectrograph." http://www.naoj.org/Observing/Instruments/FOCAS/index.html.
47. A. P. Reynolds, G. Ramsay, J. H. J. de Bruijne, M. A. C. Perryman, M. Cropper, C. M. Bridge, and A. Peacock, "High-speed, energy-resolved STJ observations of the AM Her system V2301 Oph," *Astronomy and Astrophysics* **435**, pp. 225–230, May 2005.
48. D. Steeghs, M. A. C. Perryman, A. Reynolds, J. H. J. de Bruijne, T. Marsh, V. S. Dhillon, and A. Peacock, "High-speed energy-resolved STJ photometry of the eclipsing dwarf nova IY UMa," *Monthly Notices of the Royal Astronomical Society* **339**, pp. 810–816, Mar. 2003.
49. C. M. Bridge, M. Cropper, G. Ramsay, M. A. C. Perryman, J. H. J. de Bruijne, F. Favata, A. Peacock, N. Rando, and A. P. Reynolds, "STJ observations of the eclipsing polar HU Aqr," *Monthly Notices of the Royal Astronomical Society* **336**, pp. 1129–1138, Nov 2002.
50. SAAO, "UCT photometer polarimeter SAAO." http://www.saao.ac.za/facilities/instrumentation/uct-photometerpolarimeter%/.
51. S. Potter, E. Romero-Colmenero, and D. A. H. Buckley, "Stokes imaging and Doppler mapping of the magnetic cataclysmic variable V834 Cen," *Astronomische Nachrichten* **325**, pp. 201–204, Mar. 2004.
52. P. Puxley, "Acquisition camera." http://www.gemini.edu/sciops/telescope/acqcam/acqDetector.html.
53. R. I. Hynes, P. A. Charles, J. Casares, C. A. Haswell, C. Zurita, and T. Shahbaz, "Fast photometry of quiescent soft X-ray transients with the Acquisition Camera on Gemini-South," *Monthly Notices of the Royal Astronomical Society* **340**, pp. 447–456, Apr. 2003.
54. B. Garcia-Lorenzo, S. Tulloch, and R. Bassom, "ISIS rapid spectroscopy mode." http://www.ing.iac.es/Astronomy/instruments/isis/fast_spectroscopy_mode% .html.
55. A. S. Mukadam and R. E. Nather, "Argos: An optimized time-series photometer," **26**, pp. 321–+, June 2005.

56. D. Phelan, O. Ryan, and A. Shearer, eds., *High Time Resolution Astrophysics*, ch. Transition Edge Cameras for Fast Optical Spectrophotometry by R. Romani et al. Springer, 2007.
57. R. W. Romani, A. J. Miller, B. Cabrera, S. W. Nam, and J. M. Martinis, "Phase-resolved Crab Studies with a Cryogenic Transition-Edge Sensor Spectrophotometer," *The Astrophysical Journal* **563**, pp. 221–228, Dec. 2001.
58. B. Cabrera, "Optical imaging TES array." http://hep.stanford.edu/ cabrera/optical.html.
59. J. Vallerga, A. Tremsin, J. McPhate, B. Mikulec, A. Clark, and O. Siegmund, "Photon counting arrays for AO wavefront sensors," in *Astronomical Adaptive Optics Systems and Applications II. Edited by Tyson, Robert K.; Lloyd-Hart, Michael. Proceedings of the SPIE, Volume 5903*, R. K. Tyson and M. Lloyd-Hart, eds., pp. 199–209, Aug. 2005. http://www.ssl.berkeley.edu/ mcphate/AO/SPIE2005_JVV.pdf.
60. F. Zappa, S. Tisa, S. Cova, P. Maccagnani, R. Saletti, and R. Roncella, "Single-photon imaging at 20,000 frames/s," *Optics Letters* **30**, pp. 3024–3026, November 2005.
61. F. Zappa, S. Tisa, S. Cova, P. Maccagnani, D. Bonaccini, G. Bonanno, M. Belluso, R. Saletti, and R. Roncella, "Pushing technologies: Single-photon avalanche diode arrays," in *SPIE Int. Symposium on Astronomical Telescopes & Instrumentation, Glasgow*, **5490**, 2004.
62. C. Mackay, R. Tubbs, R. Bell, D. Burt, and I. Moody, "Sub-electron read noise at MHz pixel rates," in *Proceedings of SPIE: Sensors and Camera Systems for Scientific, Industrial, and Digital Photography Applications II*, M. Blouke, J. Canosa, and N. Sampat, eds., **4306**, 2001. http://www.arcetri.astro.it/ rtubbs/papers/lllccd/sern_main.html.
63. *CCD65 series ceramic pack electron multiplying CCD sensor.* http://e2v.com.
64. D. Dravins, "Astrophysics on its shortest timescales," *ESO Messenger* (78), pp. 9–19, 1994.
65. J. Hough, "Polarimetry: a powerful diagnostic tool in astronomy," *Astronomy and Geophysics* **47**, pp. 3.31–3.35, June 2006.
66. "Galileo european satellite navigation system." http://ec.europa.eu/dgs/energy_transport/galileo/index_en.htm.
67. Niels Bohr Institute, University of Copenhagen http://www.nbi.ku.dk/page42396.htm and http://www.astro.ku.dk/ per/fosc/index.html.
68. R. Gualandi and R. Merighi, *Bologna Faint Object Spectrograph & Camera User Manual*. Bologna Astronomical Observatory, 2001.
69. M. Downing, R. Arsenault, D. Baade, P. Balard, R. Bell, D. Burt, S. Denney, P. Feautrier, T. Fusco, J.-L. Gach, J. J. Diaz Garcia, C. Guillaume, N. Hubin, J. Jorden, M. Kasper, M. Meyer, P. Pool, J. Reyes, M. Skegg, E. Stadler, W. Suske, and P. Wheeler, "Custom CCD for adaptive optics applications," in *High Energy, Optical, and Infrared Detectors for Astronomy II. Edited by David A. Dorn and Andrew D. Holland. Proceedings of the SPIE, Volume 6276, pp. 627609 (2006).*, July 2006.
70. M. Perrin, J. Graham, M. Trumpis, J. Kuhn, K. Whitman, R. Coulter, J. Lloyd, and L. Roberts Jr., "First light with the "kermit" near infrared camera." astro.berkeley.edu/ mperrin/kermit/kermit_amos.pdf.
71. Raytheon, "VIRGO-2K 2048 x 2048 SWIR HgCdTe IRFPA," tech. rep., Raytheon, 2002. http://www.sal.wisc.edu/whirc/archive/public/datasheets/raytheon/SB301U% sersGuideRevB1. pdf.
72. UK ATC, "SCUBA-2." http://www.roe.ac.uk/ukatc/projects/scubatwo/.
73. M. Fich and SCUBA-2 Team, "SCUBA-2: A Submillimeter Bolometer Array Camera for the JCMT," *American Astronomical Society Meeting Abstracts* **208**, June 2006.
74. P. Sorensen and M. Azzaro, "Staralt v2.3 observing tool." http://www.ing.iac.es/ds/staralt/index.php.
75. D. Phelan, O. Ryan, and A. Shearer, eds., *High Time Resolution Astrophysics*, ch. Geiger-mode Avalanche Photodiodes for High Time Resolution Astrophysics by D. Phelan and A. Morrison. Springer, 2007.
76. D. Phelan, O. Ryan, and A. Shearer, eds., *High Time Resolution Astrophysics*, ch. Imaging Photon Counting Detectors for High Time Resolution Astronomy by O. Siegmund et al. Springer, 2007.

77. J. Vallerga, J. McPhate, B. Mikulec, A. Tremsin, A. Clark, and O. Siegmund, "Noiseless imaging detector for adaptive optics with kHz frame rates." http://www.ssl.berkeley.edu/ mcphate/AO/SPIE_2004.pdf.
78. J.-L. Gach, O. Hernandez, J. Boulesteix, P. Amram, O. Boissin, C. Carignan, O. Garrido, M. Marcelin, G. Östlin, H. Plana, and R. Rampazzo, "Fabry-Pérot Observations Using a New GaAs Photon-counting System," *Publications of the Astronomical Society of the Pacific* **114**, pp. 1043–1050, Sept. 2002.
79. O. H. Siegmund, M. A. Gummin, J. M. Stock, G. Naletto, G. A. Gaines, R. Raffanti, J. Hull, R. Abiad, T. Rodriguez-Bell, T. Magoncelli, P. N. Jelinsky, W. Donakowski, and K. E. Kromer, "Performance of the double delay line microchannel plate detectors for the Far-Ultraviolet Spectroscopic Explorer," in *Proc. SPIE Vol. 3114, p. 283–294, EUV, X-Ray, and Gamma-Ray Instrumentation for Astronomy VIII, Oswald H. Siegmund; Mark A. Gummin; Eds.*, O. H. Siegmund and M. A. Gummin, eds., pp. 283–294, Oct. 1997.
80. O. Siegmund, J. Vallerga, P. Jelinsky, X. Michalet, and S. Weiss, "Cross delay line detectors for high time resolution astronomical polarimetry and biological fluorescence imaging," in *Nuclear Science Symposium Conference Record, IEEE*, pp. 448–452, 2005.
81. A. Tremsin, O. Siegmund, J. Vallerga, J. Hull, and R. Abiad, "Cross-strip readouts for photon counting detectors with high spatial and temporal resolution," in *IEEE Transactions on Nuclear Science*, pp. 1707–1711, 2004. http://repositories.cdlib.org/postprints/96/.
82. O. Siegmund, B. Welsh, J. Vallerga, A. Tremsin, and J. McPhate, "High-performance microchannel plate imaging photon counters for spaceborne sensing," in *Proceedings of SPIE: Spaceborne Sensors III*, R. Howard and R. Richards, eds., **6220**, 2006.
83. Sematron UK Ltd. http://www.sematron.com.
84. J. Martin and P. Hink, "Characterisation of a microchannel plate photomultiplier tube with high sensitivity GaAs photocathode." http://cddis.nasa.gov/lw13/docs/papers/detect_martin_1m.pdf.
85. J. Martin, P. Hink, C. Tomasetti, and J. Wright, "Characterisation of a microchannel plate photomultiplier tube with a high sensitivity gaas photocathode." http://cddis.nasa.gov/lw13/docs/presentations/detect_martin_1p.pdf.
86. J. Martin and P. Hink, "Single-photon detection with microchannel plate based photo multiplier tubes." physics.nist.gov/Divisions/Div844/events/ARDAworkshop/2003/ Martin_talk2% .ppt -.
87. P. Bergamini, G. Bonelli, A. Paizis, L. Tommasi, M. Uslenghi, R. Falomo, and G. Tondello, "An imaging photon counting intensified CCD for high speed photometry," *Experimental Astronomy* **10**, pp. 457–471, November 2000.
88. D. Phelan, O. Ryan, and A. Shearer, eds., *High Time Resolution Astrophysics*, ch. The Development of Avalanche Amplifying pnCCDs: A Status Report by Lothar Streuder et al. Springer, 2007.
89. D. Phelan, O. Ryan, and A. Shearer, eds., *High Time Resolution Astrophysics*, ch. EMCCD Technology in High Precision Photometry on Short Timescales by N. Smith et al. Springer, 2007.
90. D. D. E. Martin, P. Verhoeve, T. Oosterbroek, R. Hijmering, A. Peacock, and R. Schulz, "Accurate time-resolved optical photospectroscopy with superconducting tunnel junction arrays," in *Ground-based and Airborne Instrumentation for Astronomy. Edited by Ian S. McLean and Masanori Iye. Proceedings of the SPIE, Volume 6269, pp. 626909 (2006).*, July 2006.
91. E. Science and Technology, "S-CAM3." http://sci.esa.int/science-e/www/object/index.cfm? fobjectid=36685.
92. B. Mazin, P. Day, J. Zmuidzinas, and H. LeDuc, "Multiplexable kinetic inductance detectors." http://www.submm.caltech.edu/papers/pdf/2002-05-Proc-Zmuidzinas.pdf.
93. C. Cunningham, S. D'Odorico, F. Kerber, J. Blommaert, B. Brandl, J. Brinchmann, J. Cuby, T. Herbst, W. Holland, M. Iye, F. Pepe, M. Casali, A. Kaufer, M. Kasper, and A. Moorwood, "ELT working group #2: Instruments." https://ssl.roe.ac.uk/twiki/pub/ ELTInstdes/WgtwoInstruments/ELT_WG2_Ins% _report_V1_3.pdf.
94. S. Howell, *Handbook of CCD Astronomy*, ch. 4. Cambridge Observing Handbooks for Research Astronomers, Cambridge University Press, 2000.

95. G. Wirth, "Astronomical magnitude systems." http://www.astro.utoronto.ca/~patton/astro/mags.html.
96. N. Meidinger, L. Andricek, S. Bonerz, J. Englhauser, R. Hartmann, G. Hasinger, S. Herrmann, P. Holl, R. Richter, H. Soltau, and L. Strüder, "Frame store pn-CCD detector for space applications," in *Proceedings of SPIE, X-Ray and Gamma-Ray Instrumentation for Astronomy XIII*, **5165**, pp. 26–36, 2004.
97. PerkinElmer Inc., *APD Based SPCM-AQR Series Datasheet*. http://optoelectronics.perkinelmer.com/content/Datasheets/SPCM-AQR.pdf.
98. D. Phelan, O. Ryan, and A. Shearer, "Galway HTRA workshop," June 2006. http://www.htra.ie.

EMCCD Technology in High Precision Photometry on Short Timescales

Niall Smith, Alan Giltinan, Aidan O'Connor, Stephen O'Driscoll, Adrian Collins, Dylan Loughnan and Andreas Papageorgiou

Abstract We discuss the advantages and limitations of Electron Multiplying CCD technology in high precision photometry on short timescales, with special emphasis on probing the smallest structures in active galactic jets. Factors external to the EMCCD, rather than the architecture of the EMCCD itself, most often limit the precision of photometry that can be reached with groundbased observations. Although EMCCDs can be used in photon-counting mode, reliable photometry can still only be achieved when careful observing procedures are employed. Observations of a small sample of blazars taken with an EMCCD are briefly discussed. An innovative two-colour photometer, based on simultaneous observations with two EMCCDs is described and the science drivers are discussed. Novel opportunities for the use of EMCCDs in lucky photometry are outlined and it is clear that certain scientific questions cannot be addressed with groundbased observations taken with conventional CCDs.

1 The lure of Photon-Counting Detectors

The rationale to develop detectors which are capable of single photon-counting is simple—in principle they permit the highest precision photometry to be recorded in the detected flux. It is generally true that two-dimensional detectors are preferable to single element detectors, because the most efficient and most accurate way to determine if the detected flux from a stellar source is changing is to compare it to one or several other sources within the same small field of view (arcminutes) at the same time. While this can be achieved with multifibre feeds to individual photon-counting detectors, for example, it is much simpler to achieve from the end-user perspective using a 2D array detector like a CCD that can be placed directly in the

---

Niall Smith · Alan Giltinan · Aidan O'Connor · Stephen O'Driscoll · Adrian Collins · Dylan Loughnan · Andreas Papageorgiou
Department of Applied Physics & Instrumentation, Cork Institute of Technology, Rossa Avenue, Cork, Ireland
e-mails: niall.smith@cit.ie, alan.giltinan@cit.ie, aidan.oconnor@cit.ie, stephen.odriscoll@cit.ie, adrian.collins@cit.ie, dylan.loughnan@cit.ie, apapgeorgiou@cit.ie

telescope focal plane. We can estimate the photometric accuracy we might hope to achieve with a simple example: if a source covers 10 pixels of a CCD and each pixel has a well depth of 350,000 electrons, then using the simplest argument we expect the maximum integrated count to be 3,500,000 electrons, yielding a photometric error of 0.58 millimagnitudes (mmag) or ~0.06% assuming Poisson statistics. The reality is that we almost never see this precision of photometry per exposure, irrespective of how slowly we read out the CCD (and hence, irrespective of how low the readnoise is). In differential photometry of blazars, for example, experience shows that we can expect an accuracy closer to 0.5–1%, even for sources which are bright compared to the local background, and regardless of the diameter of the telescope [1], [2].

Irrespective of the scientific objectives, we probably all agree that *reliable* photometry is a crucial objective and as we push the precision limit the number of factors which affect the reliability increase and we become prone to generating lightcurves which may contain spurious datapoints. This is especially true for lightcurves of sources which do not necessarily produce periodic or quasi-periodic modulations and hence for which we cannot fold the data. We need to be mindful to differentiate between the rms precision of the lightcurve, which is sometimes quoted as an impressively small number, and the precision of individual datapoints, or small groups of datapoints, especially when we are looking for those very fast, rare events that happen unpredictably. The question arises as to how we estimate the errors on the datapoints, because we can only determine the significance of structures within our lightcurves if we have an accurate assessment of the sizes of the datapoints errors.

In this chapter we concentrate on considerations for producing high precision optical lightcurves of pointlike active galactic nuclei (AGN) with time resolutions of seconds to minutes. While this may not be considered *high time resolution* in the strict meaning of the sense of the phrase, it nevertheless represents a regime in which we have the potential to study structures on spatial scales which we cannot hope to reach using direct imaging techniques at any wavelength or with any other technique.

A few basic questions should be kept at the back of our minds: What is a reasonable precision of photometry that can be reached on a regular basis? What limits the photometric precision of a point source; of an extended source? What is the best approach to use to optimise precision? Will photon-counting detectors improve the photometric precision?

2 Electron Multiplying CCDs

2.1 The Effect of Gain on S/N

A new CCD architecture that has received much attention recently is referred to as a *Low Light Level CCD* (L3 CCD) or an *Electron Multiplying CCD* (EMCCD). The photon detecting areas for conventional CCDs and EMCCDs are identical; the

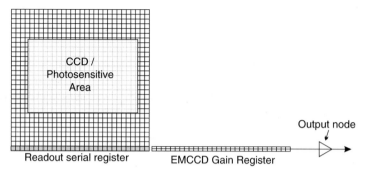

Fig. 1 EMCCDs differ from conventional CCDs in that they employ an additional register called the Gain Register

critical difference between them is the introduction of a gain register after the serial readout register in an EMCCD (Fig. 1).

This gain register consists of a set of pixels with a unique high voltage electrode (Fig. 2). This electrode is held at a low dc voltage, producing a large voltage difference between two clocking electrodes. It is the large voltage difference, which itself induces a high electric field in the smallscale silicon structure, that causes electron multiplication through impact ionization. Each multiplication process is known as a stage and corresponds to one pixel in the gain register. If the probability of multiplication per stage is g, the total effective gain, M, is related to the number of gain stages, N, by

$$M = (1 + g)^N \qquad (1)$$

The value of g is typically low at 0.01. However, currently available EMCCDs (for example, the DV887 produced by e2V) contain between 500 and 600 multiplication pixels or stages, whence gains of >1000 are easily achievable.

To see the advantage this gives the EMCCD over a conventional CCD a quick comparison of their noise statistics is useful. The conventional CCD has a typical effective noise of

$$\sigma_{eff} = \sqrt{[(\sigma_{DB}) + \sigma_{readout}^2]} \qquad (2)$$

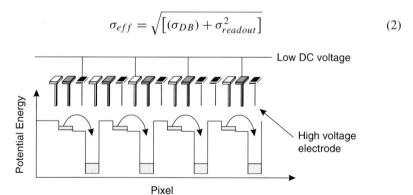

Fig. 2 Each three electrodes represents one pixel. As charge is clocked along the pixels of the gain register there is amplification in the signal through impact ionization

where σ_{DB} is the dark and background noise and $\sigma_{readout}$ is the readout noise. With sufficient cooling of the CCD, σ_{DB} reduces towards zero resulting in $\sigma_{readout}$ becoming the dominant noise factor term. This is particularly true for higher readout rates where $\sigma_{readout}$ increases substantially. (For example, for Andor Technology's DV887 EMCCD, the readout noise increases from 22e⁻ at 1MHz to 62e⁻ at 10MHz.) Reading the CCD out through a multiplication register introduces two new terms to the equation. The effective noise is now given by

$$\sigma_{eff} = \sqrt{\left[F^2 (\sigma_{DB}) + \frac{\sigma_{readout}^2}{G^2} \right]} \qquad (3)$$

where F is the multiplication noise and G is the gain. Increasing the gain sufficiently, reduces the readout noise to effectively zero. However, the multiplication register noise, which arises because the multiplication process is statistical in nature, remains. F can be estimated by noting that for N input electrons there will be a $\sqrt{N}$ error on the output number of electrons and shot noise also contributes $\sqrt{N}$ noise to the effective noise. Combining these leads to a noise contribution of $F=\sqrt{2}$. Thus, multiplication gain introduces a noise factor of 1.4 which effectively reduces the S/N by the same amount. Another way of looking at this is to assume the S/N is set by the photon shot noise but the detector quantum efficiency (QE) is reduced by half. On the face of it this would seem to suggest that EMCCDs are too inefficient to be of interest for astronomical applications where the light level is often very low. However, as we discuss below, the advantage of the EMCCD is only apparent whenever the device is operated at high frame rates (equivalent to short exposures or high time resolution) so that the detected number of photons per frame is low. If the detected number of photons is large per frame, then a conventional CCD is a better choice of detector.

There are currently a small number of players in the EMCCD market. The CCD chips themselves are fabricated by e2V and Texas Instruments (TI). Sizes of chip vary, but the standard e2V chip is a 512 × 512 array of 16 μm pixels, while the standard TI chip is a 1000 × 1000 array of 6.8 μm pixels. e2V refer to their technology as *L3 Vision* and TI refer to theirs as *Impactron*. We have used only the e2V chip packaged by Andor Technologies, Belfast, under the trade name of iXon.

2.2 Photon-Counting Mode in an EMCCD

A 50% reduction in QE is not small and, unsurprisingly, methods have been developed to retrieve the lost 50% by operating the EMCCD in a photon-counting mode. Due to the statistical nature of the multiplication process, it is not possible to use the amplitude of the output signal from a given pixel in a given frame to accurately calculate the number of photons that have been detected. Therefore, photon-counting is not possible by simply measuring the output signal amplitude. However, by working at sufficiently high gain (G>100) and using a thresholding technique one can recover the lost QE and at the same time achieve real photon-counting operation. The principle is that if the signal from a pixel at the output lies below some threshold value, the output for that pixel is set to zero for that readout. If the signal exceeds the threshold the output is set to one and it is assumed a photon has been detected. Selecting the threshold level correctly is important—if it is set too high then photons will be missed due to coincidence losses (two or more photons falling on the same pixel in a given frame); if it is set too low then random fluctuations in the readnoise or dark current will trigger false detections. Assuming no other noise source, a mean flux of 10^{-1} photon pixel^{-1} frame^{-1} should allow photon-counting operation in an EMCCD as long as the readnoise is kept below about 30e$^-$ [3].

This is good news, although there are a number of important caveats that must be considered. Firstly, since the photon flux must be kept extremely low to avoid coincidence losses, this requires that the dark current be kept exceedingly low, which means that cooling to $-100°C$ or lower is necessary. Secondly, in most astrophysical situations the highest precision photometry is achieved by intercomparisons of the fluxes of a number of sources (usually stars) that are imaged simultaneously on the same chip. The largest EMCCD chip currently available is 1k × 1k and this significantly limits the field-of-view and hence the number of intercomparison stars available to choose from. In many cases the apparent magnitudes will differ by 1 to 2 magnitudes and it is impossible to optimize the frame rate (which itself determines the flux pixel^{-1} frame^{-1} for each source) for all sources simultaneously. Thirdly, unless the EMCCD is windowed in some fashion, the volumes of data recorded become inordinately large very rapidly. And finally, because failure to use the correct thresholds results in non-linear behaviour of the CCD, there is a real danger that precision photometry may be compromised. This is particularly true of observations when there are changes in atmospheric transparency, resulting in changes in the detected flux pixel^{-1} frame^{-1}.

2.3 Clock Induced Charge

Assuming the EMCCD is operated in true photon-counting mode, a question now arises as to what limits its photometric precision. The answer lies in the degree of Clock Induced Charge (CIC) that is produced during the readout process. CIC is generated during the transfer of charges between pixels and appears as noise spikes affecting single pixels. It exists in conventional CCDs but is usually buried in the readnoise. In an EMCCD, whenever CIC originates at the front end of the gain register it undergoes a large amount of amplification and it becomes indistinguishable from the signal generated by a detected photon. Furthermore, since CIC is a function of the readout process, it scales with the frame rate, i.e., an EMCCD read out at 30 frames/s will have 30 times more CIC in 1s than the same EMCCD read out at 1fps. CIC is dependent on many instrumental parameters, including the clocking levels, pulse shape and width and operating mode of the CCD (inverted or non-inverted) and the end-user may wish to experiment with these to ensure the best response for a given experiment.

CIC is not only a limit to photometric precision in photon-counting mode. It also limits the photometric precision when thresholding is not employed. We determined the CIC for a CCD87 and a CCD97 EMCCD over a range of gains from 450 to 2400 using approximately 2000 bias frames. In Fig. 3 we show the results for the CCD97 and it is evident that the CIC reached a maximum of 0.8% at a gain of 2400, meaning that approximately 1 in every 125 pixels are affected by CIC in each readout. This value is almost five times lower than the older CCD87 EMCCD and further improvements might reasonably be expected as the technology matures.

2.4 Which Operating Mode Should One Use?

As pointed out in [4] one can think of three operating modes for an EMCCD:

1. Mode 1—unity gain. Here the EMCCD behaves identically to a conventional CCD, although the readnoise be somewhat higher as the output electronics are really designed assuming that gain will be employed. The end-user should check this if they intend to use the EMCCD in unity gain mode. The full QE is achieved, but the S/N depends on the photon shot noise and the readnoise (which increases rapidly with readout rate). The only real advantage is that the dynamic range is usually 16-bit for a single frame compared to 14-bit when EMCCD gain is used.
2. Mode 2—moderate-to-high gain. Here the gain is set sufficiently high to negate the effects of readnoise, but the QE is halved, corresponding to the S/N being reduced by a factor of 1.4. The EMCCD can be read out quickly (e.g., 34fps for Andor's 512×512 pixel CCD97 chip) without incurring any additional loss in S/N. The photometric precision is relatively insensitive to moderate changes in the

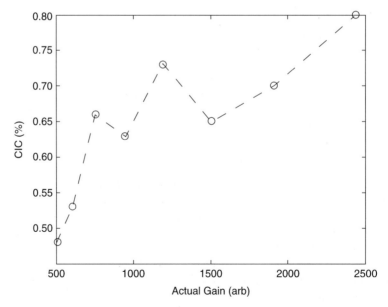

Fig. 3 The percentage of pixels affected by CIC per frame versus the actual gain of the EMCCD. The mean value is about 0.7%, rising to 0.8% at maximum gain of 2400

incident flux and so the photometry is reliable if the usual precautions are taken (section 3). The dynamic range, reduced by the gain process, can be recovered by co-adding frames and this co-addition can be done in post-processing, as required by the astrophysical timescale of interest.

3. Mode 3—high gain and thresholding. Here the EMCCD is operating in photon-counting mode with the full QE. However, the need to correctly set the threshold levels to avoid coincidence losses (which appear as non-linearity in the photometry) makes the photometry potentially less reliable under changing atmospheric conditions. Additionally, the need to keep the flux at a level of 10^{-1} photons pixel^{-1} frame^{-1} means very short integration times, and if the duty cycle is to be kept respectable very high frame rates must be continuously employed resulting in enormous data volume throughput.

2.5 Photometric Precision

We are particularly interested in the photometric performance of an EMCCD when applied to differential photometry of point sources (see section 4 for a description of the science drivers). We examined the photometric performance of two of Andor's iXon EMCCDs—data from the CCD87 was taken during two engineering runs at Calar Alto in 2003 and data from the CCD97 was taken from laboratory tests on an artificial starfield. In both experiments the EMCCDs were operated in mode 2 and in both we took many tens of thousands of frames of data at ∼1-10fps from which we were able to average the data into larger and larger time bins to plot how the scatter in the data behaves. We expect the scatter to decrease as the root of the number of data points in an integration bin. Fig. 4 shows a graph of $N^{1/2}$ against scatter for two real stars from the Calar Alto dataset, taken on 21st September 2003. Here, N is the number of bins after averaging over a given time interval, expressed as a fraction of the total number of frames. Hence, $N = 1$ corresponds to long integration times. In the absence of systematic effects, the scatter should vary as $\sigma = \sigma_0\sqrt{N}$, where σ_0 is the intrinsic error on a data point (which should be related to the flux). Hence we would expect stars of a similar magnitude to have similar slopes. In fact, the two stars shown in Fig. 4 are of comparable magnitude and have similar broadband colours and their slopes are indeed consistent within the errors.

A number of conclusions can be drawn from the data taken at Calar Alto:

1. Although the data are well determined (because of the large number of frames), the straight line fit to the data is poor, indicating some systematic effects are present.
2. The intercept is not consistent with zero scatter for most stars, implying a limit to the photometric accuracy achievable. Based upon the data from the two Calar Alto campaigns it approximates to 1.2 millimagnitude although, because of the poor straight-line fit to the data, the value is uncertain. Nevertheless, it is worthy of note that the *highest* photometric precision in our lightcurves is ∼1 millimagnitude per frame. The effect is most evident on nights when the raw data are stable, as we would expect, because of the smaller error bars.

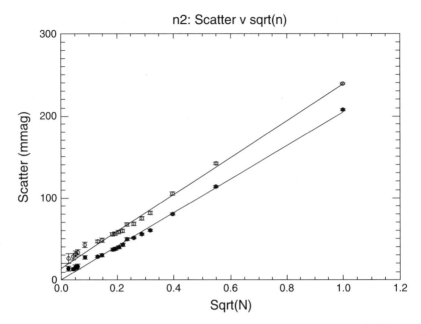

Fig. 4 A plot of the scatter in the data as a function of the parameter N which corresponds inversely to the number of individual data frames averaged to get a given data point. Small values of N imply long integration times (requiring many frames to be averaged). The two solid lines represent the best linear fit to the data for two stars. Ideally, the curves should be well fitted by a straight line, but both show evidence of significant structure, most notably at $N^{1/2} \sim 0.1$ (corresponding to integration bins of $\sim$20 s) where the two stars appear to track one another's behaviour, hinting at a common origin in this instance. Such behaviour is not always seen, however

3. There is evidence in the scatter of the data for systematic deviations from a straight line for short timescales (tens of seconds). However, plots of the lightcurves on these timescales have chi-squared values consistent with no variability. The scatter is thus larger than expected from photon statistics, but not large enough to make the stars statistically variable.
4. For longer integrations, corresponding to errorbars reaching a few millimagnitudes (usually about 300 s), many stars show evidence for apparent variability based upon the chi-squared values of their lightcurves. This may represent another type of systematic effect, perhaps uncorrelated with the short-timescale systematic noted in 3. It also generally occurs at slightly higher than the limiting photometric value implied in 2.
5. There is no correlation between the occurrence of systematic deviations and any given star.

When we perform the same analysis on the artificial starfield recorded in the laboratory we get quite similar behaviour, with three notable exceptions: the straight-line fits to the data are better, indicating the systematic effects are smaller; the intercept is consistent with an average photometric precision of 0.75 mmag for the five artificial stars, with the highest precision being 0.1 mmag; and at longer integrations the differential lightcurves show significantly less variability. The most reasonable explanation for the improved photometric behaviour in the laboratory is, primarily, the absence of atmospheric turbulence and possibly the influence of extinction effects related to variations in airmass, although we cannot rule out improvements in the CCD97 over the CCD87 as being a contributory factor. Nevertheless, what is clear is that with the latest EMCCD technology, it is possible to achieve considerably better than 1 mmag photometry in the laboratory. Since we almost never see this level of precision in field conditions, we must look elsewhere for likely problem factors.

3 Factors Affecting Photometry

Despite the fact that photometry of point sources is often allocated by telescope allocation committees during bright time, perhaps on the basis that it is "easy" to do, the reality is that photometry of point sources can only be optimized when a surprisingly large number of parameters are taken into account. We list these briefly here and suggest considerations that must be taken to lessen their negative impact when attempting photometry with precisions of better than 1 mmag.

3.1 The Signal-to-Noise Equation for Point Sources

The standard equation for estimating the S/N of a point source using aperture photometry (i.e., no PSF fitting) is [5]:

$$\frac{S}{N} = \frac{N_*}{\sqrt{N_* + n_{pix}(1 + \frac{n_{pix}}{n_B})\left(N_S + N_D + N_R^2 + G^2\sigma_f^2\right)}}. \qquad (4)$$

Here N_* is the total number of photons detected in n_{pix} pixels, n_B is the number of pixels used to estimate the background flux. Clearly, the larger the value of n_B the better, but this is actually true only if the background is well behaved. The more crowded the field, the more difficult it is to estimate n_B without making some assumptions about the PSFs of the sources. The major contributions to noise include: N_S, the total number of photons, per pixel, from the background; N_D, the total number of dark count electrons per pixel and N_R^2, the total number of electrons per pixel due to readnoise. As these are reduced the S/N increases, so we really would wish to make our observations with a noiseless detector under the darkest sky background conditions possible. Furthermore, by making n_{pix} small we increase S/N and this suggests that PSFs should be matched to the size of the pixels in the CCD, or even be smaller than them. The parameter G is the gain of the CCD (in electrons / ADU) and σ_f^2 is the 1-sigma error due to the A/D converter.

For very bright sources (4) reduces to

$$\frac{S}{N} = \frac{N_*}{\sqrt{N_*}} = \sqrt{N_*} \qquad (5)$$

with the S/N being completely dominated by photon statistics and independent of n_{pix}. In practice, however, making n_{pix} too small (*undersampling*) introduces more problems than it solves (section 3.2).

3.2 Sampling

If CCDs, conventional or electron multiplying, had a spatially uniform QE, it would be possible in principle to record all the flux from an isolated point source within a single pixel (assuming care was taken not to introduce linearity problems) and thereby obtain the maximum S/N. Unfortunately, this does not work in practice even for an EMCCD operating in photon-counting mode, due to variations in the QE within a pixel which can be as high as 10–15% for a front-illuminated chip. Movements of the source within a pixel, arising from atmospheric turbulence or telescope tracking inaccuracies result in changes in the detected flux. In the case of atmospheric turbulence there may be no simple correlation between the movements of different point sources imaged into different pixels on the same chip and the problem gets worse as the undersampling becomes more pronounced. In such a scenario, differential photometry will not allow one to extract the cause of the detected flux changes (intrinsic to the source, or a combination of atmospheric and instrumental effects) because the output from a pixel is a single value in the A/D converter and there is no way to flatfield within a pixel.

Wide-field surveys which employ small plate scales (e.g., 13" per pixel for the SuperWasp wide angle planet-finding project) sometimes defocus the image so that undersampling is avoided. This approach works well if a source is isolated, otherwise the defocusing results in complex overlapping PSFs. Photometric precisions of 1–2% per datapoint are possible only for the brightest objects with fainter objects becoming lost in the noise sources. It is hard to escape the conclusion that the highest precision photometry is achieved only with PSFs that are well-sampled from the outset.

Experience shows that the most reliable photometry is obtained when the FWHM of the PSF is 2–3 pixels, and the extraction radius is set to approximately three times the FWHM or slightly smaller. Experience also shows that it is very useful to be able to vary the parameters such as the source-extraction radius and the size of the annulus used to calculate the background—if the lightcurves derived from the data are independent of small variations in these parameters then one can be confident that the structures observed within the lightcurves are at least inherent to the raw data, though whether they are intrinsic to the source or not is another matter.

3.3 Flatfielding

Flatfielding in the true sense of the term is extremely difficult to achieve, primarily because the spectrum used to generate the flatfield is almost certainly different to the spectrum of the sources that the flatfield is being used to correct. There is no simple solution to this problem, though the optimum approach is probably to make the flatfield as spectrally flat as possible. The use of a "master flat", generated from a combination of many individual flats, does not guarantee better flatfielding, especially if dust particles have been added or removed from the optical train (notably the CCD window), or if the data are taken in bright moonlight and the scattering of photons within the telescope leads to differences in the way the dust is illuminated. Since flatfielding can *reduce* the photometric precision, it is probably wise to reduce datasets both with and without flatfielding. Interestingly, if the telescope pointing and tracking is good enough, it should not be necessary to flatfield at all. Certainly, flatfielding should never be employed as a means of correcting bad pointing or tracking, or inadequate optical design. Flatfielding also does not work well on undersampled images.

3.4 Linearity

The response of a CCD to incident light is substantially linear up to a significant fraction of the well depth (perhaps 80%, CCD dependent). In differential photometry the most insidious problems arise whenever the integrated counts change, such that an object whose flux was originally low enough to keep its PSF within the linear regime varies in such a way that this is no longer true. This could happen if, for example, the atmospheric transparency improved during an observation sequence and the detected flux either increased, or the seeing improved and the FWHM of the PSF was reduced. In either case, the peak of the stellar profile might now be in the non-linear regime of the CCD and this will result in apparent variability when the source is compared to others which are not similarly affected. Furthermore, while the behavior of CCDs can be well approximated as linear for much of the dynamic range, there are subtle structures along the linearity response curve which can vary from one CCD to another. For maximum precision these should be measured and used to correct raw data appropriately.

With EMCCDs the problems with linearity are somewhat greater than for conventional CCDs. For example, if the voltage in the gain register varies during a readout sequence the amplification will vary across the chip. There is evidence that this can happen not only because of inherent instabilities in the gain voltages, but also in circumstances where the number of electrons traveling through the gain register are high (presumably as the pixels from a bright source are being read out) as these can change the multiplication voltages. Alternatively, since EMCCD gain is highly temperature dependent, fluctuations in temperature will change the gain which can push the PSF into the non-linear regime (though this is unlikely to be significant on individual frames). As mentioned in section 2.2, non-linearities arise in photon-counting mode if the thresholds are not carefully selected or if the incident flux is too high.

3.5 Comparison Stars

The ideal comparison stars would have the same apparent magnitude and identical spectrum to the source under study. This is probably never the case, so differential photometry is always influenced by changes in both atmospheric conditions and the airmass of the field of view. (For example, if two sources have identical airmasses the flux detected from both will vary as the absolute value of the airmass changes if the two sources have non-identical spectra [6]. In general, this effect is very small, however, of the order of <1 mmag for an airmass change of 1.) It follows that the more stable the atmospheric conditions, and the smaller the change in airmass during an observing session, the higher the precision of the differential photometry is likely to be. This makes the highest precision photometry over long timescales (hours) challenging, but the corollary is that variability over timescales of minutes is almost certain to be unaffected by airmass changes at least.

A word of caution about removing comparison stars on the grounds of intrinsic variability. Since one cannot be sure *a priori* that the variability is intrinsic to the star, rather than an indication of the quality of the photometry, one should justify the star's removal from the subsequent construction of the lightcurve. In particular, if the target source varies by the same level as a comparison star, even if the variability is not correlated, one needs to be very careful about rejecting the comparison star and not the target source. This problem is particularly acute when the number of comparison stars is small.

3.6 The Value of Having Large Numbers of Datapoints

Perhaps the biggest single advantage of using an EMCCD in precision photometry is that the noiseless operation of the device, albeit at the potential expense of QE, enables one to take large numbers of frames and hence generate large numbers of photometric values for each object of interest in the field of view. These individual photometric values can be co-added to generate average photometric values for any given time bin, with the errors calculated empirically from the scatter of the data within the time bin. Thus, even without knowing where the noise in the data originates one can still measure it and generate an empirical errorbar for each averaged datapoint. In turn this enables one to empirically determine the statistical significance of any structures seen in a dataset, rebinning the data if necessary, and this gives one greater confidence that any low amplitude variations that one might observe are at least inherent to our data.

4 Observing Blazars with EMCCDs

4.1 The Importance of Variability in Blazars

The enormous energy output of active galactic nuclei (AGN) is thought to be due to accretion onto a supermassive black hole at the centre of a galaxy (Fig. 5). In about 10% of cases, the accretion results in the formation of a "jet", though the mechanism(s) responsible are not well understood. The radiation from the jet is generally considered to be *synchrotron* in the radio-X-ray regime, and *Inverse Compton* (IC) at higher energies. The apparent superluminal motion seen in high resolution

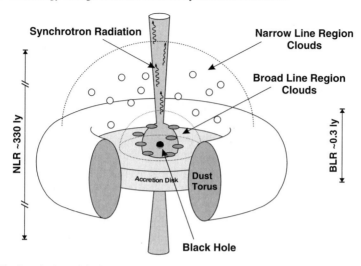

Fig. 5 The Standard Model of an AGN, showing the jet emission

radio images argues strongly for relativistic motion of the plasma constituting the jet, and the angle between the jet axis and the observer's line-of-sight determines the amount of Doppler boosting associated with this motion. In the case of *blazars*, the line-of-sight angle is believed to be small. The most extreme blazars, BL Lac objects, are thought to be viewed along the jet axis, resulting in a Doppler-boosted continuum that completely washes out spectral line emission from the broad- or narrow-line-region clouds.

Relativistic jets are expected to be highly unstable and, not surprisingly, variability is one of the defining criteria of blazars. In addition to variability over weeks and months, there is clear evidence for intraday variability in powerful AGN ([7], [8], [9]). Power density spectra indicate that variability on timescales of about a day ($\sim 10^5$ s) dominates energetically. This is true in the radio, optical and X-ray regimes ([10], [11]). The similarity in timescales of maximum power over a photon energy range of eight orders of magnitude indicates that the characteristic timescale is determined by geometrical effects (such as the width of the jet) rather than by acceleration or radiation mechanisms, which are expected to show a marked energy dependence. The true diameters of the regions where particles are accelerated must be smaller than one day and should lead to faster variations. Such rapid variations had been found in individual events (e.g., [12], [13], [14]), but have so far not been investigated in detail. The existence of rapid variations is important, since it clearly indicates the presence of discrete substructure on spatial scales given by light-travel arguments (diameter $< ct$, where c is the speed of light and t is the variability timescale). Variability measurements with high time resolution will hence determine the size scales of the individual emission regions, with the fastest variations probing the smallest spatial structures.

The objective of our work with EMCCDs is to extend studies of quasar structure to more rapid variability, and hence smaller spatial scales, than heretofore by using high-precision fast photometry of blazar variability. The work studies the rare and very fast events which may ultimately resolve the question of particle acceleration in quasars, and aims to determine the regime where variability timescales are set by acceleration and cooling processes rather than by source geometry.

Flares due to changes in the source geometry are expected to be both *temporally symmetrical* and *spectrally achromatic* (Fig. 6a). To see this, consider the output of an emitting blob moving relativistically along a curved trajectory. In this case, the viewing angle θ (i.e., the angle between the blob velocity vector and the line of sight) varies with time, and consequently the beaming or Doppler factor $\delta = [\Gamma(1 - \beta\cos\theta)]^{-1}$ changes too. Even a modest change in blob direction can cause a significant variation in flux received (e.g., a change of less than 0.5 degrees causes a flux variation of 20% in S5 0716+714, [15]). By contrast, if a flare is due to a shock in a jet [16], the particle acceleration and cooling times are expected to differ, resulting in *temporally asymmetric* flares that are *spectrally chromatic* (getting bluer as the flux rises, Fig. 6b).

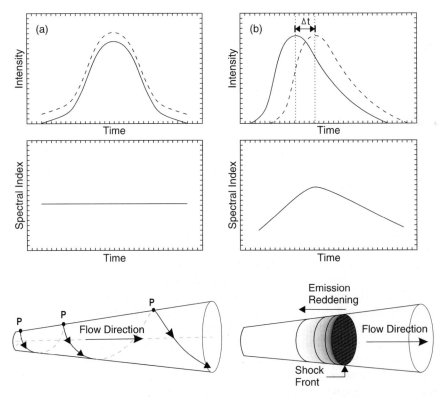

Fig. 6 The three panes on the left (**a**) show the variation in intensity of emission in red and blue filters (top pane) and the resultant spectral index (middle pane) for a blob moving in a curved trajectory (lowest pane). The intensity varies simultaneously in both filters; hence the spectral index is flat. The three panes on the right (**b**) show the equivalent behaviour for a shock propagating along a jet. The spectral index hardens as the higher energy electrons (responsible for the blue light) lose their energy first from the front of the shock

Hence, dense optical monitoring in two or more colours simultaneously acts as a powerful discriminant between geometric and acceleration/cooling mechanisms as the source of the short-timescale variability. The two different types of behaviour have been seen on timescales of hours: [17] found evidence for spectral hardening in BL Lac in the short-timescale flares, based on statistical arguments of long-term moderately-sampled lightcurves; while [14] found variability of 30% in the blazar AO 0235+164 over eight hours during which the spectrum remained unchanged.

The regions responsible for the rapid variability are crucial to our understanding of the particle acceleration mechanisms in jets, because they constrain non-thermal radiation mechanisms. **Crucially, such regions cannot be imaged directly**; even future space-borne mm-interferometry will fall short in angular resolution by three orders of magnitude.

Only with sufficiently high-precision photometry and an intensive monitoring campaign where a large number of flares are observed can one make statistically valid statements about the characteristic behaviour of blazars, whilst at the same time characterising the behaviour of individual flares with unprecedented detail.

In our work, we are attempting to reach new limits for the photometric accuracy and precision of blazar lightcurves and to push the timescale of variability that can be observed to a few minutes at the 0.5% amplitude level. This will enable us to address the following specific questions:

- what is the duration of the most rapid events that can be reliably observed?
- how frequent are they?
- what temporal shape do they have?
- how does their spectrum evolve with flare amplitude?
- on what timescale does acceleration/cooling dominate over geometric effects?
- how is flare behaviour in different objects influenced by the Lorentz factor, total luminosities and long-term variability properties?

4.2 The Significance of Rapid Optical Variability in Blazars

Despite thirty years of study of quasars, we still know relatively little about the emission mechanisms responsible for the rapid flux variations. Yet, as already mentioned, a detailed study of the flux variations can provide a considerable amount of information on the physics and dynamics of the emitting region. Although looking at cross-correlations between widely separated wavebands (e.g., optical and X-ray) has proved a powerful method for testing general theories of the radiation mechanisms in jets, the application of correlation studies breaks down in the case of rapid variability. This is because the flux at other wavelengths is too low to generate light curves on timescales of minutes (e.g., at X-ray or gamma-ray energies), or because the detectors do not have the necessary photometric precision (e.g., in the infrared), or because the sources do not vary rapidly, due to intrinsic opacity effects (e.g., in the radio). For this reason, the optical regime is the optimum to work in.

A number of well-coordinated multi-wavelength campaigns have been conducted on blazars in recent years, with the aim of gaining a better insight into the general radiation mechanisms operating from radio wavelengths through to gamma-ray energies. These campaigns generate significant amounts of optical variability data, which are intended primarily to be cross-correlated with variability at other wavelengths. However, the typical integration time for most observations is of the order of 5 minutes per data-point, which is too long to probe very-short-timescale variability, because of the need to have at least a few data-points associated with any credible variability. Furthermore, when combining the data from a number of observers, there is always the problem of systematic differences between different sets of equipment (notably due to filter response curves and CCD quantum efficiency curves), leading to typical uncertainties in the zero points of 0.08 magnitudes [17]. This large error washes out any realistic possibility of detecting small-amplitude, rapid variations when datasets are amalgamated, except on the basis of statistical arguments. This rules out analysis for individual flare events (unless they are large enough to be well above the measurement errors). The clear implication is that the most accurate short-timescale variability data are produced by a single instrument working at high time resolutions and in more than one colour simultaneously. Also, the most precise lightcurves are produced where the data is photon-flux limited, rather than limited by detector noise. This argues strongly for using an EMCCD.

The few well-sampled portions of light curves measured for intraday variability clearly demonstrate the existence of rapid variations whenever sampling and measurement accuracy are sufficient to resolve the features. However, as with the multi-wavelength campaigns, these observations have been hampered primarily by the lack of sufficient time resolution, making it difficult to draw conclusions about the temporal shapes of the fast flares and possible substructures contained within. They do, however, illustrate that variations are caused by emission regions which are no larger than 5 AU in diameter.

Intriguingly, there have been a few claims of ultra-fast variations (lasting a few minutes) which imply brightness temperatures in excess of the inverse-Compton limit, thus indicating either that the variations are *strongly* enhanced by relativistic effects, or that some other process (e.g., coherent radiation) is responsible [12]. Long time series monitoring of a sample of sources at high time resolution and high photometric precision should reveal whether such ultra-rapid events really exist, or whether they are likely to be artifacts. Importantly, if ultra-fast events do exist, an approach of performing simultaneous two-colour photometry enables one to distinguish between relativistic beaming and coherence mechanisms, since the latter should be quasi-monochromatic.

4.3 Engineering Test Observations with EMCCDs

We have already employed EMCCD technology in two campaigns (January and September 2003) on the 2.2 m telescope at Calar Alto ([1], [2]). A total of 430,000 CCD frames were recorded, amounting to 452 GB of data. We designed and fabricated the mechanical assemblies for the instrument, wrote the instrument control and diagnosis software, and the software reduction pipeline (including a realtime analysis package). The results of the two campaigns show clear evidence for fast variations on timescales of approximately 3 minutes, with photometric precisions of about 0.3%. (The best photometric precision was 0.05% rms in 15 minutes.) The campaigns reveal several intriguing features, including rapid drops in flux on timescales of less than 3 minutes (see Fig. 7c). We have also made observations with the 3-colour ULTRACAM photometer on the 4.2 m at La Palma in 2003 and 2004 (recording 134,000 frames in two hours). Interestingly, the photometric precision with ULTRACAM on the 4.2 m William Herschel Telescope at La Palma is very similar to that achieved with the EMCCD on the 2.2 m Calar Alto telescope. This result suggests that the photometry on both occassions was dominated by non-CCD effects.

4.4 TOφCAM

Based upon the results obtained in the two engineering test runs at Calar Alto we received a grant from Science Foundation Ireland to build a photometer based around EMCCD technology. The instrument, TOφCAM (**T**wo-Channel **O**ptical **P**hotometric Imaging **Cam**era, pronounced *toffee-cam*) uses two CCD97 EMCCDs

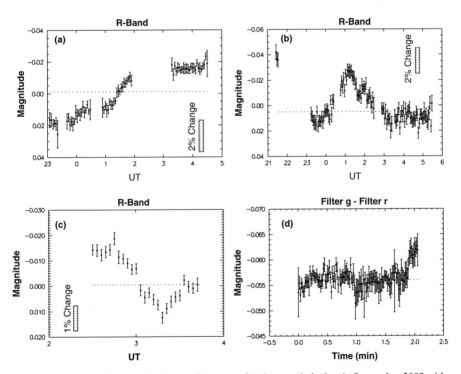

Fig. 7 (a) Variability in the blazar 0716+741 over a five-hour period taken in September 2003 with the L3 CCD. Each datapoint is a 3-minute average consisting of 1800×0.1s individual integrations. (b) An isolated, flare in the blazar 0954+51. (c) Hints of very rapid variability in the blazar *BL Lac* is seen at around 3 UT. (d) Step-like colour changes (V-R) are evident towards the end of the dataset at about 2.0 hrs in 0716+741 which can be traced to sky background colour variations approaching dawn. Data from the ULTRACAM instrument

from Andor Technologies. Its objective is high precision photometry of BL Lac objects on timescales of minutes. A schematic of the optical layout is shown in Fig. 8. The telescope output is connected to the detectors via re-imaging optics so that the plate scale is ordinarily 0.3" per pixel. The front section of TOφCAM converts the convergent telescope beam into a parallel beam and this can be changed depending on the f-ratio of the telescope. The parallel beam is split into two paths via a dichroic beamsplitter, giving a BV-band arm and an R-band arm. These two lightpaths are then passed through filters to produce effectively a B- and R-band arm, or a V- and R-band arm. One EMCCD is fixed in position and the telescope is focused onto this EMCCD in the first instance. The other EMCCD can be moved forwards or backwards along the optic axis until the focus in the second arm is identical to that of the first arm. The optics were designed at the LSW, Heidelberg and the mechanical engineering designs were generated at CIT, where most of the fabrication work was also carried out. The design paradigm calls for a compact, lightweight structure with no readily movable parts, a rigid chassis and fixed optics. Changes to the optical layout can be made, notably a change in the filters, but this cannot be done quickly. The optics are sufficiently good to image >95% of the flux into a circle with a diameter of ∼0.1 arcsec which means the image quality is limited by the local seeing.

As can be seen in Fig. 9 TOφCAM is controlled by either one or two Pentium IV PCs (each with 2GB RAM) depending on the storage capacity required and the level of realtime photometry that needs to be performed. Even with one PC the two iXons can be operated at the maximum rate of 34 frames/s (34 Mbytes/s) and the data is streamed to RAID 0 SATA hard drives. Each PC has 4×200 GB hard drives which can be removed during the observations. Control of the PC(s) is via TCP/IP over Ethernet, allowing fully remote access. GPS timing is provided to an absolute precision of 1 s, with 1 µs internal timing resolution.

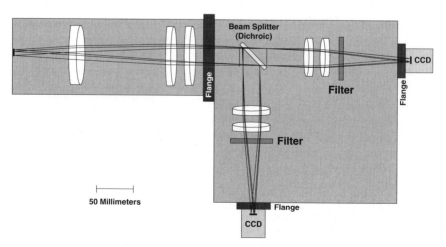

Fig. 8 The optical layout of TOφCAM

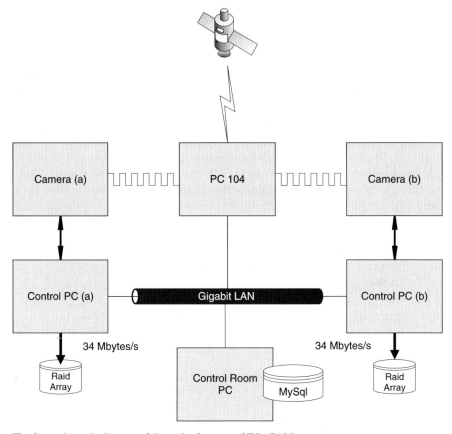

Fig. 9 A schematic diagram of the main elements of TOφCAM

The data reduction pipeline is designed to handle the large volumes of data (some 160 Tbytes in the first 18 months of operation) and to provide near-realtime photometry. This pipeline is a major advance on the one we developed to handle the 480,000 frames that we recorded during our engineering test runs at Calar Alto in 2003 and will be described in detail elsewhere. TOφCAM is expected to see first light in the middle of 2007.

4.5 High Precision Photometry of the HST-1 Knot in M87

4.5.1 Lucky Imaging

The technique of lucky imaging is now well established. It involves taking large numbers of short exposures of a "bright" stellar reference object and selecting those frames in which the reference object has effectively a FWHM which is below some specified threshold. The selected frames are shifted and co-added to build a sharpened image. In the optical, objects within ∼10" of the reference object will typically be well sharpened, and this value increases into the infrared. Diffraction-limited images can be achieved when the seeing is excellent, but in this case the final image might make use of <1% of the frames (which is rather inefficient in flux terms). Using 30–50% of the frames, lucky imaging can deliver a respectable 0.3" in average seeing. The actual exposure for each frame depends on the integrated flux from the object that is being used as the reference and hence depends on the apparent magnitude of the object and the aperture of the telescope. Conventional CCDs cannot be used for lucky imaging due to their excessive readnoise at high frame rates and this is one application where EMCCDs are necessary. From the point of view of long-term monitoring blazars, it is possible to use a percentage of the monitoring data that is used for photometry to build a lucky image. This has several interesting applications. For example, many of the TeV emitting blazars are located in optically prominent host galaxies. The photometry of the bright central core is degraded by the presence of the host galaxy, but this effect can be minimized by sharpening the image so that the two (core and host galaxy) can be deconvolved effectively before attempting the photometry. Another specific application is in finding the host galaxy of the blazar 0716+741 which currently has no known redshift. If one can detect a host galaxy it is possible to set a limit on the redshift and this in turn will allow estimates to be made of the energetics of the jet. 0716+741 is one of the best studied blazars from radio to gamma-ray energies. We plan to observe it intensively in a number of upcoming campaigns and the observations might generate sufficiently high quality lucky images to allow us to search for the elusive host galaxy with unprecedented precision.

4.5.2 TeV emission from M87

Only very recently have AGN been associated with very high energy gamma ray emission of the order of TeV (Mrk 421), [18]. Today, their number remains low with only ten sources currently known [19]. All of these sources have been identified as BL Lacertae (BL Lac) objects, with the exception of M87 (HEGRA, [20]) which is classified as an FR I radio galaxy [21].

The presence of M87 in a list of BL Lac objects represents a case of particular interest since it can be used to test the hypothesis that BL Lacs are in fact FR I radio galaxies, with their relativistic jets pointed towards the observer [22]. In addition, at a distance of 16 Mpc, M87 is one of the nearest radio galaxies. For this reason, M87's relativistic jet can be studied with unparalleled spatial resolution at all wavelengths, from radio, through optical, to X-ray. Such studies reveal that the jet consists of knots of enhanced emission (possibly due to shocks) and these knots are seen to move with apparent superluminal velocities ($\sim 6c$) that impose an upper limit to the angle between the jet direction and the line of sight ($\theta < 19°$), see Fig. 10.

Of particular interest to us is the jet knot HST-1, which is located at the base of the jet, 0.8" (60 pc projected) from the nucleus. For the past five years, the knot has been undergoing an impressive flaring event, with observed X-ray and UV fluxes increasing by a factor of ~ 50. Recently, TeV observations of M87 by HESS (S. Wagner, private communication) reveal that a significant increase in TeV flux

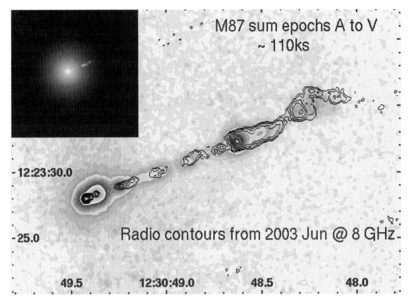

Fig. 10 The jet of M87 in X-rays (grayscale) and radio (contours) showing significant resolved structures. The knot HST-1 is closest to the core. The optical ground-based image (inset) shows how HST-1 is blended into the core due to seeing limitations. Diagram from [27]

took place contemporaneously with the X-ray and UV flux increases, opening up the possibility that TeV emission originates in the knot HST-1 in M87.

This is potentially a very exciting result because, quite simply, we presently don't know where the TeV emission in active galaxies originates. Yet identifying the origin of the TeV emission is crucial in constraining/testing models that describe either the core (e.g., [24], [25]) or the jet (e.g., [26]) as the source of the emission. Such models impose limits on the TeV fluxes, as well as on the timescales of variability, the source of the TeV photons, etc. If the TeV emission does indeed come from shocks in the jet it has important implications on modeling (unresolved) cores of BL Lacs that are further away. In this regard it is important to note that with the exception of M87 all other TeV active galaxies are BL Lac objects and because their jet axes are closely aligned to the observer's line-of-sight, the core and knots line up with one another in the observer's reference frame, making it impossible to disentangle the source of the TeV emission. M87 is truly unique amongst active galaxies and at the very least it affords us the opportunity to examine the variable optical emission from a knot in a jet, if we can resolve it from the emission from the core.

It is important to note that having precision optical photometry simultaneously in two colours enables one to draw conclusions about the most likely underlying emission mechanism. For example, simultaneous rising/falling of the fluxes at different wavelengths support a model in which the variability is due to compression and expansion within the knot itself; alternatively, if the high frequencies decay faster than the low frequencies, the likely mechanism involves injection of new relativistic particles into the knot [27].

Finally, HESS data shows evidence for regular variability within a week. There are currently no supporting data at any other wavelengths, but if the variability is due to shock fronts within HST-1 we expect correlated activity at optical wavelengths. High precision, high spatial resolution photometry with TOφCAM will test this hypothesis.

4.5.3 TOφCAM observations of M87 using Lucky Photometry

The HST-1 knot is 0.86" from the core. Given that the median seeing is typically 1" from the ground, it is not possible to perform optical photometry using long exposures (>seconds) as the core and knot become blended due to atmospheric scintillation. This perhaps explains why there is no ground-based photometry of HST-1 yet. However, diffraction-limited images equivalent to those already sparsely obtained by the Hubble Space Telescope can be regularly obtained from the ground using lucky imaging. We plan to use the core of M87, which is an unresolved point source of approx 12th magnitude, as the reference "star" for our lucky imaging. The fact that it is 0.86" from the 15th magnitude HST-1 knot actually improves the level of sharpening that can be achieved as both core and knot will almost always reside within the same non-turbulent atmospheric cell at the same time. We anticipate sharpened images with a FWHM of $\sim$0.2–0.3" under reasonable conditions. Once we have recorded the sharpened image in two colours simultaneously with TOφCAM, we will be able to perform PSF-fitting photometry to an estimated accuracy of 5% over a timeframe of about one hour. This combination of high precision and high spatial resolution is only possible with EMCCDs and represents an exciting new avenue of opportunity for blazar research.

5 Final Comments

EMCCDs offer great promise for high precision photometry in a number of application areas. They allow the observer to take large numbers of images during the observing run and to co-add these as appropriate later depending on the astrophysical timescale under examination and the photometric precision required. Laboratory tests, compared to observations in the field, suggest that the limit to photometric precision continues to lie with factors beyond the EMCCD, though poor observing strategies (notably undersampling) will always degrade the overall photometric performance.

TOφCAM is a two-channel photometer which will see first light in late 2006. It will perform high precision photometry of blazars, but it will also use the lucky imaging technique to generate photometry of the knot in M87 which cannot be done using conventional CCD imaging.

It is reasonable to suggest that EMCCD technology is here to stay. It has many applications in the lifesciences area (including fluorescence imaging and single-molecule imaging) and this will help to drive future improvements which are likely to provide moderate-cost solutions for those astronomy groups who do not have the capability to fabricate their own hardware solutions. However, EMCCDs are likely to remain physically small, probably no larger than 1k × 1k (or perhaps 2k × 2k), because they really need to be read out at high frame rates if they are to outperform conventional CCDs.

Acknowledgments TOφCAM was built with a grant from Science Foundation Ireland. A. Giltinan and A. Collins acknowledge scholarship support from a private benefactor. A. Papageorgiou is funded by the European Community's Human Potential Programme under contract HPRN-CT-2002-00321 (ENIGMA). The work of the Cork team has benefited enormously from interactions with members of the ENIGMA network. We also acknowledge the assistance of staff at Calar Alto and La Palma, and the V. Dillon and T. Marsh for assistance with the ULTRACAM observations.

References

1. Smith, N., Coates, C., Giltinan, A., Howard, J., O'Connor, A., O'Driscoll, S., Hauser, M., Wagner, S., Proc SPIE, vol.5499, pp. 162–172, 2004
2. O'Driscoll S., Smith, N., Proc SPIE, vol.5493, pp. 491–501, 2004
3. Daigle, O., Carignan, C., Blais-Ouellette, S., Proc SPIE, vol.6276, 2006
4. Mackay C., Baldwin, J., Tubbs, R., Proc SPIE, vol.4840, 2003
5. Everett, M.E. and Howell, S.B., 2001, PASP, 113, 1428
6. Young, A.T., Genet, R., Boyd, L., Borucki, W., Lockwood, G., Henry, G., Hall, D., Smith, D., Baliumas, S., Donahue, R., Epand, D., 1991, PASP, 103, 221
7. Wagner, S.J. and Witzel, A., 1995, ARAA, 33, 163
8. Wagner, S.J., 1999, in BL Lac Phenomenon, ASP Conf.Ser., Vol. 159, p.279
9. Wagner, S.J., 2000, in Scattering and Scintillations, APSS, 278, 105
10. Kataoka, J., Takahashi, T., Wagner, S., Iyomoto, N., Edwards, P., Hayashida, K., Inoue, S., Madejski, G., Takahara, F., Tanihata, C., Kawai, N., 2001, ApJ 560, 659
11. Wagner, S.J., 2002, in AGN Variability, PASA, 19, 129–137
12. Miller, H.R., Carini, M.T., Goodrich, B.D., 1989, Nature, 337, 627
13. Villata M., Raiteri, C.M., Kurtanidze, O., and 53 coauthors, 2002, A & A, 390, 407
14. Romero, G., Cellone, S., Combi, J., 2000, A&A, 360, L47–L50
15. Ghisellini, G., Villata, M., Raiteri, C.M., Bosio, S., de Francesco, G., Latini, G., Maesano, M., Massaro, E., Montagni, F., Nesci, R., and 7 coauthors, 1997, A&A, 327, 61–71
16. Marscher, A., 1996, in *Blazar Continuum Variability*, ASP Conf Series, 110, 248–259
17. Villata, M., Raiteri, C.M., Aller, H. D., 2004, A&A, 424, 497–507
18. Punch, M., Akerlof, C., Cawley, M, Chantell, M., Fegan, D., Fennell, S., Gaidos, J., Hagan, J., Hillas, A., Jiang, Y., and 13 coauthors, 1992, Nature, 358, 477
19. Horan, D., Weekes, T., 2004, NewAR, 48, 527
20. Aharonian, F., Akhperjanian, A., Beilicke, M., Bernlohr, K., Borst, H., Bojahr, H., Bolz, O., Coarasa, T., Contreras, J., Cortina, J., and 43 coauthors, 2003, A&A, 403L
21. Fanaroff, B. L., Riley, J. M., 1974, MNRAS, 167, 31
22. Urry, C. M., Padovani, P., 1995, PASP, 107, 803
23. Georganopoulos, M., Perlman, E. S., Kazanas, D. 2005, ApJ, 634, 33
24. Bai, J. M., Lee, M. G., 2001, ApJ, 549L, 173
25. Protheroe, R. J., Donea, A.C., Reimer, A., 2003, APh, 19, 559
26. Stawarz, L., Sikora, M., Ostrowski, M., Begelman, M., 2004, ApJ, 608, 95
27. Harris, D., Cheung, C., Biretta, J., Sparks, W., Junor, W., Perlman, E., Wilson, A., 2006, ApJ, 640, 211

The Development of Avalanche Amplifying pnCCDs: A Status Report

L. Strüder, G. Kanbach, N. Meidinger, F. Schopper, R. Hartmann, P. Holl, H. Soltau, R. Richter and G. Lutz

Abstract Imaging detectors with high quantum efficiency over a wide spectral range, low noise, and fast readout will prove to be very useful in many applications in astronomy and other fields. In 2005 we have started to develop pnCCDs with high quantum efficiency for fast imaging of single optical photons in the wavelength range $\sim$350–1100 nm. The concept is based on fully depleted, back-illuminated pnCCDs with a sensitive thickness of 450 μm which were originally developed for fast single photon X-ray imaging and spectroscopy. The pnCCD with column parallel readout shall be modified such that the on-chip JFET amplifiers are replaced by avalanche amplifier (AA) cells operated in the linear regime with a gain up to a few thousand. An additional on-chip n-channel MOSFET manages the suitable change of impedance such that the resulting signal can be handled by the following CAMEX readout ASICs. The amplification of the avalanche process must be sufficiently low to prevent optical crosstalk between the detector channels. The anticipated format in the imaging area will be of the order of 256 × 256 pixels with a size of $\sim$50 μm. The frame rate should be as high as 1000 frames per second.

1 Introduction

Two dimensional imaging of single photons together with accurate time tagging will open a large field of scientific investigations and applications in astronomy and other branches, where fast processes have to be recorded. Most of the 'classical' detectors with such properties are based on vacuum technology and include multi-channel plates with thin photo-electric layers and multi-anode read-out systems (so-called

---

L. Strüder · G. Kanbach · N. Meidinger · F. Schopper
Max Planck Institut für Extraterrestrische Physik, Postfach 1312, 85741 Garching, Germany
e-mail: lts@mpe.mpg.de, gok@mpe.mpg.de

R. Hartmann · P. Holl · H. Soltau · G. Lutz
PNSensor GmbH, Römerstr. 28, 808003 München, Germany

R. Richter
Max-Planck-Institut für Physik (Werner-Heisenberg-Institut) Föhringer Ring 6 80805 München, Germany

MAMA detectors, e.g. the UV STIS MAMA detectors on HST[1]). Timing can be very fast and the noise levels are low in these devices, but their quantum efficiency (<25%) and wavelength range is rather limited. Figure 1 shows a group of efficiency curves labelled 'PMT', which refers to different types of alkali photocathodes used in vacuum devices.

Solid-state detectors based on Si (shown as APD, EM-CCD, and pn-CCD in Fig. 1 offer much better quantum efficiencies over a wide spectral range. The initial difficulty with these devices was to reduce their intrinsic noise and to raise the signal amplitudes to levels where single photon detection becomes viable. Bolometric cryogenic detectors at extremely low temperatures, e.g. Superconducting Tunnel Junctions (STJ—also known as Josephson junctions [8], [9]) or transition edge sensors (TES, [4], [1]), are still quite experimental and are not considered here in comparison with the semi-conductor devices.

Single pixel photon-counting silicon detectors based on avalanche amplification in a small (few hundred μm diameter) electrically cooled diode are available from commercial suppliers. These units, called Avalanche Photo Diodes (APDs), can be selected for very low dark count rates (<100 cts/s at their operational temperature) and feature typical quantum efficiencies >20% for wavelengths in the range from 450 to 950 nm, with a maximum ∼60% at 700 nm. The big advantage of individual detectors is the very fast response, the capacity for high counting rates, and ∼ns timing accuracy. Two dimensional imaging can only be achieved in a limited way with single pixel devices, e.g. by projecting the image through a micro-lens array on

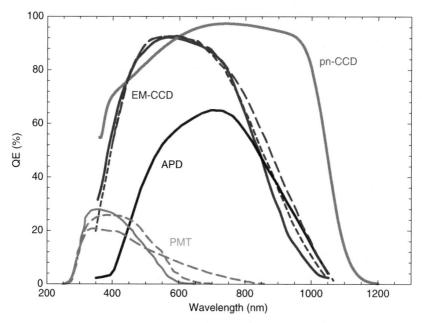

Fig. 1 Typical values for the quantum efficiency (QE) of commonly used fast timing photon detectors

---

[1] http://www.stsci.edu/hst/stis/design/detectors/

multiple detectors, or by feeding the image through a fibre array (integral field unit) to independent photon counters.

The first widely used low light-level (=L3) high time resolution CCD cameras are based on an on-chip electron multiplier (EM). The primary charges coming from the CCD are shifted sequentially through a chain of potential wells. The shift voltages are large enough to produce, with a small probability, secondary charges by impact ionisation and thus lead to a growing cascade of charges. Gains up to a factor of 1000 have been advertised for cooled EM-CCDs that also have a very low CCD dark current. The efficient wavelength range for such EM-CCDs spans 400–850 nm (QE>50%). Typical read-out rates are presently between 10 and a few 100 Hz, (depending on frame size) and these rates are limited by the serial readout scheme.

Our new design for a single-photon sensitive CCD is based on individual avalanche amplifiers integrated on the read-out nodes of every CCD column [7]. We describe in this contribution how we develop such a combination of the already highly optimized pnCCDs and the new avalanche amplifier cells. The sensitive thickness of a pnCCD is $\sim$450 µm (full depletion, back illuminated), which is much thicker than the 15–20 µm sensitive thickness of a conventional thinned back illuminated CCD. A thick pnCCD with a suitable anti-reflective coating promises therefore a much higher sensitivity (QE>80%) into the near IR up to wavelengths of $\sim$1.1 µm. The design with parallel read-out on every CCD column will also deliver fast read-out speeds (>1000 full frames/s).

Figure 1 shows a comparison of the efficiencies of the various fast photon counting devices mentioned so far. It is evident that the sensitivity of solid state devices was a substantial improvement over the photocathodes in vacuum devices and led to count-rate increases by typical an order of magnitude. If we are now able to increase and extend the sensitivity into the near-infrared range, the count-rates, especially of astronomical sources located in slightly absorbed (interstellar reddening) regions, should again improve by a factor of 2–5.

2 Technical Developments

2.1 pnCCD

pnCCDs were already produced with pixel sizes of 36 µm and 51 µm ([3]) in a fabrication process that is fully compatible with the process needed to integrate avalanche amplifier cells. The devices were characterized with respect to their imaging capabilities, stability, and ease of operation. Figure 2 shows an X-ray pnCCD with split field readout. The Figure caption gives typical dimensions and operational parameters.

The thermally generated DC leakage current in the pnCCD was typically 300 pA cm^{-2} at room temperature. This current is equivalent to about 2×10^9 e$^-$ s^{-1} cm^{-2}, and, after assuming pixels of 50×50 µm^2 and a read-out rate of 10^3 frames s^{-1} the rate reduces to 50 e$^-$ frame^{-1} pixel^{-1}. Cooling the device down to $-40°$C reduces the dark current by a factor of $\sim 10^3$ and the dark rate, equating one electron to one

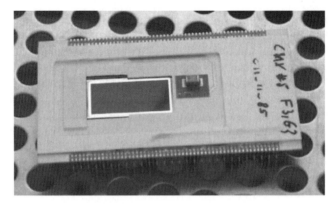

Fig. 2 pnCCD for X-ray measurements mounted on a ceramic carrier board. The radiation entrance window is covered with an anti-reflective coating. The overall detector size is 27 × 13.5 mm and contains an imaging area and two frame store areas. The 264 × 264 imaging pixels and the storage area pixels are of size 51 × 51 μm^2. The frame store is arranged in two halves on both sides of the image field. Image transfer is achieved in 20 μs and the read-out through on-chip FET's occurs in parallel chains (ASIC) for all CCD columns. The readout noise ranges from 1.8 e^- for readout rates slower than 400 frames per second (fps) to 2.3 e^- for readout rates up to 1100 fps. These parameters were measured at an operational temperature of −40°C [3]

count, will be about 5×10^{-2} counts frame^{-1} pixel^{-1} or 50 counts s^{-1} pixel^{-1}. From our experience with single-photon APD counters in an astronomical setting we estimate this pnCCD dark-rate to be nearly an order of magnitude below the expected rate from the dark night sky.

2.2 Avalanche Amplification

In order to reduce the read-out noise of currently available pnCCDs and to reach a 'single-photon' sensitivity, a quasi 'noise-free' amplification of the primary charges generated in the CCD has to be developed. Such an amplifier must be integrated directly on the detector chip to ensure the low noise characteristics. The 'electron multiplication (EM)' concept of on-chip amplification, that is now available in commercial low-light-level CCD cameras, operates in serial read-out mode, which limits the overall output speed. We have chosen another approach which is shown schematically in Fig. 3: attached to the read-out nodes of the CCD columns individual avalanche cells multiply the primary charges by a factor that can be adjusted between ∼100–1000. The amplified charge is then used as input to on-chip preamplifiers (FET) and their output is then connected to the external parallel read-out ASIC. Full frame read-out speeds exceeding 1000 fps have already been achieved with non AA pnCCDs.

In a first step the avalanche amplification cells were developed through simulation and experimental verification. Figure 4 shows the basic design of the AA cell with its dopant profile inside the Si substrate ([6]). Charges from the pnCCD are shifted into the AA cell, which is biased through a high ohmic resistor. The AA cell is coupled to a FET pre-amplifier (possible types MOSFET, JFET, or DEP-

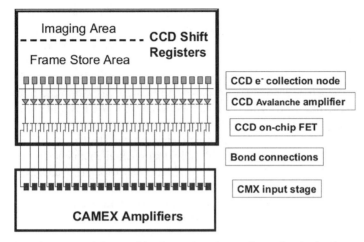

Fig. 3 Schematic layout of the amplification and read-out scheme for single photon sensitive pnCCDs with on-chip avalanche cells

MOS) for output to the external ASIC array (type Camex, CMX). We performed a technology study and detailed experimental evaluations of dopant profiles to verify the anticipated spatial electrical field distributions under defined electrical boundary conditions. About 50 different profiles were analyzed including samples submitted to different temperature treatments. The dopant profile results were convoluted with

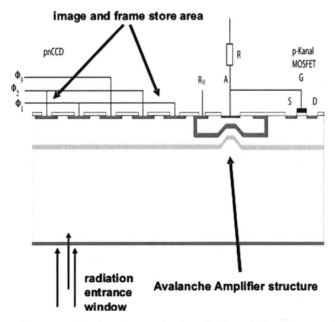

Fig. 4 Schematic cross-section through an avalanche cell. The radiation entrance window (bottom) and the shift registers are realized through rectifying p^+ implants. In this design a p-channel MOSFET is used to couple to the external amplifier ASIC

the simulation tools to get a realistic prediction of the performance of the avalanche cells. Comparison of such simulations, with the recently measured full AA device characteristics show very good agreement—a goal of our first programmatic milestone.

The onset of an avalanche happened within 2 Volts of the expected value of about –45 V. There was no breakdown in neighboring areas which could perturb the avalanche amplifier activity by cross-talk between cells. We have measured around 100 different structures on a single wafer. They show an excellent homogeneity. The transition between no avalanche and Geiger type avalanche is sufficiently soft to establish a common voltage for all individual readout channels in a monolithic device. Connected to a fast preamplifier, we have observed the avalanche type amplification of single electrons just compatible with the previously measured amount of thermally generated leakage current. The rise time was as fast as 2 ns. All signals had the same amplitude with a spread of ~10%. Visible light was injected in the AA cell by projecting a pulsed LED through a microscope. The increase of counts caused by the optical photons was clearly detected, and the correspondingly pulsed output on the avalanche amplifier was proof for the detection of the LED photons. The n-channel MOSFET was also characterized in detail: it had perfect output characteristics just as desired and simulated. The trans-conductance g_m is close to 100 µS, the pinch-off voltage is of the order of 2 V and the output resistance was above 250 $k\Omega$. More details and test measurements from the development of the avalanche cells can be found in [3].

2.3 Efficiency and Radiation Entrance Window: Anti-Reflective Coating

The internal efficiency of an indirect semiconductor like Silicon, is not easy to determine theoretically. Measurements [2] indicate that the intrinsic efficiency in the pnCCD detector material is nearly constant at one electron per photon for radiation with wavelengths longer than ~300 nm. Above 950 nm Si starts to become transparent and the efficiency drops toward the band gap of 1.12 eV. This good internal efficiency of silicon (~100%) can only be exploited if the surface reflection is suppressed by an anti-reflective coating (ARC). Suitable 2 or 3 layer coatings have been developed for various applications, e.g. readout of scintillators or good efficiency in the near IR, and their optical properties were modelled and verified experimentally. Figure 5 shows theoretical curves and experimental data for some of the available coatings. The new broadband ARC, which will be applied to the AApnCCDs is shown as a band. It has so far only been modelled, but we are confident, that the good predictability of the previous ARCs assures the quality of the new design.

2.4 Read-out System and ASIC

The read-out electronics for the optical AApnCCD will be very similar to the systems used with conventional X-ray pnCCDs, which were developed for space

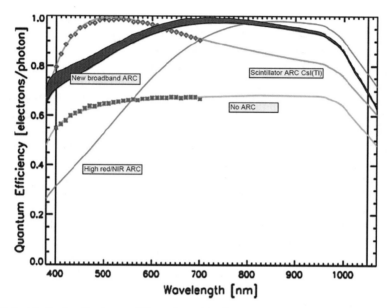

Fig. 5 Anti-Reflective Coatings on Silicon: Several available ARCs are displayed with experimental verification plotted as symbols. New designs for a red/near IR coating and for the envisaged broad-band ARC for the AApnCCDs are shown as a band

missions (e.g. XMM-Newton, eRosita) as well as for laboratory applications. Both detector types share the concept of pre-amplifier FETs integrated on the CCD wafer and connected in parallel to front-end multi-channel ASICs. This ASIC, known as CAMEX, performs further amplification, gain selection, signal filtering, sample and hold, and multiplexing onto an output line. Figure 6 shows the layout and functionality of the CMX chip. The current CMX features 128 parallel read-out channels and can be operated with a nominal speed of $\sim$8 μs per row. Full AApnCCD frame (= 256 × 256 pixels) readout at a rate of 1000 frames/sec can thus be achieved with 4 CMX nodes attached to frame store zones with a depth of 128 and a width of 256 pixels arranged on both sides of the imaging region. This layout was realized in the detector shown in Fig. 2.

The ASIC development for the further processing of the pre-amplified signals is already close to the specified performance figures. The chip has been characterized with respect to readout speed, noise, power dissipation, stability, linearity, output compatibility to the subsequent ADC etc. All functions are within the specifications but details like linearity need to and will be improved in the next versions.

2.5 Data Acquisition Electronics, DAQ

The DAQ system for the readout of the new AApnCCDs is also based on a long heritage of readout electronics developed for the X-ray pnCCDs. Figure 7 shows the block diagram of the planned DAQ.

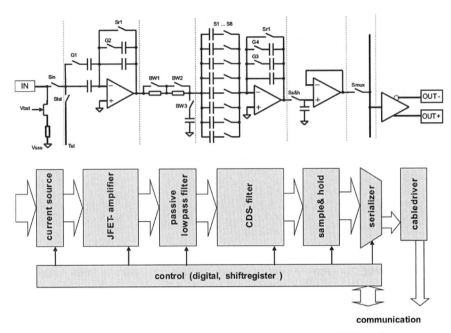

Fig. 6 Schematic design and functions of the readout ASIC of type CAMEX. Note that parameters like gain and bandwidth can be set by external command

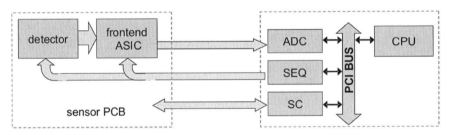

Fig. 7 Block diagram for the data acquisition system for the AApnCCDs

3 Perspectives

The device concept of the AApnCCDs seems to be sufficiently robust to be implemented in larger systems without a loss of homogeneity. No technical constraints were found so far in the fabrication of developmental and pre-production samples—only "good surprises" were encountered. The system work, which includes the sensors, the ASIC qualification, the mechanical and thermal components, the experimental set-ups and the physics program, is going on smoothly.

In the time between now and the availability of the first large AApnCCDs we intend to perform astronomical measurements at 1–2 m telescopes with an existing 'conventional' optical pnCCD, i.e. readout through FETs and no on-chip avalanche amplification. This camera will be operated at high readout speeds at a noise level below 2 electrons (rms). The results will be used to confirm our expectations on efficiency from 0.3 to 1.2 μm and the stability and ease of use of the system as an astronomical instrument.

References

1. Bay, T.J., Burney, J.A., Barral, J., et al.: *NIM A*, **559**, 506 (2006)
2. Hartmann, R., et al.: *Proc. SPIE* 5903, pp. N1–N9, 2005
3. Holl, P., Andritschke, R., Eckhardt, R. et al.: *IEEE Conference, San Diego, Oct. 2006*, Workshop N44 'Astrophysics and Space Instrumentation II', paper available from http://www.hll.mpg.de/Hauptnavigation/publications/2006/N44-1.pdf
4. Irwin, K.: *App. Phys. Lett.*, **66**, 1998 (1995)
5. Lutz, G., et al.: *IEEE Trans.Nucl.Sci.*, **NS-52**, 1156 (2005)
6. Lutz, G., Holl, P., Laatiaoui, M., Merck, C., Moser, H.G., Otte, N., Richter, R.H., Strüder, L.: *NIM A*, **567**, 129 (2006)
7. Meidinger, N., Andritschkea, R., Hälker, O., et al.: *NIM A*, **568**, 141 (2006)
8. Perryman M.A.C., Foden C.L., Peacock A.: *NIM A*, **325**, 319 (1993)
9. Rando N., Vervear J., Andersson S., Verhoeve P., Peacock A., Reynolds A., Perryman M.A.C., Favata F.: *Rev. Sci. Inst.*, **71**, 4582 (2000)

Geiger-mode Avalanche Photodiodes for High Time Resolution Astrophysics

Don Phelan and Alan P. Morrison

Abstract Geiger-mode Avalanche Photodiodes (GM-APDs) are establishing themselves as potential candidates for the broad temporal range covered in high time resolution astrophysics (HTRA). These detectors have already been employed in astronomical instrumentation and significant results have been obtained to date. Their high time resolution and quantum efficiency make these single photon event counting detectors ideal for observations of stochastic phenomena, and ultimately for extreme HTRA observations. In this chapter, we review the technology and to illustrate their potential we briefly touch on specific science goals and astronomical applications. We then focus on the fabrication and characterisation of GM-APDs, and discuss the development and challenges posed in designing array devices.

1 Introduction

In HTRA, conventional high speed charge coupled devices (CCDs) can now make measurements of bright sources down to a time of a few tens of milliseconds—as in, for example, ULTRACAM [3]. Fainter sources in the same time regime can be detected by the use of L3 (E2V) or Impactron (TI) electron-multiplying CCDs (EM-CCDs). Of future astronomical interest will be the investigation of higher speed phenomena, including signals from stochastic processes which require time resolutions far exceeding those available from CCDs, in any operating mode. This is due to the fact that CCDs are not event counting devices, and require the readout of whole sub-frames at very high clock speeds, with the resultant data pipelining and data storage difficulties. For future large telescopes the time resolution requirements become more severe, because large photon fluxes will enable shorter timescales to be investigated, even down to nanosecond timescales, where second and third order

---

Don Phelan
Dept. of Experimental Physics, National University of Ireland, Galway, Ireland
e-mail: don.phelan@nuigalway.ie

Alan P. Morrison
Dept. of Electrical and Electronic Engineering, University College Cork, Ireland
e-mail: a.morrison@ucc.ie

photon correlations may be expected [4]. GM-APDs are ideal for such astronomical applications, with characteristics such as high quantum efficiency, high timing resolution, low noise, high counting rates and event counting.

2 Science Case

An extensive discussion of the science case of high time resolution observations is given by Dravins [4], and by other authors in this volume. Many examples are given of sources of high speed observational astrophysics, including optical and X-ray pulsars; lunar and planetary-ring occultations; rotation of cometary nuclei; cataclysmic variable stars; pulsating white dwarfs; flickering high luminosity stars; X-ray binaries; accretion-disk instabilities; gamma-ray burst afterglows, and many more.

To expand further on a significant astronomical phenomena to illustrate the potential of GM-APDs, pulsars have been observed for over thirty years and there is still no concensus as to their structure and emission mechanism. Recent observations indicate that simultaneous multiwavelength observations—radio, IR, optical, X-ray, Gamma Ray—are showing promise in solving the pulsar problem. A solution to this problem will have implications to other areas where there is an electron-positron plasma and strong magnetic fields. In the radio, the emission it is probably coherent synchrotron whilst at higher energies it is likely to be incoherent synchrotron or curvature radiation. Joint optical–radio observations of the giant radio pulse (GRP) phenomena indicated that during GRPs a similar energy is liberated in the optical and radio regions [19]. Similar studies on these and other GRP pulsars are required to strengthen this link.

To make these observations it is essential to measure both a periodic signal—at a resolution down to a few microseconds and stochastic variations at times scales of 10–100 μs. Pulsars have rotation periods ranging from <2 milliseconds to >5 seconds, and consequently they require detectors which have time resolutions of at least 50 μs, or preferably higher. The time domain from 10 ms+ is covered by ULTRACAM, although this instrument has significant read noise problems at the shortest time scales. Currently only avalanche photodies (APDs), transition edge sensors (TES), and superconducting tunnel junctions (STJs) can cover the region less than 10 ms for faint non-periodic signals (this includes stochastic variability in periodic signals).

The recent observations of rapid radio transients (RRATs) [9] show that there are other pulsar phenomena which require a similar temporal resolution. RRATs typically show burst activity of duration 10–20 ms with a random repetition rate ranging from 200 seconds to several hours. Optical detections will only be possible using large telescopes and low noise detectors. For example, RRAT J1819−1458 has a burst duration of 20 ms equivalent to or around 5 photons incident on the VLT in 20 ms. It is clear that systems with low read–noise per resolution element will have to be required. For all of these observations a further dissection of the light for spectral and polarisation will be needed.

GM-APDs, in particular PerkinElmer's SLiK APD modules, have already been employed sucessfully to observe optical pulsars and other highly variable sources in instruments such as OPTIMA [20], and TRIFFID [19]. These single photon counting APD modules have quite a large active area, with a diameter of 180 μm, photon detection efficiencies typically as high as 60% at 650 nm, and dark count rates as low as 25 counts per second. The prototype QuantEYE astronomical photometer [16] is also using GM-APDs, the Single Photon Avalanche Diodes (SPADs) [2], with better than 45% quantum efficiency, timing accuracy better than 50 ps, dark count rate on the order of 50 counts per second, and a dead time of less than 70 ns.

2.1 Requirement for Arrays

It has been shown [6] that aperture photometry of faint sources in the presence of background and variable atmospheric properties (seeing) is best achieved by using an imaging detector capable of resolving a small region encompassing both the source (an aperture defined in software), and a comparison background region, under all conditions of seeing, telescope wobble, etc., rather than a physical aperture. One is able, post-exposure, to determine the variable optimum aperture size to maximise the signal-to-noise adaptively during the observation, and to track movements of the aperture centroid (due to seeing and also wobble) and to select from the data stream only photons falling within this aperture. A physical (rather than software) aperture cannot be optimized and must be sufficiently large to ensure that the target is easily acquired, and that 100% of the target photons are intercepted – otherwise, seeing-induced variations in brightness will occur, which are damaging to high precision time series analysis. A software-defined, adaptive aperture can therefore be smaller than a fixed, physical aperture, and can consistently give optimum signal to noise. One is able to select, post-exposure, background photons from a comparable region close to the adaptive aperture. This is always essential in order to achieve accurate photometry, but is even more important when background polarization may be variable across the field.

For these reasons, it is desirable to have an array detector, at least 4×4, or preferably 16×16 pixels, with temporal resolution in the microsecond domain, but down to nanoseconds for future instruments on future large telescopes. It is also desirable to have event counting in order to minimize data pipeling and storage problems, and extremely high detector quantum efficiency, even under conditions where there are a mixture of low and high peak rates.

3 Geiger-Mode and Active Quenching

There are two mechanisms employed in the operation of an APD, the more commonly used linear mode and Geiger–mode. In the former method, when the APD is biased below the breakdown voltage, the current is proportional to the incident photon flux. In Geiger–mode the device is biased above the breakdown voltage, leaving the device in a metastable state. An incident photon can then cause a catastrophic breakdown with an infinite and self-sustained avalanche, that can only be quenched with external electronics. Figure 1 illustrates the multiplication process within the APD, and a schematic illustrating the operation of Geiger-mode in terms of the V-I characteristic of the diode.

In order to operate an APD in photon counting mode, careful consideration must be given to the design of the external operating circuit. Two methods are currently employed and are referred to as passive quenching and active quenching [1]. Passive quenching, while relatively simple, has a number of drawbacks in its operation. Active quenching, Fig. 2, on the other hand, is the preferred method of operation but the design is critical for the efficient operation of an APD in photon counting or Geiger-mode.

3.1 Passive Quenching

When an APD is reversed biased above the breakdown voltage, V_a, an incident photon can trigger an avalanche in the depletion region and the device will start to breakdown. The large current that subsequently flows can be quenched by passive elements, in this case resistors. A large ballast resistor, on the order of 200 kΩ in series with the APD, can be used to limit the current flowing through the device. The increase in voltage at the anode of the APD will drop the voltage across the device to below breakdown. The APD will then start to slowly recharge until the voltage across the device has returned to the original bias voltage.

During this long recharge period, the APD is less sensitive to photons. The recharge time, or dead time, can be in the order of 5–10 µs, which limits the overall maximum counting rate to approximately 2×10^5 counts per second. Another drawback of using this passive quench arrangement is the dead time of the detector itself which is not constant nor well defined.

The dark count of the detector is also affected by this passive quench arrangement. The initial large current that is allowed to flow before being quenched will

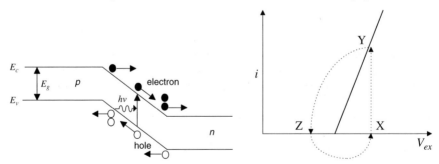

Fig. 1 Left: The multiplication process in an Avalanche Photodiode. **Right:** V-I characteristics illustrating the operation of an Avalanche Photodiode in Geiger-mode. The device is initially biased with an excess reverse bias voltage to point X. Once a photon is absorbed, the device breaks down to Y. The quench circuit will then bring the bias voltage below the breakdown voltage to Z, and then reset the bias voltage back to the original value at X

Geiger-mode Avalanche Photodiodes

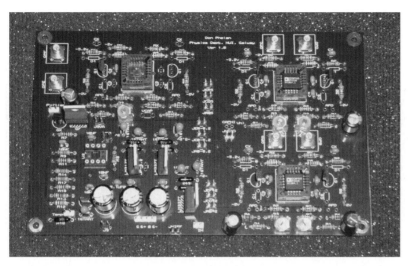

Fig. 2 An example of an active quench circuit (AQC) for operating a GM-APD. This board has a numer of AQCs, along with full temperature control for operating a Peltier cooler with the devices

dissipate a large amount of energy in the device. The resulting increase in temperature is responsible for an increase in the dark count rate.

A further drawback is after-pulsing, which can have a considerable effect on the dark count rate. During an avalanche, some electrons and holes are trapped in the depletion layer of the device, only to be released later to cause a further avalanche. This again is a major result of the large current flowing because of the passive quench arrangement. If the current could be limited quicker, fewer carriers would fill these traps and result in less after-pulsing and therefore fewer dark counts. These limitations make a passive quench arrangement less desirable for HTRA applications.

3.2 Active Quenching

To overcome the drawbacks of passive quenching, active quench circuits have been developed [5] [14]. When the device breaks down after an incident photon has been absorbed, an active quench circuit will quickly sense and reduce the bias voltage across the APD. After a certain hold off time, t_d, the voltage will be subsequently restored. Previous designs have been developed to operate APDs with high breakdown voltages (250–400 V) [5], and many different circuit designs have been developed by the authors of this chapter. These have proved very reliable with numerous different APD designs and are capable of detecting well in excess of 10^7 counts per second for long periods. Some of the benefits of the design include reduced power dissipation, lower dark counts and less after-pulsing. This after-pulsing, which manifests itself as an excess of correlated counts, is particularly important when dealing with second and third order correlations in very high time resolution observations.

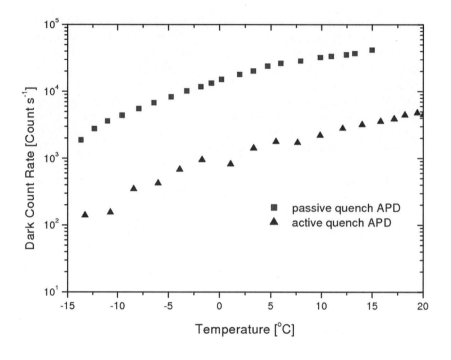

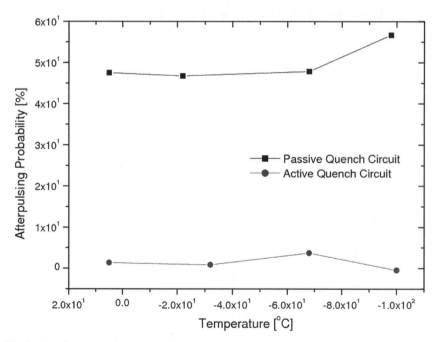

Fig. 3 The advantages of using an AQC over the traditional passive arrangement, results in lower dark counts and reduced after-pulsing

In the latest designs, the dead time of the APD can be adjusted between 10 ns and 1 µs by an ECL monostable multivibrator. The particular circuit used for these devices relies upon the ADCMP565 analog devices comparator, with an input/ouput delay of 300 ps, 50 ps propagation delay dispersion and 5 GHz equivalent bandwidth. The current board also includes temperature control circuitry with an AD590 temperature sensor providing stability of ± 1.0°C. Current efforts are also under way to develop the active quench circuit into a hybrid IC. The APD devices have been custom packaged and wire bonded on ceramic boards and mounted on a double stage Peltier cooler and contained within a heat sink enclosure.

The performance of the active quench circuit was compared to passive quenching and the results are shown in Fig. 3. The slope of the top graph, and hence the activation energy, was consistent with the result from active quenching.

4 Design and Fabrication of Shallow Junction Avalanche Photodiodes

4.1 General Device Architecture

Early attempts at photon counting using solid-state detectors involved the use of existing APDs operating above breakdown. Various conventional avalanche photodiode architectures were employed with varying degrees of success. To optimize the performance of APDs for use in photon counting applications the basic device architecture was modified to engineer a suitable electric field profile that would accomplish several objectives, namely:

- sufficiently wide depletion region to achieve high quantum efficiency
- peak electric field approaching the critical electric field of the material making the device
- optimize the carrier collection time at the contacts
- minimize the response time of the detector to incoming photons

The commercial leader in photon counting detectors at visible wavelengths is presently the reach-through avalanche photodiode developed by McIntyre [22, 21]. This device is schematically illustrated in Fig. 4, and is characterized by a large active area (~180 µm), wide depletion region (~30 µm), high photon detection probability in excess of 50% and good timing response less than 300 ps. The main disadvantages of this device include the high breakdown voltage (in excess of 300 V), and the proprietary process required to thin the device structure to allow the depletion region *reach through* the entire device volume. The proprietary fabrication process contributes to the expense of these devices and more importantly limits their usefulness in fabricating arrays of devices, which potentially opens up new astronomical applications.

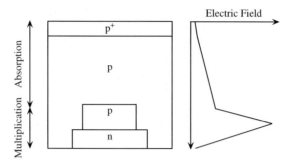

Fig. 4 Schematic illustration of the general architecture of a reach-through APD showing the electric field profile through the device

To address some of the short-comings of the reach-through avalanche photodiode, other device architectures were optimised for photon-counting HTRA applications. One such architecture is the shallow junction GM-APD [10, 12, 13], illustrated schematically in Fig. 5.

There are several aspects to the architecture of the shallow junction GM-APD that are worth noting. In particular the shallow junction can be appropriately placed to optimize the response to particular wavelengths; placing the junction closer to the surface improves the blue light response, while placing the junction deeper enhances the red light response of the device. Although the depletion region is only a fraction of the width obtained in the reach-through structure, the doping profile ensures that the peak electric field is almost at the breakdown field of silicon, thereby requiring relatively low reverse bias voltage ($\sim$30 V) to achieve a self-sustaining avalanche current from only a single charged carrier. The n-type doped region overlaps the p-type doped region creating what is known as a *virtual guard-ring*, the purpose of which is to ensure uniform breakdown of the central junction region, without edge breakdown effects that can normally occur due to electric field crowding at the edges of the device. The process steps used to fabricate the device are compatible

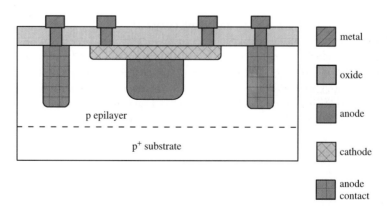

Fig. 5 The shallow-junction Geiger-mode Avalanche Photodiode architecture, a planar device structure employing a virtual guard-ring to ensure uniform junction breakdown

with standard CMOS processing steps [11], thereby facilitating the monolithic integration of these devices with the control electronics that make the device useable in photon counting mode. The planar nature of the device structure opens up the possibility of fabricating arrays of devices for imaging purposes [8, 12, 15], something that is not possible with the reach-through structure. The proximity of the junction to the surface, and hence the contacts, ensures a high speed of response, largely limited by the diffusion of carriers created by absorption deep in the semiconductor. This *diffusion tail* can be eliminated by using a silicon-on-insulator substrate and limiting the overall thickness of the epilayer. Finally, the oxide layer over the active region of the device can be used in conjunction with a silicon nitride passivation layer to form an anti-reflection coating that will improve the quantum efficiency of the device.

Ultimately the advantages of the shallow-junction GM-APD come at a price. The main disadvantages of these devices include: limited active area diameters (50 μm typical) due to the rapid increase in the dark count noise with increasing active area diameter. The main cause of this noise increase is the presence of process-induced defects that act as carrier generation centres within the device. There is a reduced quantum efficiency due to the smaller detection volume limited by the depletion region having a width typically less than 1 μm.

4.2 Device Fabrication

The fabrication of shallow-junction GM-APDs utilizes process steps common to the fabrication of CMOS transistors. The main process options include:

- choice of substrate; either standard epitaxial wafer or silicon-on-insulator (SOI)
- choice of doping technique; either implantation or diffusion
- choice of contacts; either planar or vertical—planar places both anode and cathode contacts at the surface, while the vertical structure places the anode contact at the back surface

The basic device process flow is summarized in Table 1. The critical factors for successful device fabrication include the execution order of the process steps to minimize process induced defects (particularly oxidation induced stacking faults) and the engineering of the doping profile to satisfy the requirements of a peak electric field close to the ionisation field for silicon and an electric field distribution that optimizes the speed of response and maximizes the avalanche initiation probability. A typical electric field profile for these devices is shown in Fig. 6.

4.3 Typical Device Performance Characteristics

A number of important parameters are identified which are necessary to fully characterize and efficiently operate avalanche photodiodes in Geiger-mode. These include quantifying the dark count rate, the excess reverse bias voltage, the optimum

Table 1 Outline process steps for fabrication of silicon shallow-junction GM-APDs

Step Number	Description
1	Select appropriate start material (epitaxial or SOI)
2	Define alignment marks
3	Oxidise surface
4	Define anode and anode contact regions lithographically
5	Implant boron to define anode and anode contacts
6	Drive-in anneal to activate boron
7	Define cathode lithographically
8	Implant phosphorous to form cathode
9	Drive-in anneal to activate dopants
10	Define metal contacts lithographically
11	Deposit metal and anneal

operating temperature, the photon detection probability, the after-pulsing probability and the hold-off time.

Several of these parameters are interrelated and a trade-off is necessary between the different variables for optimum performance, depending on the application. For example, decreasing the temperature will decrease the dark count, but increase the after-pulsing probability. In fact the dark count rate of the GM-APD will drop

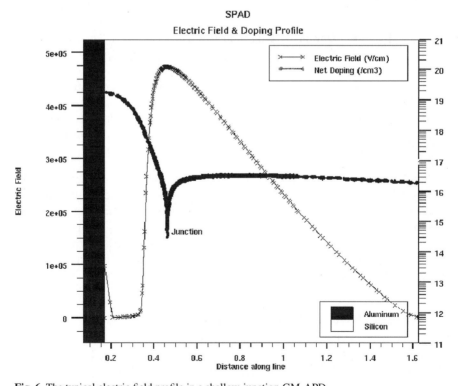

Fig. 6 The typical electric field profile in a shallow-junction GM-APD

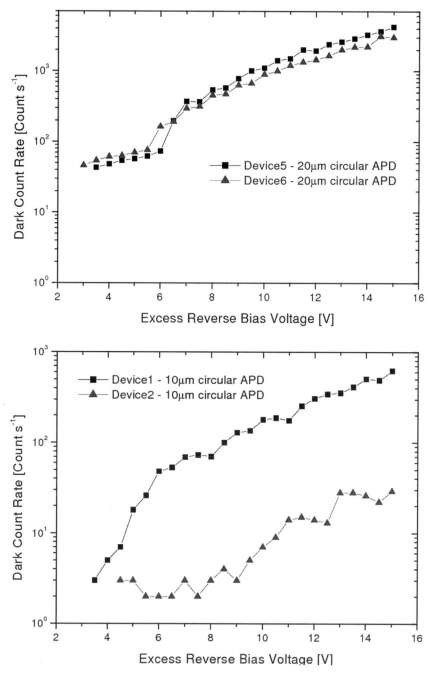

Fig. 7 The top figure shows the dark count rate versus excess reverse bias voltage for 20 μm diameter APD, while the bottom figure is for a 10 μm diameter device. Both APDs are at room temperature with a hold-off time, $\tau_d = 250$ ns

exponentially with temperature and while it is desirable to have a low dark count to obtain a highest signal to noise ratio and lowest statistical uncertainty, a trade-off must be made with the other parameters which affect it. These include the excess reverse bias voltage, V_{ex}, and the active area of the detector.

As the excess reverse bias voltage increases, the photon detection probability increases. While it would be desirable to have a high photon detection probability by increasing V_{ex}, it is to the detriment of the dark count rate, which also increases.

In the case of the active area, both the dark count rate and after-pulsing will increase significantly with an increase in the APD active area. This dependence has limited the size of APDs to small active areas. Fig. 7, while showing the exponential increase in dark count rate with excess reverse bias voltage for both a 10 and 20 μm diameter APD, also illustrates the difference between two similar size devices fabricated on the same wafer. Fig. 8 gives an example of the performance of both a 20 and 100 μm APD with respect to temperature.

The main performance characteristics for the shallow-junction GM-APD are summarised in Table 2.

In contrast to the reach-though APD the GM-APD has a much lower operating voltage that permits integration with CMOS electronics provided suitable electrical isolation elements, such as a trench isolation, are employed. The device size is obviously one of the main limiting factors for these devices in applications requiring the detection of diffuse radiation. The size limitation is only imposed due to the rapid increase in dark count noise with increasing device active area. Reducing this rapid increase in dark count rate with device area remains one of the goals of GM-APD design, although solutions have been demonstrated that limit the rate of increase in dark count by employing a parallel array of smaller devices to give a larger effective area [7]. One such device is illustrated in Fig. 9.

The spectral sensitivity achievable is typical for silicon based devices, but the blue response can be further enhanced by appropriate placement of the device junction closer to the semiconductor surface. The dark count rate for GM-APDs appears optimal around a diameter of 20 μm, with 100 counts per second typical at room temperature for bias voltages up to 5% in excess of the device breakdown voltage. The dark count rate is typically improved by cooling the devices using either a Peltier element or a cryogenic system. The photon detection probability for these devices comes close to matching that for reach-through devices despite not being able to achieve the same quantum efficiency due to the limited depletion region width. The key to the large photon detection probability is the avalanche initiation probability in these devices approaches unity, in effect requiring only a single charge carrier to create a self-sustaining macroscopic avalanche current.

5 Development of Photon Counting Arrays

For many astronomical applications, single pixel photon counting devices have limited potential. The shallow junction GM-APD already described provides an excellent opportunity to produce arrays of photon counting detectors by virtue of

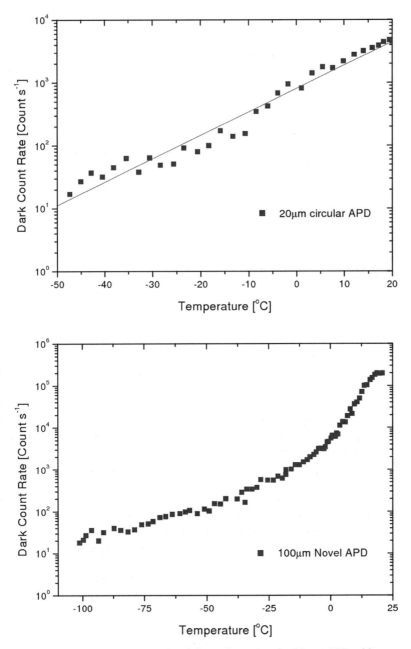

Fig. 8 The top graph shows an example of the performance of a 20 μm APD with respect to temperature, while the bottom graph shows the performance of a large 100 μm APD device. The latter device, at the operating temperature of a double stage Peltier cooler, −60°C, has a dark count rate of 100 counts per second, which is comparable to a commercial APD detector. At liquid nitrogen temperature, 77 K, no dark counts were observed at all for numerous devices that were tested

Table 2 Performance characteristics of silicon shallow-junction GM-APDs

Parameter	Min.	Max.	Typical
Diameter	5 μm	50 μm	20 μm
Operating Voltage	−25 V	−35 V	−30 V
Spectral Sensitivity	400 nm	900 nm	650 nm
Dark count rate (20 μm device)	10 Hz	1 kHz	100 Hz
Photon detection probability (650 nm)	30%	60%	50%
Timing response	100 ps	300 ps	150 ps

their planar architecture and CMOS compatible process steps. There are several considerations to be aware of in the pursuit of these arrays and these are outlined in this section.

5.1 Optical Crosstalk

One of the fundamental limitations in the development of arrays of avalanche photodiodes is the creation of photons by an APD when it is avalanching. This light is easily coupled to neighbouring pixels in an array leading to the generation of false counts in these pixels. There are several approaches that may be taken to overcome this fundamental difficulty. The first of these techniques involves sensing the onset of the avalanche breakdown and quenching it rapidly before significant light generation takes place. Rochas [18] have demonstrated that this is possible when the active quench and reset circuit is monolithically integrated with the APD. Another

Fig. 9 Novel device structures employing large arrays of smaller devices are used to limit the rate at which the dark count rate increases with active device area

approach is to optically isolate the neighbouring pixels from each other by incorporating features in the architecture such as filled trench isolation or by employing mesa structures to limit light coupling between devices. Trench isolation provides the added benefit of electrically isolating the pixels from each other.

5.2 Connectivity

Unlike CCD devices each pixel in an APD array requires its own electrical connection to a control circuit. This makes it extremely difficult to fabricate large two dimensional arrays, because of the difficulty in connecting to pixels near the centre of the array. Large one-dimensional arrays pose no such difficulty. The connectivity issue can be overcome by using clever routing of the connecting wires and making use of multilayer metal interconnect, thereby allowing crossover of metal tracks without short-circuits. The requirement for at least one unique connection to each pixel results in at least the same number of bond-pads as pixels. This makes it difficult to package large arrays and solutions such as flip-chip packaging to preprepared windowed boards becomes essential. An example of a 4 × 4 array of quad cells in development is shown in Fig. 10, where the large number of bond-pads obviously takes up the largest amount of the silicon area, thereby limiting the number of arrays that can be produced on a single wafer.

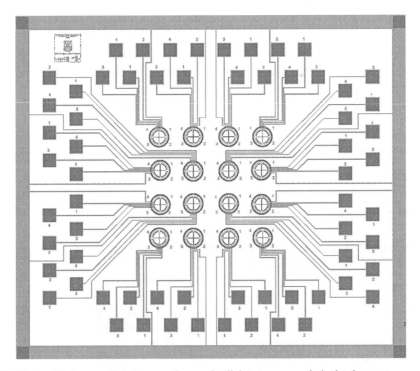

Fig. 10 4 × 4 2-D array of photon counting quad-cell detectors currently in development

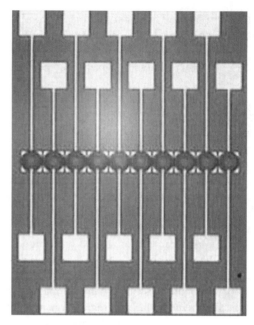

Fig. 11 1-D array of photon counting detectors

5.3 Fill-Factor

With the dilemma of optical crosstalk and the inevitable requirement to separate the pixels in an array the question of the array fill-factor becomes important. In other words, for a given array area how much of this is sensitive to light? For arrays of photon counting detectors this turns out to be less than 50% in most cases and a poor as 20% in some cases. Apart from including trench isolation between pixels, which is the main limitation for linear arrays as shown in Fig. 11, there is a requirement to leave room for the interconnect to the pixels in the case of 2-D arrays, as shown in Fig. 12.

6 Astronomical Performance

In astronomical observations, in order to increase the fill-factor it will be necessary to use a monolithic lenslet module with the GM-APD array . Custom lenslet arrays are available with a pitch of 100 µm and a focal length of 300 µm. These desirable low f/number lenslets are compatible with the current larger APD devices under development, which also have pitch of 100 µm. The GM-APD array with a lenslet array in front of it (at the focal plane of the telescope) forms a pupil image on each of the APD array elements. While the f/number of the lenslet is f/3, the small diameter of the APD pixel limits the f/number of the system. For example, a 25 µm active area device with the f/3 lenslet array will have an f/number of f/12.

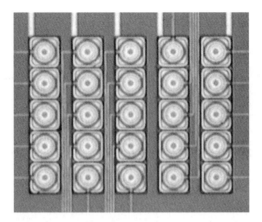

Fig. 12 2-D array of photon counting detectors

This potential mismatch between the f/number of the telescope and the lenslet/APD system can cause a significant loss of light, proportional to the square of the ratio of the f/numbers. To overcome this problem a focal extender could be placed before the focal plane, to increase the f/number of the telescope, as shown in the schematic in Fig. 13.

Simulations of the performance of an APD array on a 42 m future telescope have been carried out by Ryan [17] in the V Band with a 1 arcsecond point source. The calculations assume a SPAD array [24, 23], with a quantum efficiency of 53% and 50 dark counts per pixel per second. A number of parameters were assumed, including a 20% obscuration with atmospheric extinction of 0.21 mag in the light collected by the 42 m telescope. A 81% telescope throughput (four reflections of 95%) and 70% detector optics thoughput were also assumed. Using flux at $m_v(0) =$

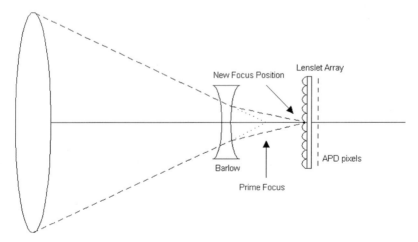

Fig. 13 A potential design of a physical scheme where the f/number of the telescope is increased (in the case where the mismatch of the f/number of the telescope and the lenslet/APD system can cause a significant loss of light)

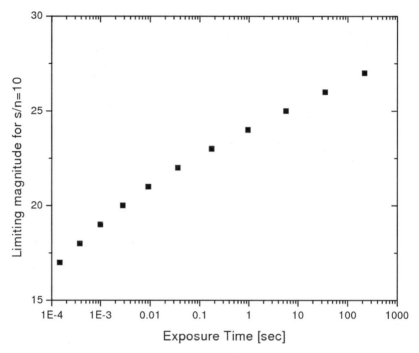

Fig. 14 An example of predicted astronomical performance showing the limiting magnitude of an APD array on a future large 42 m telescope, from data simulated by Ryan [17], e.g. a star of magnitude ∼18 needs ∼1E-3 seconds to reach a signal-to-noise of 10

3460 Jy, $\lambda_c = 550$ nm, $d\lambda/\lambda = 0.16$, 1 Jy = 1.51×10^7 photons s^{-1}m$^{-2}(d\lambda/\lambda)^{-1}$, the integrating times for SPAD detectors was calculated for a S/N of 10. This data is plotted in Fig. 14.

7 Conclusions

Geiger-mode Avalanche Photodiodes (GM-APDs) provide a number of unique advantages as detectors having the potential to solve numerous astrophysical problems when incorporated with future instrumentation in HTRA. While they offer many advantages over other detectors, substantial further research is necessary to overcome the challenges posed in their development. Future large telescopes will open up an exciting new domain in astronomy, and given further development the potential of GM-APDs could be realised to meet the highest time resolution demands of detectors in astrophysical exploration.

Acknowledgments The authors would like to acknowledge the support of OPTICON, which has received research funding from the European Community's Sixth Framework Programme under contract number RII3-CT-001566.

References

1. S. Cova, M. Ghioni, A. Lacaita, C. Somari, and F. Zappa. Avalanche photodiodes and quenching circuits for single-photon detection. *Applied Optics*, 35, 1996.
2. S. Cova, M. Ghioni, A. Lotito, I. Rech, and F. Zappa. Evolution and prospects for single-photon avalanche diodes and quenching circuits. *Journal of Modern Optics*, 51:9, 2004.
3. V. Dhillon and T. Marsh. ULTRACAM - studying astrophysics on the fastest timescales. *New Astronomy Review*, 45:91–95, January 2001.
4. D. Dravins, C. Barbieri, V. Da Deppo, D. Faria, S. Fornasier, R. Fos-bury, L. Lindegren, G. Naletto, R. Nilsson, T. Occhipinti, F. Tamburini, H. Uthas, and L. Zampieri. Quanteye - quantum optics instrumentation for astronomy. *ESO*, OWL Instrument Concept Study, 2005.
5. M. Ghioni, S. Cova, F. Zappa, and C. Somari. Compact active quenching circuit for fast photon counting with avalanche photodiodes. *Review of Scientific Instruments*, 67, 1996.
6. A. Golden, A. Shearer, R. M. Redfern, G. M. Beskin, S. I. Neizvestny, V. V. Neustroev, V. L. Plokhotnichenko, and M. Cullum. High speed phase-resolved 2-d ubv photometry of the crab pulsar. *Astronomy and Astrophysics*, 363:617, 2000.
7. J. C. Jackson, A. P. Morrison, D. Phelan, and A. Mathewson. A novel silicon geiger-mode avalanche photodiode. In *IEEE-International Electron Devices Meeting (IEDM)*, volume 32.2, December 2002.
8. J. C. Jackson, D. Phelan, A. P. Morrison, M. Redfern, and A. Mathewson. Towards integrated single photon counting arrays. *Optical Engineering*, 42(1):112–118, January 2003.
9. M. A. McLaughlin, A. G. Lyne, D. R. Lorimer, M. Kramer, A. J. Faulkner, R. N. Manchester, J. M. Cordes, F. Camilo, A. Possenti, I. H. Stairs, G. Hobbs, N. D'Amico, M. Burgay, and J. T. O'Brien. Transient radio bursts from rotating neutron stars. *Nature*, 439, 2006.
10. A. M. Moloney, A. P. Morrison, C. J. Jackson, A. Mathewson, and P. J. Murphy. Large-area Geiger-mode avalanche photodiodes for short-haul plastic optical fiber communication. In *SPIE, Opto Ireland, Optoelectronic and Photonic Devices*, volume 4876–83, September 2002.
11. A. M. Moloney, A. P. Morrison, J. C. Jackson, A. Mathewson, and P. J. Murphy. Geiger mode avalanche photodiode with CMOS transimpedance amplifier receiver for optical data link applications. In *IT&T Annual Conference, Transmission Technologies*, October 2002.
12. A. P. Morrison, V. S. Sinnis, A. Mathewson, F. Zappa, L. Variscoand M. Ghioni, and S. Cova. Single-photon avalanche detectors for low-light level imaging. In *Proceedings of SPIE, EUV, X-Ray, and Gamma-Ray Instrumentation for Astronomy VIII*, volume 3114, pages 333–340, 1997.
13. A. P. Morrison, V. S. Sinnis, L. Varisco, F. Zappa, and M. Ghioni. Single photon avalanche detectors for fluorescence imaging applications. In *Proceedings of SPIE, Chemical, Biochemical and Environmental Fiber Sensors IX*, volume 3105, pages 122–128, 1997.
14. N. S. Nightingale. A new silicon avalanche photodiode photon counting detector module for astronomy. *Experimental Astronomy*, 1, 1991.
15. D. Phelan, J. C. Jackson, R. M. Redfern, A. P. Morrison, and A.Mathewson. Geiger mode avalanche photodiodes for microarray systems. In *SPIE, the International Society for Optical Engineering, Biomedical Nanotechnology Architectures and Applications, San Jose, CA*, volume 4626, pages 88–97, January 2002.
16. D. Phelan, O. Ryan, and A. Shearer(eds.). High time resolution astrophysics, ch. From QuantEYE to Aqueye – Instrumentation for astrophysics on its shortest timescales by C. Barbieri. *Springer*, 2007.
17. D. Phelan, O. Ryan, and A. Shearer(eds.). High time resolution astrophysics, ch. Use of an Extremely Large Telescope for HTRA by O. Ryan et al. *Springer*, 2007.
18. Alexis Rochas, Alexandre R. Pauchard, Pierre-A. Besse, Dragan Pantic, Zoran Prijic, and Rade S. Popovic. Low-noise silicon avalanche photodiodes fabricated in conventional CMOS technologies. *IEEE Transactions on Electron Devices*, 49:387–394, 2002.
19. A. Shearer, B. Stappers, P. OConnor, A. Golden, R. Strom, M. Redfern, and O. Ryan. Enhanced optical emission during crab giant radio pulses. *Science*, 301, 2003.

20. C. Straubmeier, G. Kanbach, and F. Schrey. OPTIMA: a photon counting high-speed photometer. *Experimental Astronomy*, 11, 2001.
21. P. P. Webb, R. J. McIntyre, and J. Conradi. Properties of avalanche photodiodes. *RCA Review*, 35:234–278, 1974.
22. P. P. Webb and R. J. McIntyre. Large area reach-through avalanche diodes for x-ray spectroscopy. *IEEE Trans Nucl. Sci.*, 23:138–144, 1976.
23. F. Zappa, S. Tisa, S. Cova, P. Maccagnani, D. Bonaccini, G. Bonanno, M. Belluso, R. Saletti, and R. Roncella. Pushing technologies: Single-photon avalanche diode arrays. In *SPIE Int. Symposium on Astronom- ical Telescopes and Instrumentation, Glasgow*, volume 5490, 2004.
24. F. Zappa, S. Tisa, S. Cova, P. Maccagnani, R. Saletti, and R. Roncella. Single-photon imaging at 20,000 frames/s. *Optics Letters*, 30:3024–3026, February 2005.

Transition Edge Cameras for Fast Optical Spectrophotometry

Roger W. Romani, Thomas J. Bay, Jennifer Burney and Blas Cabrera

Abstract When one spends $10^{8-8.5}$ euros on a large aperture telescope, it behooves one to make maximal use of the photons concentrated at the focus. One popular path is to use very large focal plane arrays, for a high multiplex advantage; this is a natural driver in extragalactic survey science. The other avenue to efficiency applies to single source science, where a large field of view is not necessary, but energy resolution over a broad wavelength range, high time resolution and, possibly, polarization measurements wring maximum benefit from the collected photons.

Cryogenic energy-resolving sensors are proving to be promising detectors for many areas of astronomy. Significant astronomical observations have, in fact, already been made using two detector schemes: Superconducting Tunnel Junction (STJ) and Transition-Edge Sensor (TES). TES instruments, in particular, have found application as bolometric detectors from the sub-mm to the hard X-ray range. We focus here on TES applications in the near IR through UV, reviewing the basic technology and describing current work toward effective imaging arrays. A TES camera using such arrays can address the need for broad-band single source efficiency at large telescopes noted above. A review of early test observations and a description of a few science goals are presented to illustrate this potential.

1 Transition Edge Sensor Technology in the IR-Opt-UV

TES are calorimeters measuring the temperature rise in an isolated heat capacity. The thermistor monitors the superconducting-normal transition of a thin film (typically elemental tungsten W, or bi-layers, e.g. Mo/Au). To keep the heat capacity low and the energy resolution high, the critical transition temperature is generally set by ion implantation to be $\leq$ 100 mK [1]. In fact, the energy resolution scales with device saturation energy $\Delta E_{FWHM} \propto E_{Sat}^{1/2}$ and with $T_c \sim 0.1$ K practical devices have $\Delta E_{FWHM}(3eV) \approx 0.1$ eV. Accordingly, with a fast readout the devices are photon counting, energy-resolving detectors sensitive down through the near-IR

Roger W. Romani · Thomas J. Bay · Jennifer Burney · Blas Cabrera
Dept of Physics, Stanford University, CA 94305–4060, USA
e-mails: rwr@astro.stanford.edu, tbay@stanford.edu, burney@stanford.edu, cabrera@stanford.edu

range. At longer wavelengths TES can be run as total-power bolometers, with very good sensitivity and noise performance.

The other principal technology uses tunnel junctions and essentially counts quasi-particles with excitation energy close to the ∼1 meV superconductor energy gap. For these devices the energy resolution scales as $\Delta E_{FWHM} \propto E_\gamma^{1/2}$. As it happens, present STJ and TES devices can achieve comparable energy resolutions at ∼2–3 eV photon energies. Both techniques have now been used in this optical range with small arrays instrumented with full readout channels for each pixel. As for which device proves most attractive for fast energy-resolved spectrophotometry, much depends on their ability to approach the theoretical resolution limits in practical astronomical packages with a modest number of active pixels. To illustrate the potential and challenges we focus on our recent TES work.

A schematic and photo of an optical TES pixel are shown in Fig. 1. In the optical regime, the photons are absorbed directly into the ∼25 μm × 25 μm superconducting W (tungsten) film (with ∼50% optical quantum efficiency QE for bare W). The film is biased at a fixed voltage so that Joule heating raises the W temperature above the ∼60 mK substrate temperature to the middle of the steep superconducting-normal transition. Thereafter, when an absorbed photon deposits energy into the tungsten electron system, the resistance rises and the Joule heating current drops. This current pulse is read with a low noise, low impedance 100 × SQUID series amplifier which is controlled, shaped and recorded by a room-temperature digital feedback system (one channel for each pixel). The net effect of this 'negative electro-thermal feedback' (negative ETF) is a substantial speed-up in the pulse recovery and improved linearity and stability, since it ensures that the system makes modest excursions around the bias temperature [2].

The fundamental resolution [3] is given by

$$\Delta E_{FWHM} = 2.355\sqrt{4k_B C T^2 \sqrt{n/2}/\alpha}, \quad (1)$$

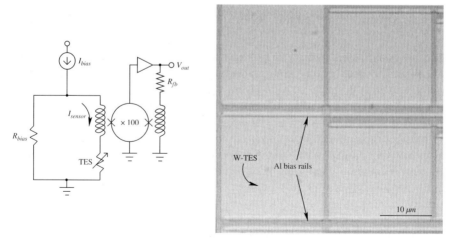

Fig. 1 The electrical circuit (left) of the TES showing the voltage bias scheme and SQUID current readout. A photograph (right) of a W TES sensor array with Al rails

where T and C are the temperature and heat capacity of the electron system (very near T_c), respectively, $n = 5$ for electron-phonon limited conductance, and $\alpha = (d \ln R / d \ln T)_{V=const}$. This resolution is independent of energy in the limit of a small temperature excursion around T_c; however, the optimum detector design gives

$$\Delta E_{FWHM} \approx 2.355\sqrt{4\sqrt{n/2}k_B T_c E_{max}}, \qquad (2)$$

with $E_{max} \approx CT_c/\alpha$, which scales as $\sqrt{T_c}$. For example, for $T_c = 80$ mK and $E_{max} = 10$ eV we obtain $\Delta E_{FWHM} \approx 0.05$ eV. This intrinsic resolution is limited by thermodynamic fluctuations. The W TES sensors are intrinsically fast, with a pulse rise-time constant of ~ 0.3 μs and pulse fall-time of ~ 10 μs.

In practice, some of the photon energy is lost as high-energy athermal phonons of the initial energy cascade penetrate from the W electron system to the substrate. Monte Carlo simulations and direct measurement of the electro-thermal feedback integral show that in our present devices 58% of the initial energy is lost [4]. With this correction, the limiting resolution for a $E_{max} = 10$ eV pixel (size ~ 20 μm) is $\Delta E_{FWHM} = 0.09$ eV. Our best energy resolution to date is 0.12 eV. More typical resolutions in array applications are ~ 0.15–0.2 eV. The source of the extra broadening, which likely includes far-IR thermal loading, antenna pick-up and some positional-dependence in the W response, is not yet fully understood, but we describe below our work to isolate the dominant effects in the instrument applications and ideas for further improvement in energy resolution. For example, patterning the TES devices on a SiN membrane instead of a bulk Si substrate should increase the efficiency of energy retention in the TES pixel to near 100%.

To obtain good energy resolution, while taking advantage of the high count rates allowed by the pulse fall time, fast dedicated read-out electronics are required. At present we use custom FPGA boards, with controllable algorithms to correct for background variations and to shape the pulses. The system assigns photon energies based on pulse height, while monitoring pile-up, and can operate at rates >100 kHz/pixel. However, full pulse digitization confirms that somewhat greater performance should be achievable with more sophisticated energy measurement schemes. To explore these effects, we have monitored the response to calibrated heat pulsed injected into the pixels, which can be averaged to produce high S/N pulse profiles (Fig. 2). This work is backed up by detailed device simulations. As the Figure demonstrates, as the deposited energy approaches saturation, the pulse shape changes. Of course, the area integral remains nearly proportional to the photon energy.

An optimal filter can be used to measure the energy of the full digitized pulse shape [5]. In the linear regime $E < E_{Sat}$ the energy resolution is constant, as noted above. At higher energies, $\Delta E_{FWHM} \propto E^{1/2}$, as the longer pulse is affected by lower frequency noise. The models suggest that one can operate out to 3–4 × E_{Sat} with good energy recovery. Thus ground-based optical-UV applications should be able to run with smaller $E_{Sat} \approx 1$ eV pixels. resulting in a factor of 2 increase in energy resolution. Fast digitization FPGA measurement of the pulse integral has, however, not yet been implemented.

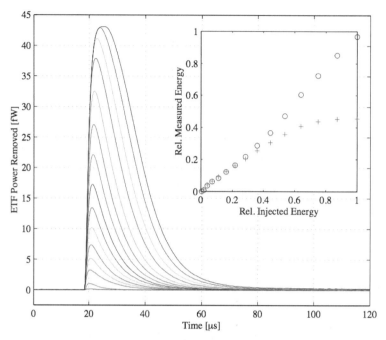

Fig. 2 Heat pulse (0.5 eV to 8 eV photon equivalent) calibrations of a 25 μm TES. Each curve averages 1000 pulses and shows amplitude saturation at 4 eV. The decay time constants are ∼8 μs and pulse flattening is apparent for the highest energies, with a saturation power of ∼43 fW. The inset shows the sensor linearity for peak-height (+) and pulse integral (o)

2 Initial Astronomical Applications

The first demonstration of individual TES sensors and few pixel arrays in the optical were made using a KelvinOx-15 He dilution refrigerator. Light was brought to the 20 μm pixels operating at ∼70 mK via optical fiber. To achieve superconductivity and TES operation, it is essential to exclude the large heat load from the 300 K thermal photons. This is a critical task for all cryogenic optical detector technology. A simple solution is to employ small-core ∼5 μm single mode fibers, which transmit only the acceptable near-IR. However, for astronomical applications at reasonable telescope focal lengths, a larger entrance aperture is needed. Our solution is to use 50–200 μm core 'wet' (high OH) fibers, with 5–10 m spooled at 4 K in the fridge [6]. At these lengths, the OH absorption bands in the fiber are opaque, providing 4 K cold stops, while allowing good transmission in the optical and portions of the near-IR J and H bands. The output end of the fiber is focused with a (Gradient index) GRIN and ball lens assembly to the TES pixels (Fig. 3).

Astronomical first light was achieved with a commercial 0.2 m telescope [7], followed by the first scientifically interesting measurements with a 0.6 m telescope [6]. This series of experiments culminated in a run at the 2.7 m Harlan J. Smith telescope at McDonald observatory [8]. This system had a 400 μm (=1.7″) entrance

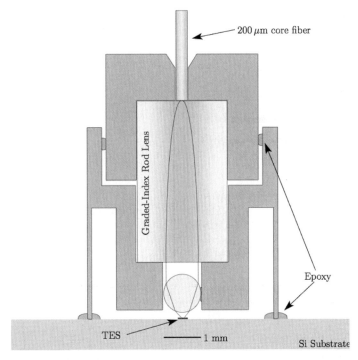

Fig. 3 (left) Diagram of a machined Al optics mount for coupling a 200 μm fiber to a small (40 × 40 μm) set of four TES pixels (right). Photograph of the assembly mounted on the sample stage of the dilution refrigerator

fiber, tapered to 200 μm and (inefficiently) coupled to the dewar internal fiber. The internal fibers acted as a cold filter and were focused onto a sub-set of a 6 × 6 array, where a four-channel prototype DAC read out the central pixels. Measurements of the Crab pulsar detected phase-resolved spectral variations from the H through U bands for the first time (Fig. 4, Left). During this run, we also observed several CVs and other compact object binaries, as well as other pulsars and emission line sources. For example Fig. 4 (Right) shows the spectrum of the bright 114 s orbital period polar (magnetic CV) ST LMi, showing above the difference between phases where the white dwarf and accretion column were visible ('On') and eclipsed ('Off'). Below is plotted the 1–2 eV light curve (Med band) along with the color ratios to high (>2 eV) and low (<1 eV) bands. Crude spectro-photopolarimetry was also demonstrated for the Crab pulsar. While on-sky efficiencies as high as 20% have been achieved with these fiber TES systems (including the ∼50% bare W absortivity), the limitation of a small entrance aperture makes it increasingly difficult to work at large focal length. Also, aperture losses, variable seeing and atmospheric dispersion make absolute spectrophotometry very difficult. Finally, most of the most interesting sources are at $V>25$, so that acquisition and sky subtraction are nearly impossible. Clearly an effective instrument must have a larger aperture and an imaging array.

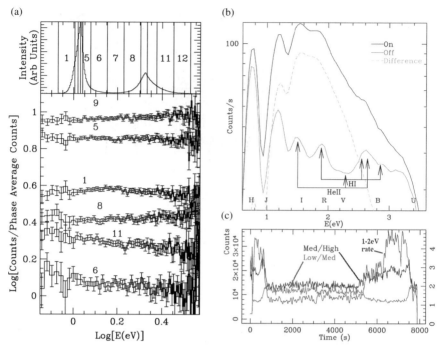

Fig. 4 Left: Calibrated phase spectra for selected phase bins based on 3500 s of Crab 2.7 m data. The spectra extend into the near-IR and show significant variations, especially around the main peak. Right: Observations covering 1.25 orbits of the polar ST LMi. Above are spectra with (on) and without (off) the white dwarf and accretion column visible. Below are shown the light curve and colors, with flux ratios showing spectral hardening at white dwarf egress

3 TES Array Camera Technology

We have focused on a focal plane imaging solution employing small arrays of uniform TES W pixels. Each pixel has a sensitive area of $24 \times 24\,\mu\mathrm{m}$, 35 nm thick W and is connected to read-out pads with 1–1.5 μm wide Al rails, which are superconducting at the base temperature. We have made 4×8, 6×6 and 8×8 arrays, all with the pixels on a 36 μm pitch (Fig. 5). Each pixel has independent wiring and the traces are carried out of the array in the 12 μm gaps.

We found during the initial experiments that if photons are absorbed on the portion of the W pad overlapping the superconducting rails, more energy is retained in the W e^- system. This gives a monochromatic source energy response function with a higher energy satellite peaks containing several % of the counts. Similarly, photons absorbed on the Si substrate can transfer heat to adjacent W pads with low efficiency, depending on distance: these leads to a low energy continuum in the energy response function. Moreover, the effective loss of the Si and Al area leads to a low 44% fill factor. We have combatted both of these effects with a reflective grid mask, suspended above the array (Fig. 6). Using a reactive ion etch, we form tapered wells in a 60 μm thinned Si wafer which end in 22 μm square apertures. This mask is aluminized to reflect the masked photons to the nearest pixel. The result has an effec-

Fig. 5 Three generations of TES imaging arrays, starting with 4×8, then 6×6, and finally the 8×8 array. Each pixel has a sensitive area of 24×24 μm and the array has 36×36 μm center-to-center spacing. The space between pixels is covered with reflection mask (shown in Fig. 6)

tive fill factor of ∼92%, and completely removes energy response function satellites while providing 12 μm of covered real estate on the chip to bring out the Al traces. Thus our arrays have a dense-packed 36 μm pitch better matched to large aperture telescopes. The smaller active W area decreases the heat capacity and increases the energy resolution, while the single peak energy PSF greatly simplifies the spectral analysis. The array focal plane lies at the top of the Al-coated reflection mask.

The minimum fraction of the array in active TES pixels depends on the speed of the beam hitting the array. For our imaging camera solution, the fraction is ≥ 0.3, meaning that the amount of area available for read-out channels is limited. Thus while our present lithography can be used for array sizes of up to $\sim 10 \times 10$, much larger arrays cannot use this simple single plane solution. Possibilities for larger formats include e-beam Al rail lithography, allowing much larger trace density in the pixels gaps or multi-plane fabrication, with read-out traces running below the active W plane.

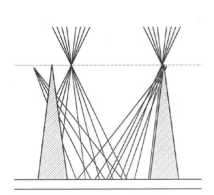 

Fig. 6 Schematic (left) showing the image plane at the top of the mask and the reflected rays which avoid the rails. An SEM (right) of 36 μm center-to-center by 60 μm thick reflection mask, which is not yet coated with Al

3.1 Refrigerator, Imaging package

The real challenge is to bring the desired near-IR through UV photons from a large aperture telescope to a TES array at $T \sim 70$ mK at reasonable efficiency, while maintaining thermal, electrical and mechanical isolation from the outside environment. In particular, the incident power on each TES pixel needs to be well below the ~ 10 eV/10 µs ≈ 150 fW that saturates the device. Note that a 4 K load is already a not insignificant 19 fW/steradian.

The original KelvinOx-15 dilution refrigerator system was bulky and required pumping systems that made close coupling to a telescope difficult. We have instead based our imaging system on a compact two-stage GGG/FAA adiabatic demagnetization refrigerator (ADR), standing about 1m tall and weighing under 60 kg. The system was manufactured by Janis Research Company after an original design by our NIST collaborators. It achieves base temperatures of ≤ 60 mK and has a hold time of ~ 12 h at $T \leq 80$ mK, depending on radiation load. We have configured the dewar with a side-looking aperture, with windows at the room temperature (300 K), lN$_2$ (77 K), and lHe (4 K) walls. The base temperature mount and array are contained within a 1 K shield at the GGG stage.

Expected applications will be at the broken Cassegrain or, better, Nasmyth foci of large aperture telescopes where the plate scale will be large. For example, the f/15 Nasmyth foci of the VLT have a plate scale of 580 µm/arcsecond. Accordingly, we incorporate an all-reflective Schwarzschild focal-reducing optic (a 15× microscope objective) at the 1 K stage to more effectively couple a slow telescope beam to our small array (Fig. 7). The microscope objective has a long working length and thus we can relay the focal surface from a convenient distance outside the RT window to provide optimal $\sim$f/1 coupling to the array at an effective magnification of $\sim 13 \times$.

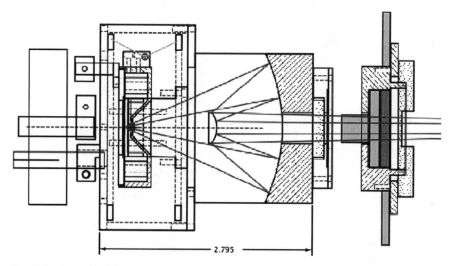

Fig. 7 A schematic of the camera interior. From right to left, the 4 K quartz filter and acrylic plug, the 1 K grid filter and reflection optic, the 1 K enclosure and the suspended 60 mK base stage with the reflection mask and focal plane array

Thus at a 120 m focal length, our array covers ~0.7 arcsec/pixel. This design has several additional advantages. With the slow beam suffering little distortion in the entrance windows, the all reflective on-axis optic provides a nearly achromatic focus that couples well to the grid mask. Also, this system introduces no additional instrumental polarization, beyond that produced at the Cass. or Nasmyth fold. The large solid angle of the optic and its support structure are also carried at 1K, meaning that the TES system only views higher temperatures along its active optical path.

The optical system allows us to produce ~2 μm spots at the focal plane. This is very convenient for both imaging the array in situ and for probing the array light response. For example, a laser line source can be conveniently scanned across the array to measure flat field uniformity, as well as probe position dependence of the pixel performance, including mask efficiency, pulse shape and QE (Fig. 8).

Further, the focal reduction is also helpful in reducing the importance of internal flexure. Since the array must be carried on the base-stage which is suspended on a Kevlar thermal isolation system, some flexure associated with gravity changes and with magnet thermal control is inevitable. However, the overall motion is small and is easily contained within the entrance windows, while the focal-reduced displacement at the array can easily be guided out. Indeed, since most applications will feature point source targets, real-time feedback from the TES system itself can be used to compute the small guiding corrections.

The ADR system can also be modified to have an on-axis, downward-looking port. This configuration could be of interest for use without the internal re-imaging

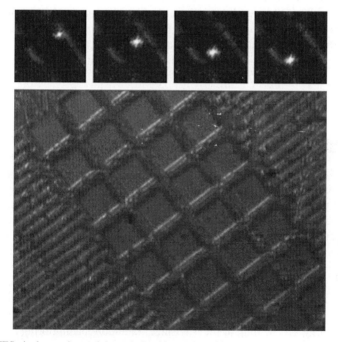

Fig. 8 A TES pixel array imaged through the microscope objective. Above: images of a laser spot focused by the objective being scanned across a pixel

optic at the fast prime focus of a large telescope. At a focal length of 15 m (eg. at Subaru) the array covers 0.5 arcsec/pixel. While this arrangement maximizes delivered light and minimizes instrumental polarization, is suffers the disadvantage of some chromatic aberration in the entrance windows, as well as access difficulty.

3.2 Shutter and Filters

To control the light delivery to the focal plane, we have installed a cryogenic shutter at the 4 K wall. This allows us to test pixel performance and probe the effect of environmental heat loading [9]. Fiber optic cables can be used to introduce optical calibration signals into the dewar even when the shutter is closed.

The greatest challenge to a cryogenic imaging solution is the extreme $\sim 10^8 \times$ suppression of the unacceptable 10 µm peak photons from the room temperature environment while allowing good transparency for the resolvable photons [10]. In practice this means very efficient suppression for $\lambda \geq 1.7\,\mu$m. The solution that we employed with the initial dilution fridge system worked quite well and can, in principle, be adapted to the ADR system. Since ≥ 5 m of cold fiber are required, a long coherent bundle of high-OH fiber would need to be spooled inside the dewar. This would also require vacuum feed-though for the bundle or a cryogenic 'snout' at the windows, to put the fiber entrance aperture in the focal plane. This remains a fall-back solution for a TES camera system, but several factors make this less desirable. Fill factor losses (limits on fiber packing) and IR and especially UV cutoffs to the system response will limit the imaging performance. Further, because of the extraordinarily broad-band sensitivity of the device ($\sim$0.1–10 eV, $\sim$10–0.1 µm), a true imaging solution adaptable to eventual space applications is highly desirable.

Even in our present system, we would like to pass photons from $\sim 1.7\,\mu$m to the $\sim 0.3\,\mu$m atmospheric limit. This is well over an octave in energy so most traditional filter solutions are completely inadequate. One traditional way of rejecting infrared is heat-absorbing glasses (e.g. Schott KG glasses). The ESA S-Cam team, for example employ KG and BK7 glasses in their filter stack [11]. The result is a ~ 0.7–$0.4\,\mu$m transmission window with a modest (30–40%) peak transmission efficiency. One serious problem with heat absorbing glasses is a significant $\sim 2\%$ leak at 2.7 µm which adds substantial thermal load at the focal plan. In addition, the challenges of employing glasses at cryogenic temperatures, where poor thermal conductivity makes cooling difficult and far-IR leaks can let through additional thermal load, are substantial.

We have adopted a more complex solution (Figs. 9 & 10). At room temperature a simple sapphire window seals the dewar. Then quartz filters are employed at 77 K and 4 K; at cryogenic temperatures these provide effective blocking between 6 µm and $\sim 27\,\mu$m. We filter the shorter wavelength radiation with a thick optical grade, UV transmitting acrylic plug at the 4 K stage; radiation loading brings the acrylic itself to somewhat higher T_{eff}. The last element is a reflective metal grid filter at the 1 K stage, which reflects the longest wavelength radiation. In Fig. 9 we show

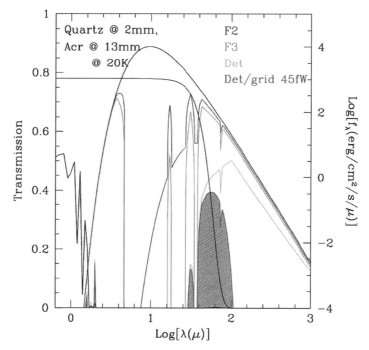

Fig. 9 Filter Transmission in wavelength space at successive stack planes: 300 K entrance flux, 77 K quartz, 4 K quartz, Acrylic at ∼20 K. This emphasizes the many components needed to block at large λ. The shaded region gives the energy flux through the 1 K grid filter onto the detector. The transmission (left scale) peaks at ∼50% for optics without AR coatings

the computed filter performance, first emphasizing the challenges by plotting the thermal load at each stage and overall optical transmission as a function of Log(λ). Fig. 10 emphasizes the science opportunities by plotting in linear eV (i.e. TES energy resolution steps). This shows reasonable sensitivity from 0.85 eV (1.5 μm) to beyond the atmospheric limit. For contrast the approximate filter pass band of the S-Cam glass system is shown. The model residual power into each detector pixel (45 fW) is in reasonable agreement with the measured result.

3.3 Data Acquisition System

At present 32 channels are instrumented, each with a dedicated SQUID amplifier and digital feedback system. These are organized in banks of eight cards in a small crate and are slaved to a time signal from a GPS receiver[12]. The entire system is controlled by a MatLab GUI, which allows for control of the device biasing and calibration and allows real-time quick-look analysis of the imaging data stream. The individual events are time-tagged, energy resolved, assigned to a pixel and flagged for pile-up and other system problems and are dumped to disk on a dedicated PC. Calibration signals (pulses of monochromatic photons and/or heat pulses) can be injected into the data stream for improved running calibrations.

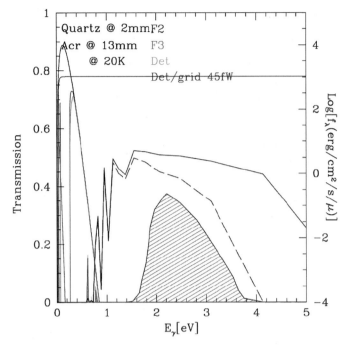

Fig. 10 Filter Transmission in energy space, emphasizing the sensitive band ($\Delta E \sim$const in eV). The black solid line shows stack transmission for uncoated optics, dashed line shows response including atmospheric extinction. The shaded region is the transmission of a KG5/BK7 heat absorbing filter stack, as used in the STJ system S-CAM3

Figure 11 shows the result of an imaging test demonstrating time, space and energy resolution. Here the array is illuminated with a diffuse beam from an IR/green laser and a focused red laser spot. The spot was scanned across an array column. As shown in Fig. 11 an energy cut isolates the moving component from the stationary diffuse photons.

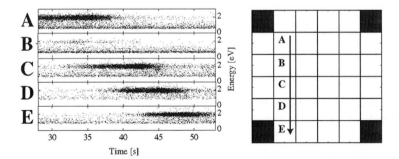

Fig. 11 Left: a time series of optical photon events from one column of the TES array. A red laser spot is scanned down a column (pixel B had a partly blocked focusing mask). The background counts come from diffuse illumination from a frequency doubled Nd-YAG laser with IR and green peaks

4 What Next?

At present the TES camera is still in the laboratory testing stage, and more work is needed to make the system robust and optimize performance. However prospects look good for a 32 pixel system with modest ($R \sim 15$–25) energy resolution and usable 10–20% on-sky efficiency. This should be sufficient for PSF-contained imaging spectrophotometry of pulsars and other faint time variable sources from the H through U bands. Brighter sources may be defocused for study at total count rates up to ~ 500 kHz, but with more limited absolute photometry. Many of the sort of experiments that one might wish to make have in fact already had proof-of-principal demonstrations with S-CAM observations of bright targets by the ESA STJ group, e.g. [13, 14, 15]. The science reach of a TES camera, depends critically on the robustness and efficiency of the system achieved under routine observations. Only demonstration observations can tell that tale.

This does not however prevent us from speculating on further TES camera development. The simplest step is to convert the existing system to a imaging dual polarimetry camera. We have designed a (room temperature) Wollaston prism plus rotating super-achromatic 1/2 waveplate system that splits the present 4×8 array into two 4×4 subfields with E and O images. The energy resolution of the detector allows one to finely calibrate the residual chromatics in the polarization response giving high precision near-IR through UV spectro-photopolarimetry. Observations of spin-powered pulsars with their high polarization are very interesting. Since the emission zone geometry of these objects is still controversial, polarization measurements can be a strong driver for improving magnetospheric emission models. Polars (accreting magnetized white dwarfs) are also attractive targets, with polarized broad cyclotron features dominating the continuum; the strongest superimposed atomic emission lines can also be measured. The cyclotron lines should vary in both in intensity and polarization as the system rotates, allowing novel diagnostics. Higher frequency $\sim$Hz QPOs in the accretion shock should also exhibit interesting polarization fluctuations.

A related but more ambitious scheme is to use the the TES array as an order-sorter in an Echelle spectrograph. We have blocked out low resolution configurations that can provide good wavelength coverage with ~ 100 pixel linear arrays. If arrays can grow to the kilo-pixel scale more ambitious spectrograph arrangements could be pursued, as demonstrated in designs by the STJ group [16].

This leads naturally to the question of array size limits. The S-CAM3 STJ system has already demonstrated a 10×12 array, with individual amplifier chains for each channel [11]. Manufacture of TES arrays with single plane wiring can reach $\sim 10 \times 10$ scales with our present fabrication facilities; the limit with more sophisticated lithography might reach $\sim 16 \times 16$ formats. Beyond this e-beam trace writing or multi-plane solutions may be required. However the rapidly growing heat load associated with wiring to independent SQUID channels for each pixel rapidly makes such 'brute force' schemes impractical. STJ systems share this problems, although with 0.3 K base stages it is not so severe. Happily, TES arrays have a demonstrated time division multiplex (MUX) solution [17]. Here, first stage SQUIDs rapidly switch a bias current along a pixel row whose signal is measured by a single SQUID amplifier channel. The challenge is that high (~2 MHz/channel) switching rates are required to sample the optical TES pulses with their short time constants. Present SQUID bandwidths should allow ~10 pixels/amplifier in feedback mode and 2–3 × more in open loop mode. One issue is that the 20 kHz/amplifier count rate limit now applies to an entire row rather than a single pixel. This is a serious limit for ground based applications, with the sky background supplying a large fraction of this rate unless the system is narrow banded or the pixels subtend a small angle on the sky. MUXed TES cameras can, however be very interesting for AO corrected beams on large telescopes. We should note that the optical high time resolution application of SQUID MUX is particularly challenging, with high count rates and short pulses. TES MUX schemes have in contrast already allowed $\geq$ 10 kilopixel arrays in the sub-mm (SCUBA-2) where the TES operate in total power mode and the MUX cycle is slow[18]. There is also good promise of instrumenting large MUX arrays in the X-ray, where appropriate adjustment of thermal time constants in the absorbers can lead to ms fall times and simple accommodation of low X-ray count rates.

Finally, what about increased energy resolution and bandwidth? We have already described how smaller pixels, operated well into saturation should give a modest decrease in ΔE_{FWHM}. Smaller pixels and smaller E_{sat} can be particularly interesting for an instrument focused on near-IR applications at high angular resolution AO ports on large telescopes. Realization of such improvement with require both more sophisticated pulse sampling and improved suppression of thermal and pick-up noise. However, with fixed ΔE, TES systems will always have the most energy bins and the highest energy resolution at short wavelengths. Thus balloon and space platforms, which can allow access to the 0.1–0.3 μm range will provide the widest application of TES technology. Interestingly, with colder optics and no atmospheric background, effective photon counting out to 5 μm or even 10 μm becomes possible. We have kept this future potential in mind while developing our filtering scheme. Many challenges remain to realization of such a system. However, while we are working toward this broad-band energy-resolved view of the universe, existing TES systems provide some exciting opportunities for unique, albeit specialized, science at high time- and low energy- resolution.

This work was supported in part by NASA grants NAG5-13344 and NAG5-9087. We wish to thanks our many colleagues at Stanford and NIST Boulder who have assisted with the TES research program, including J. Barral, P. Brink, J.P. Castle, A. Miller, S.W. Nam and A. Tomada. The devices described here were fabricated in the Stanford Nanofabrication facility.

References

1. B.A. Young, T. Saab, B. Cabrera, J.J. Cross, R.A. Abusaidi, NIMPA, 444, 296 (2000).
2. Cabrera, B.; Clarke, R.M.; Cooling, P.; Miller, A.J.; Nam, S.W.; and Romani, R.W., Appl. Phys. Lett., 73, 735 (1998).
3. K. D. Irwin, Appl. Phys. Lett. 66, 1998 (1995).
4. Nam, S.W.; et al., IEEE Trans. Appl. Supercond., 9(2), 4209 (1999).
5. Cabrera, B.; and Romani, R.W., *Crogenic Particle Detection*, (ed. C. Enss), Topics in Appl. Phys. 99, 417 (2005).

6. R.W. Romani, A.J. Miller, B. Cabrera, E. Figueroa-Feliciano, and S.W. Nam, ApJ 521, L153 (1999).
7. R.W. Romani, B. Cabrera, E. Figueroa, A.J. Miller, and S. W. Nam, Bull. Am. Astron. Soc. 193, 11.12 (1998).
8. R.W. Romani, A.J. Miller, B. Cabrera, S.W. Nam, J.M. Martinis, ApJ 563, 221 (2001).
9. J. Burney, T.J. Bay, P.L. Brink, B. Cabrera, J.P. Castle, R.W. Romani, A. Tomada, S.W. Nam, A.J. Miller, J.M. Martinis, E. Wangc, T. Kennyc, and B. Young, Nucl. Instr. and Meth. A 520, 533 (2004).
10. Brammertz, G.; Peacock, A.; Verhoeve, P.; Kozorezov, A.; den Hartog, R.; Rando, N.; Venn, R.; AIP Conference Proceedings; 2002; no.605, p.59–62.
11. Martin, D.D.E; et al., SPIE Proc. 6269 22 (2006).
12. Thomas J. Bay, et al.; Proceedings of SPIE Vol. 5209 Materials for Infrared Detectors III, edited by Randolph E. Longshore, Sivalingam Sivananthan (SPIE, Bellingham, WA, 2003).
13. M.A.C. Perryman, M. Cropper, G. Ramsay, F. Favata, A. Peacock, N. Rando, A. Reynolds, Mon. Not. R. Astron. Soc., 324 899 (2001).
14. J.H.J. De Bruijne, A.P. Reynolds, A.M.C. Perryman, A. Peacock, F. Favata, N. Rando, D. Martin, P. Verhoeve, N. Christlieb, Astron. Astrophys., 381, L57 (2002).
15. Reynolds, A.P., Ramsay, G., de Bruinje, J.H.J., Perryman, M.A.C., Cropper, M., Bridge, C.M., and Peacock, A., A & A, 435, 225(2005).
16. Cropper, M.; et al., MNRAS, 334, 33 (2003).
17. J. A. Chervanek, K.D. Irwin, E.N. Grossman, J.M. Martinins, C.D. Reintsema, and M.E. Huber, Appl. Phys. Lett., 74, 4043 (1999).
18. Irwin, K.D.; et al., SPIE Proc. 5498 (2004).

Imaging Photon Counting Detectors for High Time Resolution Astronomy

O. H. W. Siegmund, J. V. Vallerga, B. Welsh, A. S. Tremsin and J. B. McPhate

Abstract Ground based high time resolution astronomical (HTRA) observations require specialised detectors. Photon counting detectors that combine high spatial resolution imaging with fast event timing capability have been a significant technical challenge. We have developed a high-temporal and -spatial resolution, high-throughput sealed tube microchannel plate detector with electronic readout as a tool for HTRA applications. The design is based on a 25 mm diameter S20 photocathode followed by a microchannel plate stack, read out by a cross delay line anode with timing and imaging electronics. The detector supports 500 kHz global count rate, 10 kHz /100 μm^2/sec local count rate, and 100 ps timing resolution. We also describe high time resolution astronomical observations currently being obtained with the NASA *GALEX* UV satellite, as well as preliminary visible wavelength imaging results obtained on the Crab pulsar at the Lick Observatory.

1 Photon Counting Imaging Detectors

The motivation for our current detector development [1]–[3] stems from advances in adaptive optics, laser imaging and the emerging discipline of high time-resolution astronomy which are challenging the limitations of current detector performance capability [4]–[10]. We have used the cross delay line (XDL) anode [1]–[3] in a number of microchannel plate (MCP) detectors for spaceflight imaging and spectroscopy missions. The cross delay line detector could open new windows in high time-resolution astronomy, with a sufficient photon flux and temporal resolution it will be possible to investigate very small emission regions of astrophysical objects in great detail [11]. When coupled to the additional diagnostic tool of polarization, whose angle and degree constrains the geometry of the local magnetic field, it will thus be possible to investigate the restricted emission zone geometry of a wide variety of astronomical objects such as pulsars, isolated neutron stars, gamma-ray

O. H. W. Siegmund · J. V. Vallerga · B.Welsh · A. S. Tremsin · J. B. McPhate
Space Sciences Laboratory, University of California, Berkeley, CA, 94720, USA
e-mail: ossy@ssl.berkeley.edu

bursters and the emitting regions of accretion disks associated with active galactic nuclei (AGN) and cataclysmic variable stars.

This type of detector, with a suitably designed polarimeter mounted on a large aperture ground-based telescope, will perform high-speed (µs) optical photopolarimetry of all these types of objects. A major scientific goal is to use the detector to simultaneously measure all the Stokes' parameters of an incoming signal from an astronomical object by recording simultaneous position/time data for a set of polarimetric split images. Additional advances in photocathode materials and readout technologies also show promise for enhancement of the performance of photon counting imaging detectors. Such instrumentation will be the trailblazers for future general-user high-speed detection devices that are required to fully exploit the scientific potential of the next generation of extremely large aperture telescopes.

2 Cross Delay Line Detector

The implementation of a monolithic-ceramic XDL detector in a sealed vacuum tube was first accomplished in our development of the GALEX mission MCP detectors [1] that are currently operating in-orbit. The basic configuration for current work is the same, however the anode and detector are specifically optimized for the high counting rates expected for the time resolved applications.

We have employed a multi-alkali semitransparent proximity-focused photocathode (Figs. 1, 2) as the photo-conversion medium. A stack of three low resistance (25 MΩ) MCPs were used to accommodate high local event rates at modest gain (6×10^6). The XDL anode was designed with a cross-serpentine configuration (Fig. 3)

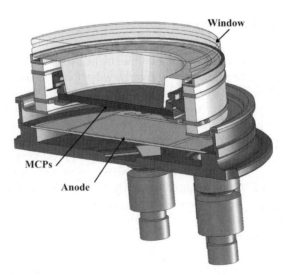

Fig. 1 Cutaway view of a cross delay line imaging microchannel plate sealed tube detector. Incident photons pass through the entrance window and are converted to photo-electrons by the S20 photocathode on the inner window surface. These are proximity focused onto a set of three MCPs in a back to back Z configuration. After amplification in the MCP stack (gain $\sim 5 \times 10^6$) the charge cloud falls onto a cross delay line anode that performs event position encoding

Fig. 2 25 mm sealed tube detector with S20 photocathode and 29 mm^2 XDL anode. High voltage and fast timing connections are radial, the anode signal connections are positioned on the base

to minimize the overall footprint and establish a fast propagation delay (20 ns) appropriate for high global event rate capacity. All the materials used were metal or ceramic for compatibility with the ultra-high vacuum sealed tube configuration.

The XDL anode was mounted in a vacuum header with SMA coaxial connectors for the anode signal outputs, and welded to a hermetic metal/ceramic brazed assembly that holds the MCPs. Verification tests were accomplished with

Fig. 3 29 mm cross delay line anode. The anode consists of a serpentine conductor for each axis, with a winding period of 0.6 mm. The end to end delay produced is approximately 20 ns. Axes are isolated from each other by a refractory layer

the detector in an open vacuum tank prior to processing and sealing the detector. Processing consisted of a vacuum bakeout, high flux charge scrubbing of the MCP stack, followed by cathode deposition and final sealing of the tube. The vacuum seal was a hot indium seal between the window and the brazed assembly.

3 Cross Delay Line Detector Performance

3.1 Photocathode Efficiency

The photocathode that we have employed for our current development is a conventional S20 type multialkali. This was deposited onto the inside of the fused silica window using a standardized process, producing a quantum efficiency comparable with commercial S20 devices (Fig. 4.) [12, 13]. The background rate for this cathode was approximately 50 events/sec with the detector un-cooled. Since the counting rates of the planned tests were orders of magnitude above this level it was deemed unnecessary to use a cooler for the detector.

3.2 Electronic Position and Time Encoding Scheme

The electronic system for fast two dimensional imaging and timing is shown and described in Fig. 5. The position encoding for the cross delay line anode is based on encoding the arrival time differences of the event signals observed at the two ends of the delay lines. In this case we have added a new circuit to accomplish time encoding for each event that relies on a GPS standard reference.

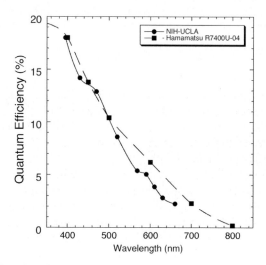

Fig. 4 Quantum detection efficiency for the cross delay line detector S20 photocathode (NIH-UCLA) compared with a commercial S20 photocathode (Hamamatsu)

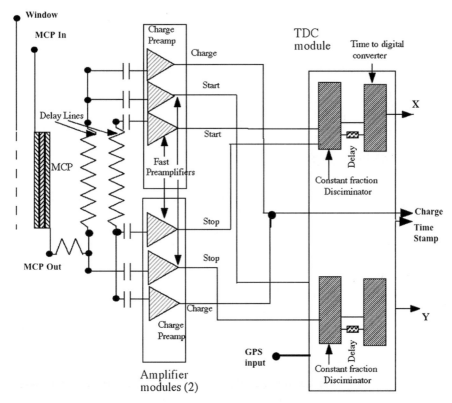

Fig. 5 Electronic processing chain for imaging with the cross delay line detector. Fast amplifiers on each delay line end amplify the signals, and one end of each delay line is delayed by slightly more than the overall delay of the anode. All fast signals are discriminated by constant fraction discriminators, and the output time differences are converted to analogue values by a time to amplitude converter. Analogue to digital conversion of the result gives a digital event centroid position for each axis. Charge signals are also amplified and digitized to monitor the overall gain of the detector

The delay-line encoding electronics consists of a fast amplifier for each end of the delay line, followed by time-to-digital converters (TDCs) (Fig. 5). Ultimate resolution of delay line anodes in this scheme is determined by the event timing error which is dominated by the constant fraction discriminator (CFD) jitter and walk, and noise in the time-to-amplitude converter (TAC) part of the TDC, which is typically of the order ~10 ps FWHM total. The overall dead time is ~0.4 μs allowing rates of 500 kHz to be achieved with acceptable (<15%) (Fig. 6) deadtime losses. For the applications in astronomical polarimetry the MCP output time signal is simply correlated to the local GPS clock to provide a time stamp (25 ns accuracy) associated with each valid event. For even higher time resolution applications a second TDC can be used to encode event time to ~100 ps. This uses the signal directly from the MCP output as the stop signal and the start signal can be obtained from a GPS, laser, or trigger depending on the specific application.

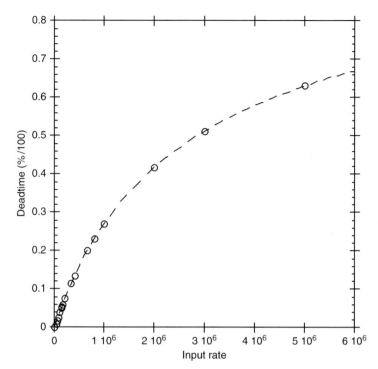

Fig. 6 Dead time event loss as a function of the input counting rate for the 400 ns dead time of the detector electronics (Fig. 5)

3.3 Imaging Performance

The detector was extensively tested under various conditions. The response of the sealed tube to uniform illumination is shown in Fig. 7a. There is little fixed pattern noise, and no evidence of hexagonal structures that are often seen for MCPs due to the pore multi-fiber stacking structures. This is attributable to better MCP manufacturing processes established by Photonis recently. The edge enhancement is due to fringe fields resulting from the metal ring supporting the MCPs perturbing the field between the MCP stack and the XDL readout.

Resolution tests were performed both before and after the detector was sealed. Images of pinhole mask arrays (Fig. 8) allow both the point spread function and image linearity to be evaluated. The average width distributions of the pinhole images were measured as a function of MCP gain (Fig. 9) demonstrating that the resolution is better than 40 μm at modest gain (8×10^6). The resolution of the completed tube was measured by focusing a small spot (<50 μm) onto the cathode. This gave a ∼80–90 μm image, showing that the proximity focus spread from cathode to MCPs is dominating the final resolution performance. The linearity over the central portion of the image is good with <1 resolution element position distortion.

The stability of the image and the resolution at high counting rates was also assessed. Uniformly illuminated images taken at 10 kHz, and 450 kHz showed no

Imaging Photon Counting Detectors 333

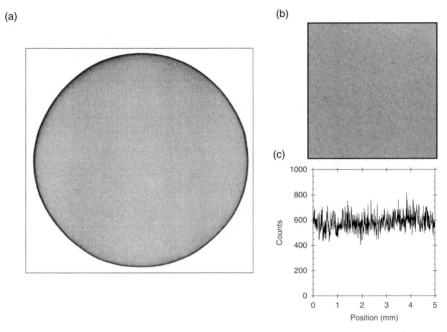

Fig. 7 Uniformly illuminated 25mm image using visible light, 1k × 1k pixels, 7×10^8 events at event rate 500 kHz, showing the uniformity of the response, and the lack of MCP multi-fiber modulation. **Fig. 7b.** 5 mm × 5 mm area showing detail. **Fig. 7b.** Single row intensity histogram

Fig. 8 Pinhole mask image taken before the final tube seal. A thin mask is in contact with the top MCP with 1mm hole spacing, and 10 μm hole width

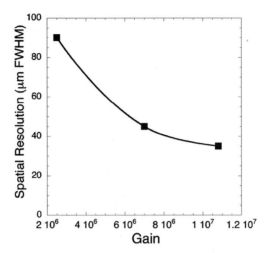

Fig. 9 The measured average width of the mask pinhole images shown in Fig. 7 for various gain MCP settings. This resolution is dominated by the electronics timing errors. The resolution of the final tube is closer to 80 μm FWHM due primarily to the large window to MCP proximity gap (~600 μm) causing photoelectron spread. A smaller proximity gap (<250 μm) can be achieved, but was not necessary for the current application

significant differences, and local spot images (90 μm spot) did not change position or width up to rates of 10 kHz in the spot (~100 events pore^{-1} sec^{-1}).

3.4 Timing Results

The timing accuracy of the entire system was evaluated by illuminating the detector with a fast pulsed laser and measuring the difference between the laser trigger and the fast MCP pulse signal (Fig. 5). The overall timing jitter was ~100 ps FWHM (Fig. 10) as compared to the trigger jitter of the laser which is specified to be ~80 ps FWHM. The transit time spread for this type of detector due to the photoelectron trajectory spread in the proximity gap and MCP stack is ~25 ps, so the main contribution to the system jitter is provided by the timing electronics.

4 High Time Resolution Observations with GALEX

Our recent implementation of cross delay line detectors on the GALEX satellite has provided some key results on time variable sources observed in the UV. The primary scientific mission of NASA's *GALEX* satellite mission is to carry out imaging sky-surveys in both the NUV and FUV wavelength bands, with the scientific goal of understanding star formation in galaxies and how its evolution changes with time (see reference 14 and references therein).

During the course of these sky-survey observations *GALEX* has serendipitously detected numerous variable and transient ultraviolet sources, many of which exhibit

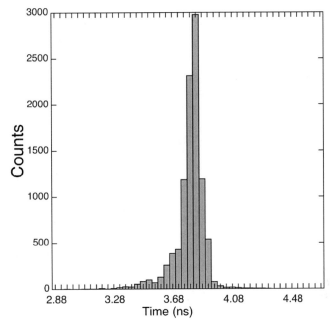

Fig. 10 Measured transit time spread of events using a pulsed laser with ∼80 ps jitter produced by comparing the laser trigger with the MCP output fast pulse (Fig. 5). The detector system jitter is probably better than 80 ps FWHM and is controlled by the electronics jitter, the proximity gap and MCP transit time spread

much larger amplitudes of flux variation in the ultraviolet region than that recorded at visible wavelengths. Such stellar sources include M-dwarf flare stars [15], RR Lyrae variable stars [16] and variability associated with the accretion surrounding black holes in AGN. The unique capability of UV time-variability studies by *GALEX* is due to the photon counting nature of its detectors, which allow time-tagging of individual UV photons with a precision of <0.05 seconds. Such timing accuracy cannot be obtained with CCD sensors, and in Figs. 11 and 12 we show the UV light-curve and image of a stellar flare recorded on the star GJ 3685A in which an overall brightness increase of >4 magnitudes was observed in both the FUV and NUV bands in a time period of only 20 seconds! In addition to the detection of variable celestial sources, *GALEX* has also serendipitously observed the passage of both asteroids and man-made satellites across its 1.2° field-of-view. Such objects are detectable in the UV due to their inherent reflective properties of the incident solar (Fraunhofer) spectrum. Since *GALEX* has the post priori ability to reconstruct images as a function of time, if several orbital parameters such as time of observation and satellite viewing angle are known with precision, then it is possible to determine the orbital path of the observed man-made satellite. This is an issue which has been discussed in several previous publications [17, 18]. The detection of such objects by GALEX shows that this is possible in the UV regime and stimulates the evaluation of photon counting detectors in this type of application.

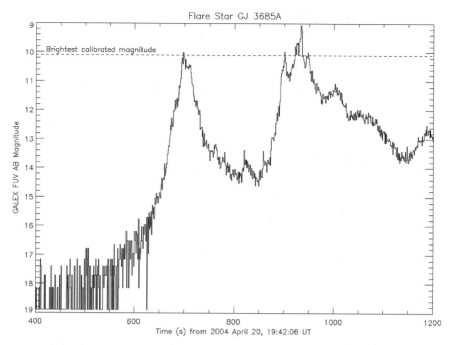

Fig. 11 Galex NUV detector channel light-curve of a flare on the star GJ 3685A, the detector can accommodate the highest flux without saturation efects

Fig. 12 Image of GJ 3685A at peak emission intensity, the flare star is the brightest object in the image, but is essentially undetectable in the image prior to the flare outburst

5 High Time Resolution Visible Astronomy

Visible radiation from celestial sources such as pulsars and white dwarf binaries displaying fast temporal variations in their emitted high-energy flux, is often associated with the non-thermal particle populations that generate X-ray and gamma-ray photons. Thus, photons emitted at visible wavelengths possess timing and polarization information that is linked to their origin and these phenomena can potentially be observed from large ground-based telescopes. On time-scales of a tenth of a second, echo mapping of the geometry of cataclysmic variable and X-ray binary stars becomes possible, and at the hundredth of a second time-scale direct speckle imaging of nearby stars is achievable. At time-scales of milliseconds the optical emission of pulsars and the detection of the optical analogue of the kilohertz quasi-periodic oscillations found in X-ray binary accretion disks also becomes possible. However, recording these various phenomena with good signal-to-noise and good spatial resolution, a visible wavelength detector that records the arrival time of each emitted photon is required. Conventional CCDs are presently incapable of such observations due to their inherent limitations with respect to read-out noise on short time scales.

A particularly important scientific goal for high time-resolution astronomy is the study of pulsar polarization. For objects such as the Crab pulsar, with a period of 33 ms, simultaneous measurements of the polarization parameters are required to determine how the polarization varies across the entire pulsar period. It has been established that a large polarization is associated with the "off" phase of the Crab pulsar's rotational phase. This suggests that the emission during the "off" phase of the visible light-curve is consistent with some form of non-thermal (synchrotron-related) origin. This issue can only be investigated by recording all of the Stokes polarization parameters for each individual optical pulse. This can potentially be achieved using a photon counting detector to simultaneously image and record the time of photon events in separate images with different polarizations.

As a conceptual test, an optically sensitive imaging tube was built and attached to the prime focus of the 1 m Nickel telescope at Lick Observatory on Mt. Hamilton, CA. The detector consisted of the imaging tube (S20 photocathode, MCPs for charge amplification, and a cross delay-line anode for charge collection and imaging), fast amplifiers, and a TDC for position and arrival time encoding. The TDC time reference was reset once per second via a GPS one pulse per second signal, resulting in an absolute timing accuracy of $\pm 1\mu s$. However, the TDC provides 25 ns time stamping precision for each photon recorded. The dead time of this TDC is about 400 ns giving <30% dead time at 1 Mhz. A picture of the optical tube and amplifiers mounted to an adapter flange at the telescope Cassegrain focus is shown in Fig. 13.

This setup was used to observe the Crab pulsar on two nights. Observing conditions were relatively poor on both nights, the seeing was approximately 4–5 arcsec FWHM, and the sky background was dominant source of counts on the detector. We observed between 7 and 10 events /sec from the Crab pulsar, in agreement with our expectations based on source brightness, telescope size, and detector QE. The pulsar light curve shown in Fig. 14 was derived from 1800 seconds of data by extracting the photons that fell within a 3 arcsec radius of the pulsar and making a histogram

Fig. 13 25mm MCP sealed tube cross delay line detector mounted at the Cassegrain focus of the Lick Observatory Nickel 1m telescope

of their arrival times binned to the expected period of the pulsar for the date of the observation. The characteristic light curve shape [19] for the Crab is clearly visible. By extracting the photons arriving within any desired time range of the light curve we can produce the corresponding image of the Crab field of view during that time from the stored data set (Fig. 15). Our current plan is to employ a polarimeter to project several images of the Crab at different polarizations onto a single detector. Then the recorded data will allow these to be optimally analyzed after the fact, without foreknowledge of their position, seeing conditions, or pointing constraints.

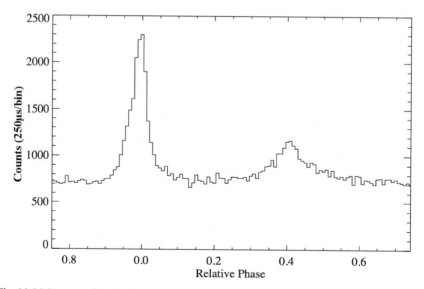

Fig. 14 Light curve of the Crab 33 ms period pulsar. 250 μs time binning for an 1800 sec exposure time

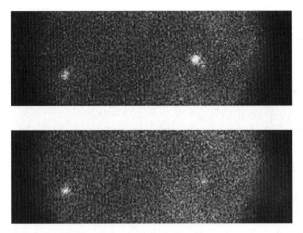

Fig. 15 Image of the Crab pulsar at the primary emission peak (250 µs window, upper) and the low phase (lower)

New sealed tubes with higher efficiency (Super GenII, GaAs) are in development, which in combination with larger telescope apertures will provide the sensitivity required for these studies.

6 Detector Development

Our goal for the implementation of photon counting MCP sensors in high time resolution applications is to dramatically improve the detection efficiency and rate handling capability, while maintaining the imaging performance. For sealed tube applications we have proposed to use extended red SuperGenII photocathodes as provided by Photonis, and ultimately GaAs or GaAsP photocathodes, which will increase the efficiency (Fig. 16.) greatly over the standard S20 optical photocathodes. The increased efficiency will also result in the need for some form of device cooling, but this need only be to $\sim 0°C$ to achieve low background counting rates. In addition the sky background rates for high time resolution astronomy will also increase, thus promoting the need for higher event rates.

To increase the readout rate, we propose to use a cross strip readout concept [20]. The cross strip (XS) looks very similar to the XDL anodes. However the XS anode works completely differently than delay-line anodes, to achieve much higher resolution and speed, while using gain that is several orders of magnitude lower. The XS anode is composed of two coarse (~ 0.5 mm) sets of fingers arranged orthogonally as a multi-layer metal and ceramic pattern on an alumina substrate. The top and bottom layers of fingers are used to collect the charge from the MCPs with equal charge sharing between the axes. The charge cloud is matched to the anode period so that it is collected on several neighbouring fingers to ensure an accurate event centroid can be determined. The anodes made in this fashion are low outgassing, and thus can be put into sealed tube devices. Each finger of the anode is connected to a low noise charge sensitive amplifier (Fig. 17) and followed by subsequent analogue to digital

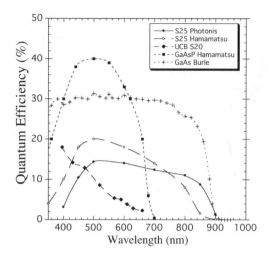

Fig. 16 Measured quantum efficiencies of S20, S25 (SupergenII) and GaAs type photocathodes. (Photonis and Hamamatsu catalog data, Burle data courtesy P. Hink, Burle Inc.)

Fig. 17 32 mm×32 mm cross strip anode read out with an amplifier board for each axis. (IDEAS, VaTagP3 ASIC)

conversion of individual strip charge values and a software, or hardware, centroid determination. The centre peak of the charge cloud determines the coarse position of the registered photon. The charge cloud centroid is calculated with a resolution equivalent to a fraction of the strip width. Provided the noise and the number of bits used for signal digitization are sufficient, centroids of better than 1/100 of a strip are possible.

XS anodes are mounted close to ($\sim$3 mm) the MCPs and each strip is connected through hermetic vias to the backside of the anode, allowing mounting of all the detector electronics outside the vacuum tube. This allows the amplifier ASIC boards to be easily mounted or exchanged. A 32 mm$\times$ 32 mm XS open face laboratory detector (Fig. 17) has demonstrated excellent resolution (<7 μm FWHM) using low MCP gain ($\sim$5 $\times 10^5$) [21]. Since the spatial resolution achieved is much better than most readouts, we can also choose to relax the resolution by lowering the gain further in favor of higher local counting rate performance and longer overall device lifetime. We estimate a device running at 5 MHz event rate and $<10^5$ gain would have a 10,000 hour lifetime.

Since our current electronics for XS anodes is relatively slow (<10 kHz), high speed parallel electronics is being developed to digitize the strip signals, and process the events to generate the x,y position with MHz rates and time stamp the events with sub-nanosecond accuracy. We are developing a full parallel ASIC amplifier system using our standard centroiding algorithms which assume processing of a single event at a time on the detector. In this parallel scheme each strip on the anode (64 X and 64 Y) is connected directly to a preamp board using a fully parallel 32 channel ASIC amplifier (RD-20). The output of these amps are shaped unipolar pulses of $\sim$40 ns rise time, with a noise floor of $\sim$800 electrons. The 64 parallel outputs (32 $\times$ 2 per axis) are amplified again before being continuously digitized by 64 discrete analogue to digital converters operating at >50 mega-samples per

Fig. 18 18 mm cross delay line anode read out sealed tube detector

second. These digital samples are fed into an FPGA (Xilinx) where they are digitally filtered to extract pulse peak information. This is used to derive the event centroid for both X and Y axes which is then combined with the local digital timing signal which can be syncronized with an external clock (GPS). The events are buffered to await transfer to a downstream computer as an event list of X, Y and T.

Sealed tube XS anode detectors are being developed to accompany the new electronic readout scheme. These include the 32 mm XS anode we have already comissioned and a smaller 18 mm version that is suitable for a more standard size 18 mm tube such as the one currently employed for our cross delay line anodes (Fig. 18). These will provide significant performance improvements for high time resolution astronomy, with the ability to utilize many of the available cathodes and readout schemes. The compact size and modest cooling requirements should enable easy adaptation to many ground based instruments.

Acknowledgments GALEX (Galaxy Evolution Explorer) is a NASA Small Explorer, launched in April 2003. We acknowledge the dedicated team of engineers, technicians, and administrative staff from JPL/Caltech, Orbital Sciences Corporation, University of California, Berkeley, Laboratory Astrophysique Marseille, and the other institutions who made this mission possible. We acknowledge the efforts of J.Hull, J. Malloy, R. Raffanti, and S. Brown for their work in accomplishing these studies. This work was supported by NSF grant AST0352980, and NASA grants NAG5-12710 and NNG05GC79G.

References

1. O.H.W. Siegmund, P. Jelinsky, S. Jelinsky, J. Stock, J. Hull, D. Doliber, J. Zaninovich, A.S. Tremsin and K. Kromer, Proc. SPIE 3765 (1999) 429.
2. O.H.W. Siegmund, M.A. Gummin, J. Stock, D. Marsh, R. Raffanti, T. Sasseen, J. Tom, B. Welsh, G. Gaines, P. Jelinsky, and J. Hull, *Proc. SPIE*, 2280 (1994) 89.
3. O. H. W. Siegmund, Proc. SPIE . 4854 (2003) 181
4. S. Weiss, Science 283 (1999) 1676.
5. S. E. D. Webb, Y. Gu, S. Leveque-Fort, J. Siegel, M. J. Cole, K. Dowling, R. Jones, P. M. W. French, M. A. A. Neil, R. Juskaitis, L. O. D. Sucharov, T. Wilson, M. J. Lever, Rev. Sci. Instr. 73 (2002) 1898.
6. O. Krichevsky, G. Bonnet, Rep. Prog. Phys. 65 (2002) 251.
7. X. Michalet, F. F. Pinaud, L. A. Bentolila, J. M. Tsay, S. Doose, J. J. Li, G. Sundaresan, A. M. Wu, S. S. Gambhir, S. Weiss, Science 307 (2005) 538.
8. L. A. Kelly, J. G. Trunk, K. Polewski, J. C. Sutherland, Rev. Sci. Instr. 66 (1995) 1496
9. L. A. Kelly, J. G. Trunk, J. C. Sutherland, Rev. Sci. Instr. 68 (1997) 2279.
10. V. Emiliani, D. Sanvitto, M. Tramier, T. Piolot, Z. Petrasek, K. Kemnitz, C. Durieux, M. Coppey-Moisan, Appl. Phys. Lett. 83 (2003) 2471.
11. A. Shearer, and Golden, A., ApJ, 547 (2001) 967
12. Hamamatsu, www.hamamatsu.com
13. Burle, www.burle.com
14. D. C. Martin, J. Fanson, D. Schiminovich, et al, Astrophysical Journal **619**, (2005) L1.
15. R. D. Robinson, J. M. Wheatley, B. Y Welsh, et al, Astrophysical Journal **633**, (2005) 447.
16. J. M. Wheatley JM, Welsh BY, Siegmund OHW, et al., Astrophysical Journal **619**, (2005) L123.

17. C. Ho, W. C. Priedhorsky, M. H. Baron, SPIE **1951**, (2003) 75.
18. W. Priedhorsky, J.J. Bloch, **44,** (2005), 433.
19. G. Kanbach, S. Kellner, F. Z. Schrey, H. Steinle, C. Straubmeier, H. C. Spruit, Proc. SPIE **4841**, (2003) 82,
20. A. S. Tremsin, O. H. W. Siegmund, J. V. Vallerga, J. Hull, IEEE Trans. Nucl. Sci. **51**, (2004) 1707.
21. O.H.W. Siegmund, A. S. Tremsin, R. Abiad, J. V. Vallerga, Proc. SPIE, **4498**, (2001) 131.

Index

A0620–00, 42, 47
AApnCCDs, 286, 287, 288
AAT, Anglo-Australian telescope, 64, 76, 82, 188
Accreting binary stars, 78–80
Accretion disc, 21, 22, 25, 27, 28, 29, 31, 33, 37, 39, 42, 45, 80, 135
AcqCam, 240, 251
Active quenching, 295–297
ADAF, 44, 46, 47, 49, 50, 51
Adiabatic demagnetization refrigerator, 318
ADR, 318, 319, 320
AFOSC, 180, 181
AGILE, 168
AGN, 258, 268, 269, 276, 328, 335
AM Her, 164, 207, 208
AM Her binaries, 164, 207, 208
Anomalous X-Ray Pulsar, 9–10
APDs, 11, 16, 153, 154, 157, 161, 183, 243, 282, 292, 295, 297, 302
Aperture, 44, 51, 61, 71, 77, 97, 118, 121, 136, 141, 153, 154, 155, 161, 172, 173, 192, 249, 265, 276, 293, 311, 314, 317, 318, 320, 328
Apogee, 154
AquEYE, 171, 180, 181, 183, 242
Argos, 235
Aristarchos Telescope, 138
Asiago Quantum Eye, 179
Asteroseismology, 30–33, 53, 56, 57–58, 61, 62, 64, 71, 72, 77
ATNF catalog, 4
Autocorrelation, 167, 172
Avalanche amplifier, 245, 281, 283, 286
Avalanche multiplication, 284–286
Avalanche Photodiodes, 1, 153, 154, 157, 172, 243, 291, 299, 304
AXPs, 13–16

Balloon 090100001, 68, 70, 71
Balmer, 42, 65, 201
Bias subtract, 200
BK7, 320, 322
Blazar 0716+741, 273, 276
Blazars, 257, 258, 268, 269, 271, 276, 278
BL Lac, 271, 274, 276, 277
Bolometric, 40, 282, 311
Bose-Einstein, 97, 98, 101, 102, 120
Boundary layer, 21
BX Cir, 56

CAHA, 164
Calar Alto Spain, 154
CAMEX, 281, 285, 288
Cassegrain, 143, 155, 156, 157, 174, 180, 236, 239, 318, 337, 338
Cataclysmic variable, 21, 30, 166, 206, 207, 292, 328, 337
CCDs, 9, 13, 23, 25, 38, 51, 56, 76, 82, 83, 85, 86, 87, 88, 89, 133, 134, 137, 140, 141, 142, 144, 146, 150, 168, 187, 197, 234, 245, 250, 257, 258, 261, 266, 267, 276, 279, 283, 291
Cen X-3, 206
CIC, 87, 88, 89, 92, 93, 261, 262
Clock Induced Charge, 87, 92, 261
Close binary systems, 107
CMOS, 299, 302, 304
Crab, 2, 5, 7, 8, 13, 111, 112, 145, 158, 160, 161, 162, 163, 164, 168, 191, 205, 209, 210, 211, 212, 213, 214, 215, 220, 221, 226, 236, 240, 242, 315, 327, 337, 338, 339
Crab Pulsar, 7–8
CRGO, 9
Cross Delay Line Detector, 328, 330, 331, 334, 338
Cross strip, 244, 339, 340
Cryogenic Imaging Spectrophotometers, 246
CV, 21, 22, 29, 31, 315

345

DAQ, 157, 159, 287–288
DEPMOS, 285
Derotator, 241
DNOs, 23–24
DOAP, 215
Doppler boosting, 269
DOWP, 215
DQ Her, 23
Drift mode, 64, 65, 142, 148, 188
DROIDs, 247
Dwarf Nova Oscillations, 23, 78
Dwarfs, 22, 30, 32, 33, 53, 56, 57, 62, 76, 77, 78, 80, 82, 84, 125, 134, 135, 172, 187, 213, 231, 292, 323

E2V, 137, 140, 141, 221, 245, 291
Eclipse map, 25
ELT, 1, 17, 79, 183, 203, 229, 230, 231, 233, 234, 236, 239, 240, 241, 242, 250
Emission theory, 7
ENIGMA, 279
EPICS, 241
ESA, 320, 323
Eta Carinae, 106, 172
Ex Dra, 27
EXO 0748–676, 39, 40
Extrasolar Planets, 233

FFT, 159
Fibre bundle, 163
Flatfield, 200, 266, 267
FO Aqr, 199–200, 201
FOCAS, 235
FORS-1, 187
FORS-2, 187, 188, 189
FORS2, 187, 191, 193, 197, 203, 251
FOS, 28
4U 1626–67, 38
4U 1636–536, 39
4U 1735–444, 39
FPGA, 178, 244, 313, 342
FUSP, 235
FWHM, 7, 50, 140, 190, 191, 197, 198, 243, 244, 247, 266, 267, 275, 331, 334, 335, 337

G29-38, 77, 78
GaAs, 89, 208, 221, 244, 249, 339, 340
GaAsP, 339
Galactic halo, 62
Galactic latitude, 136, 167
Galileo network, 231
Gamma-ray, 9, 96, 153, 168, 271, 272, 276, 327, 337
GASP, 214–220

Geiger, 157, 243, 286, 291, 293, 294, 298
Geiger-mode Avalanche Photodiode, 291–308
Giant radio pulse, 8, 112, 209, 292
GJ 3685A, 335, 336
GLAST, 168
Globular clusters, 4, 62, 234, 250
GM-APD, 291, 292, 293, 298, 299, 300, 302, 306, 308
GPS, 38, 142, 145, 153, 158, 159, 179, 183, 195, 201, 239, 244, 246, 321, 330, 331, 337, 342
GRO J0422+32, 41
GRP, 2, 7, 8, 13, 210, 211, 220, 292
GS 1826–24, 39
GU Mus, 49–51
GX 1+4, 38

Hale, 134
Hanbury Brown-Twiss, 120, 172, 183
He^4, 98
Helium-4, 98
Her X-1, 38, 206
Himalaya Chandra Telescope, 70
HIT modes, 192
HIT-MS, 188, 189, 192, 193, 196, 197, 200
HIT-OS, 192, 193, 194, 197, 198
HiTRI, 229, 243
Horizontal clocking, 146, 147
HST, 28, 163, 277, 278
HU Aqr, 164, 166
HW Vir, 70
Hydrogen, 29, 56, 62, 77, 105, 107, 108, 109, 118, 201

ICCD, 244
Image Photon Counting System, 75
Imaging arrays, 311, 317
IMPOL, 241
INTEGRAL, 168
Intensified CCD, 244
Intensity interferometer, 99–102
Intermediate Polars, 21, 199
Inverse Compton, 7, 111, 268, 272
IPCS, 75, 87
IP Peg, 26, 79, 80
Isaac Newton Telescope, 209
ISIS, 65, 188

JFET, 281, 284
Jitter, 183, 196, 210, 331, 334, 335
JOSE Camera, 134

Keck, 77, 78
Keplerian, 21, 26, 42, 46, 79, 80, 108
KG5, 322

Index

KPD 1930+2752, 65, 66, 67
KPD 2109+4401, 63, 65, 66, 67, 68, 69
KT Per, 24
KV UMa, 167, 168

L3CCD, 13, 29, 33, 89, 240, 245, 249, 250
La Silla, 154
LED, 155, 286
Lick Observatory, 327, 337, 338
Lightcurve, 24, 25, 38, 39, 42, 45, 46, 47, 50, 163, 165, 190, 197, 198, 199, 258, 268
LISA, 81, 82
Lithography, 317, 323
Lomb-Scargle, 24
Long-wave pass, 139
LRIS, 188, 251
LSST, 75, 76
LuckyCam, 242
Lucky Photometry, 257, 278
LWP, 139

M31, 2
M4V, 166
M87, 275, 276, 278
Magneto-hydrodynamics, 22
Magnetospheric, 5, 10, 113, 164, 323
MAMA, 154, 282
MatLab, 321
Mayall, 134
MCPs, 243, 244, 328, 329, 332, 337, 339, 341
MCVs, 206, 207, 208, 214
MHD, 22, 25, 33
MIDAS, 193
MIDIR, 241
MIT, 187, 197
MOMSI, 241
Montreal Lapoune photometer, 70
MOS, 187, 188, 190
MOSFET, 281, 284, 286
MPPP, 235
Mt. Skinakas, 154, 155, 157
Mt. Stromlo, 154
MXU, 187, 193

NASA GALEX satellite, 334
Nasmyth, 220, 318, 319
Neutron stars, 3, 7, 53, 54, 55, 57, 78, 106, 113, 125, 134, 135, 168, 172, 206, 231, 327
NIST, 318, 324
Non-LTE departure coefficient, 101
Normal modes, 53, 54, 55, 57
NOT, 154, 163, 164

OPTIMA, 153, 154, 155, 157, 158, 160, 161, 162, 163, 164, 168, 183, 213, 293
OPTIMA-Burst, 154, 156, 157
Oscillations, 23–24, 37, 53, 54, 55, 56, 57, 58, 63, 68, 70, 72, 78, 135, 337
OSSE, 9
OverWhelming Large Telescope, 171
OWL, 172, 173, 174, 175, 176, 177
OY Car, 27

Passive quenching, 294–295
PB 8783, 63, 65, 66, 67
PDS, 37, 38, 44, 45, 46, 47, 49, 50, 51
Peltier, 141, 157, 195, 197, 302, 303
PG 0014+067, 64, 68, 77
PG 1047+003, 63
PG 1219+534, 64, 77
PG 1336–018, 65, 66, 68, 70
PG 1338+481, 64
PG 1605+072, 65, 66
PG 1627+017, 64
PG 1716, 63, 64, 70
PG 1716+426, 56, 63
Phase spectra, 316
Photocathode, 120, 157, 244, 282, 283, 327, 328, 329, 330, 337, 339, 340
Photocathode imager, 244
Photometry, 5, 6, 22, 23, 29, 33, 41, 51, 53, 58, 59, 68, 70, 71, 72, 75, 76, 79, 83, 133, 134, 135, 139, 146, 153, 154, 174, 190, 192, 199, 211, 213, 214, 217, 218, 234, 244, 250, 257, 258, 261, 263, 265, 266, 267, 268, 269, 271, 273, 274, 275, 276, 277, 278, 293, 323
Photon Correlation Spectroscopy, 121–122
pnCCDs, 245
Poisson noise, 85
Polarimetry, 8, 17, 96, 153, 160, 162, 168, 203, 205, 206, 207, 209, 211, 234, 236, 250, 323, 331
Polarization, 96, 97, 115, 116, 120, 125, 153, 160, 162, 164, 165, 168, 205, 206, 207, 208, 209, 210, 211, 212, 215, 219, 220, 222, 223, 224, 225, 226, 236, 241, 246, 250, 311, 319, 320, 323, 337
Polars, 21, 24, 199, 207, 234, 250, 323
Power density spectrum, 37, 38, 39, 47, 48, 50
Prism, 153, 155, 161, 162, 205, 212, 215, 216, 217, 218, 222, 323
Prism spectroscopy, 162
PSR B0540–69, 4
PSR B0656+14, 4, 5, 11
PSR B0950+08, 2
PSR B1055–52, 2
PSR B1509–58, 2

PSR B1929+10, 2
PSR J1824–2452, 9
PSR J1939+2134, 9
Pulsar, 1–17

QuantEYE, 124, 125, 171–184
QFT, 179
QPOs, 23, 24, 37, 42, 44, 50, 51, 323
Quantum efficiency, 1, 13, 23, 76, 117, 137, 140, 154, 157, 168, 172, 173, 174, 180, 183, 213, 220, 230, 245, 260, 272, 281, 282, 291, 292, 293, 297, 299, 302, 307, 312, 313
Quantum Fourier Transform, 179
Quantum Optics, 95, 96, 97, 109, 115, 116, 117, 118, 124, 125, 135, 174, 177, 183

Rotating polaroid filter, 160–161
RPF, 160–161
RRAT, 2, 10, 11, 16, 17, 292
RR Lyrae, 335
RR Tel, 105
RSS, 235
RX J0806+1527, 84
S20, 327, 328, 329, 330, 337, 339, 340

S5 0716+714, 270
SAAO, 39, 63, 208
ST LMi, 236, 315, 316
SALTICAM, 240, 241
S-Cam3, 154, 242, 322, 323
SCELT, 241, 248
Schwarzschild radii, 46, 49
SCSI, 144, 146
SCUBA-2, 241, 247, 324
sdB, 57, 59, 63, 64, 67, 70, 71, 76, 77, 78
SDSS1602–0102, 33
SDSS J171722.08+58055.8, 70
SDSU, 142, 144, 145
SEM, 317
Ser X–1, 39
Single photon, 98, 115, 116, 125, 153, 157, 168, 172, 211, 229, 243, 247, 257, 281, 282, 283, 284, 291, 293
SKA, 109, 118
Sloan, 45, 46, 48, 49, 50, 140
SPAD, 172, 173, 174, 176, 178, 180, 181, 182, 243, 249, 293, 307, 308
Spectra, 5, 8, 17, 25, 28, 29, 33, 56, 63, 65, 68, 71, 75, 76, 77, 78, 80, 82, 97, 106, 107, 109, 116, 117, 123, 125, 135, 140, 187, 188, 193, 195, 200, 201, 268, 269
Spectrograph, 80, 83, 84, 153, 155, 180, 187, 188, 240, 241, 323

Spectroscopy, 22, 23, 33, 53, 62, 64, 66, 67, 68, 70, 71, 75, 76, 78, 80, 81, 82, 83, 84, 89, 95, 96, 100, 121, 123, 124, 125, 135, 153, 160, 162, 168, 172, 187, 192, 193, 194, 195, 199, 201, 203, 242, 250, 281, 327
Spectrum, 5, 7, 25, 37, 38, 40, 42, 57, 59, 63, 65, 66, 68, 70, 77, 80, 82, 96, 97, 101, 107, 115, 122, 134, 135, 158, 161, 162, 168, 187, 193, 200, 201, 210, 243, 266, 267, 271, 315, 335
Square Kilometer Array, 109, 118
SQUID, 247, 312, 321, 323, 324
SS Cyg, 25, 26
Stars, 3, 7, 17, 22, 25, 29, 30, 33, 53, 54, 55, 56, 57, 58, 59, 60, 62, 63, 64, 68, 71, 76, 77, 78, 79, 80, 81, 82, 102, 104, 105, 106, 107, 112, 113, 114, 119, 124, 125, 134, 135, 136, 141, 142, 145, 147, 148, 154, 162, 168, 172, 183, 191, 194, 206, 207
Stellar flares, 2, 234
Stellar Pulsation, 53, 54, 62, 72, 231
Stellar radius, 54
STJ, 11, 246, 247, 248, 249, 282, 292, 311, 312, 323
Stokes, 3, 5, 206, 207, 208, 209, 212, 213, 214, 217, 219, 220, 221, 222, 223, 226, 242, 328, 337
Subdwarf B, 53, 54, 62, 70, 76
Superconducting Tunnel Junction, 246–247, 282, 292
SWIFT, 168
Synchrotron, 3, 5, 7, 8, 11, 46, 49, 96, 103, 206, 268, 269, 292

TDCs, 331
TES, 247
Throughput, 213
Time-to-digital converters, 331
Transition Edge Camera, 311–325
TRIFFID, 154, 183, 213
Triple-beam CCD, 133–150

UCTPol, 208, 212, 251
UTC, 145, 158, 159, 179, 183, 196, 240
UV, 5, 13, 37, 38, 40, 47, 139, 140, 244, 247, 276, 277, 282, 311, 313, 318, 320, 323, 327, 334, 335
UX Uma, 28

V361 Hya, 63, 64, 70s
V404 Cyg, 38, 41, 42, 44, 46, 47, 49, 50
V652 Her, 56, 63
Vela, 4, 210, 242

Vertical clocking, 146, 147
Virtual guard ring, 298
VLT, 13, 33, 42, 68, 70, 79, 84, 85, 139, 149, 150, 183, 187, 192, 212, 292, 318

WD0957−666, 81, 82
White dwarf, 21, 25, 29, 31, 33, 46, 56, 62, 63, 67, 76, 78, 79, 80, 81, 82, 166, 199, 202, 206, 207, 208, 315, 316, 337
Whole Earth Telescope, 70
WHT, William Herschel Telescope, 8, 11, 24, 43, 44, 46, 47, 64, 65, 66, 67, 68, 70, 134, 137, 142, 143, 149, 150
Wollaston, 154, 155, 160, 161, 212, 215, 216, 218, 323

XDL, 327, 328, 329, 332, 339
X-ray, 4, 8, 9, 11, 37–44, 47, 49, 50, 57, 79, 104, 106, 113, 145, 167, 168, 234, 248, 250, 268, 269, 276, 277, 281, 283, 286, 287, 292, 311, 324, 337, 199, 231
X-ray binaries, 37, 38, 39, 41, 44, 51, 79, 234, 250, 292
X-ray transients, 37, 41, 43, 44
XRT, 38, 41, 44, 49, 51
XS, 339, 341, 342
XTE J1118+48, 167
XZ Eri, 31

ZIMPOL, 241
ZZ Ceti, 56, 72, 77, 78